Information Mechanics

Information Mechanics

FREDERICK W. KANTOR

A WILEY-INTERSCIENCE PUBLICATION

JOHN WILEY & SONS, New York • London • Sydney • Toronto

Published by John Wiley & Sons, Inc.

Library of Congress Cataloging in Publication Data:

Kantor, Frederick W.
Information mechanics.

"A Wiley-Interscience publication."
Includes bibliographical references and index.
1. Physics. 2. Information theory. I. Title.
QC28.K36 530.1 77-6747
ISBN 0-471-02968-8

Printed in the United States of America

10 9 8 7 6 5 4 3 2 1

NOTICE

Preface

This monograph is a report on work in progress and not yet complete, intended to provide to the reader much of the conceptual basis and formal methods to date of what appears so far to be an interesting new path of inquiry. I do not know whether, in what ways, or to what extent material presented herein might be changed or discarded as these ideas are pursued further.

In this work, I have explored some aspects of a conceptual picture, in which information represented by physical systems appears fundamental. In the informal introduction, I summarize what I found, referencing detailed discussions in the main text. The main text presents investigation of whether a physical picture in terms of information is allowed by prior knowledge and may be formalized. It seems that it is and may be, and further development and experiments are suggested. Many interesting discussions in the main text are not repeated in the summary, and the casual reader may wish to skim through the formal parts and read these (for your convenience, abbreviations are listed in Appendix 2).

For the more thorough reader, I have included lists of brief descriptions of postulates, principles, definitions, lemmas, theorems, corollaries, chain, links, and conjectures. (You might find it helpful to duplicate this material and keep it handy while reading the main text, to help with references in various steps.)

In developing this approach, I started with part of nonrelativistic quantum mechanics and electromagnetic theory, related to information and self-consistent, compatible with extension by those information arguments which I then used. By this route, I came to develop a more nearly formal treatment, in which concepts relating to information are taken as basic and the previous starting theories appear as derived special cases. Often in physics one sees an effort to say as much as possible at each step. In this work, much of the development is pursued with an effort to say as little as necessary to proceed. Thus, models are depicted for only a few steps of the formal reasoning.

If the ideas presented herein prove helpful, I would like to thank, for both content discussion and human context, my brother P. B. Kantor, and D. J. Kanter, W. J. Kober, J. E. Lucas, and D. I. Sverdlik; and for human context, R. H. Laguardia and E. B. Ulman; but if the ideas presented herein prove worthless, it would be a reflection on my own lack of understanding.

Part of my interest in this direction of inquiry came from wondering about transportation for colonizing space. Transportation might be thought of as changing the relative position of objects.

What is position...?...

FREDERICK W. KANTOR

New York, New York
June 1977

Contents

1. INFORMAL INTRODUCTION 1

1.1 Conventions for Order of Mathematical Operations and Labeling, 3
1.2 Compatability with Prior Knowledge (Section 2) (2.1–2.4), 4
1.3 Special Relativity and a Limitation on Its Domain of Applicability (2.4.1–2.4.2), 7
1.4 Inertia (2.5–2.11), 10
1.5 Momentum and Position (2.5), 10
1.6 Detection and Transcription (2.6), 11
1.7 Position, 'Red Shift', Nearness, and Gravity (2.7–2.8), 11
1.8 Discreteness of Mass, Charge, and Statistics (2.9–2.11)B_0 and 'Elementary Particles' (2.9) 15
1.8.1 *Fast* B_0s (2.10), 19
1.8.2 *Information and Inertia* (2.11) *and Compatibility with Prior Knowledge*, 21
1.9 Axiomatization (Section 3), 22
1.10 Looking For Postulates (3.1), 22
1.11 Postulates (3.2), 22

1.12 Testing Postulates (3.3), 23
1.13 Describing Communication (3.3.1), 23
1.14 3-Space and Apparently Unidirectional Time (3.3.2), 24
1.15 Separability of Internal and External Information (3.3.3), 25
1.16 Spectrum of *w* States in *U* (3.3.4), 27
1.17 Describing Information Representation by *ws* (3.3.5), 27
1.18 Adseparability for Systems (3.3.6), 28
1.19 Coding (3.3.7), 29
1.20 Quantized Electromagnetic Theory and Relativistic Quantum Mechanics (3.3.8), 30
1.21 Sufficiency of Postulates (3.3.9), 30
1.22 Axiomatic Form (3.4), 30
1.23 Partial Test of Axiomatic Form (3.5), 31
1.24 Comments (Section 4), 33
1.25 Some Predictions and Experiments which Might Be Interesting (Section 5), 33
1.26 Questions (Section 6), and Appendix 1, 33
1.27 Lists (Appendix 2) and Map (Appendix 3), 33

2. COMPATIBILITY WITH PRIOR KNOWLEDGE **35**

2.1 Information Representations, 35
2.2 Motion of a Material Particle, 39
2.3 Information Conservation, 39
2.4 Separability of Internal and External Information, 52
  2.4.1 *Special Relativistic Covariance of Theory*, 71
  2.4.2 *A Limitation on Applicability of PODIC*, 81
2.5 Momentum, 96
2.6 Detection and Transcription, 99
2.7 Position, 100
  2.7.1 *Cosmological Red Shift*, 102
2.8 Gravity, 105
  2.8.1 *Nearness*, 111
  2.8.2 *Gravity*, 112
2.9 B_0 and 'Elementary Particles', 124
  Table 1. *Comparisons of F and* α^{-1} *for Various Sets of Recommended Values*, 150
  Table 2. *For Our Universe U as 'Box' B, some collected results*, 155

2.10 Fast B_0s, 156
2.11 Information and Inertia, 172
2.12 Compatibility with Prior Knowledge, 173

3. AXIOMATIZATION **174**

3.1 Looking for Postulates, 174
3.2 Postulates, 179
3.3 Testing Postulates, 179
- 3.3.1 *Describing Communication*, 179
- 3.3.2 3-*Space and Apparently Unidirectional Time*, 181
- 3.3.3 *Separability of Internal and External Information*, 188
- 3.3.4 *w Spectrum in U*, 206
- 3.3.5 *Describing Information Representation by ws*, 207
- 3.3.6 *Adseparability for Systems*, 217
- 3.3.7 *Coding*, 226
- 3.3.8 *Quantized Electromagnetic Theory and Relativistic Quantum Mechanics*, 227
- 3.3.9 *Sufficiency of Postulates*, 229

3.4 Axiomatic Form, 230
3.5 Partial Test of Axiomatic Form, 232
- 3.5.1 *Photon*, 235
- 3.5.2 *Electron Neutrino and Antineutrino*, 236
- 3.5.3 *Electron and Positron*, 238
- 3.5.4 *Muons*, 240
- 3.5.5 *Muon Neutrino and Antineutrino*, 244
- 3.5.6 *Pions*, 246
- 3.5.7 L17.10 *Interaction*, 258
- 3.5.8 *Comparisons with Reports of Observations*, 265
  - Table 3. *Some Comparisons with Reports of Observations of Various Particle Properties*, 268
  - 3.5.8.1 Table 3, *Items* 13–20, *and* 22 (*Pions*), 267
  - 3.5.8.2 Table 3, *Items* 9–12, *and* 21 (*Muons*), 288
  - 3.5.8.3 Table 3, *Items* 7 *and* 8 (*Muon Neutrino and Antineutrino*), 290
  - 3.5.8.4 Table 3, *Items* 4–6 (*Positron, Electron, and* α^{-1}), 290
  - 3.5.8.5 Table 3, *Items* 2 *and* 3 (*Electron Neutrino and Antineutrino*), *and 'Weak Interaction'*, 292

3.5.8.6 Table 3, *Item* 1 (*Photon*), 294
3.5.8.7 Table 3, *Columns* 5 *and* 6, *and 'Lepton Number'*, 294
3.5.8.8 Table 3, *Columns* 11–14, *and 'Strong Interaction'*, 294
3.5.8.9 Table 3, *Columns* 15 *and* 16, 296
3.5.9 *Apparent Correspondence*, 296
3.6 Break of Development, 297

4. COMMENTS **298**

4.1 Relativistic and Nonrelativistic Theories, 298
4.2 Treatment of Gravity, 303
4.3 Inertia, 304
4.4 Reality, 304
4.5 Introspection, 305
4.6 Free Will, 306
4.7 Planck's Constant and Photons, 307
4.8 Representing Measurements, 309
4.9 Finite Accessibility of Information, 312
4.10 Tunneling and Virtual States, 312
4.11 Hubble's Constant, 313
4.12 Phononic and Photonic Systems, 313
4.13 Postulates and Constants, 314
4.14 'Why Do Things Fall?', 315

5. SOME PREDICTIONS AND EXPERIMENTS WHICH MIGHT BE INTERESTING **318**

5.1 Predictions, 318
5.2 Experiments, 321
Table 6. *Some Experiments which Might Be Interesting*, 322

6. QUESTIONS **326**

7. CONCLUSION **349**

APPENDIX 1: CONJECTURES ON FAST B_0s IN GRAVITY **350**

APPENDIX 2: REFERENCE LISTS 368

Appearing in Section 1, 368
Appearing in Section 2, 369
Conditions, Lemmas, Theorems, and Corollaries, 369
Definitions, 374
Principles, 375
Appearing in Section 3, 375
Postulates, 375
'A-' Prefix Conditions, Lemmas, Theorems, and Corollaries, 376
A-Definitions, 377
Chain 17, Links, and Definitions 21 and 22, 378
Abbreviations, 379

APPENDIX 3: MAP 382

REFERENCES AND NOTE 384

INDEX 387

Section One

Informal Introduction

This introduction is intended to state conventions for, and as an informal summary of, the much longer main text, in which I present derivations of results presented here. Parenthetical references in this Section 1 indicate where in the main text various derivations appear, and parenthetical numbers after subheadings are used to reference headings in the main text discussed. This summary omits many detailed results presented in the main text.

This report includes material written over a period of several years, much of it originally for my own use in exploring various of the ideas presented. I find that how I say things, and what I choose to say and/or emphasize, have changed during that period, and have been in the more recent part of that period more nearly as presented in Section 3 hereof. At least partly to offer this report of ongoing inquiry now, I offer it (including this Section 1) as not of uniform kind and/or quality; and caution that it might contain typographical, cross reference, and/or other error(s). Please look to the substance offered, and excuse awkwardness, sloppiness, and/or error, herein. And, please note the warning in the first paragraph of the Preface.

For this presentation (not just this Section 1), the following interpretation applies:

Every statement herein referring, and/or implying reference, to prior (to the time of this writing) and/or present knowledge and/or theory(s) shall be construed as referring to only that part thereof which I have (with both knowledge and awareness) correctly used in such statement.

In Section 2, Compatibility with Prior Knowledge, the first part of the main text, compatibility of this approach with prior knowledge is examined by deriving it from a part of prior theory. In Section 3, Axiomatization, postulates are used to starkly display the nature of the physical picture presented, and sufficiency of such postulates is tested.

This work took me a long time to do, so don't be surprised if you find it slow reading. Especially, you may wish to keep track of those parts of generally accepted physical theory which are *not* used in this derivation.

Underlying all of the following is a ('locally') new approach to physics, in which concepts related to information provide a basis for new physical insights. By information, I mean specification of one out of more than one possibilities. The amount of information, in 'bits', is defined as $\log_2$ of the total number of alternatives from which one is designated (Def 1). For convenience, 'bit' is occasionally used herein as meaning 'bit of information'. 1 of 4 is 2 bits, 1 of 1024 is 10 bits, etc. I do *not* mean some thermodynamic concept related to entropy. Many concepts for entropy are based on statistical arguments which do not apply to finite closed systems, and many beliefs about information derived from such concepts of entropy have seemed to me either erroneous and misleading, or merely uninteresting. As an example of this difference, the amount of information represented by a system does not depend on which of its possible states it is in, because the number of alternatives is the same, but the entropy does.

Information appears encoded in a representation, and in receiving and transcribing such a 'signal' one needs other suitably represented information as a 'code book' for the old and new representations. For suitable amounts of information in the signal and 'code book', and suitable codings, each can be used as a 'code book' to decode the other, so that all information in both would be available for encoding in other representation with signal and 'code book'. Such a signal and 'code book' are said to be 'complementary' information representations, and their codes are also said to be 'complementary' (Def 2). Together, a pair of complementary information representations may be regarded as 'pure' information, fully available for encoding for representation in other ways.

By way of getting into the spirit of this approach, consider the following puzzle. In both classical and quantum mechanics, an object's position is described by three real numbers. A real number is one of an infinite number of alternatives, and so represents infinite information. Presumably, when an object is moved from one place to another, its position information goes with it. In quantum mechanics, all detectors have quantum 'uncertainty' noise. From communication theory, transporting infinite information in finite time in the presence of such noise would require infinite energy. So, if an object's position really required three real numbers to describe it, the object would be immovable.

A wrong comment which I have occasionally received is that one might define a real number by a rule, such as 'divide the circumference of a circle by its diameter', which does not require infinitely many letters to write, and so a real number might represent finite information. However, the number of different sentences which can be written depends on the number of symbols in the alphabet and the number used in the sentence. Thus, each sentence is like a number with a specific number of digits. The alternatives include longer numbers, and there are infinitely many for the set of real numbers, so by designation of one from that set, by any means, infinite information would be represented.

By this simple argument about the position information of an object, one can see that prior theory of nonrelativistic quantum mechanics ('NRQM') is internally inconsistent. For the nonaxiomatic development of this theory, I have used part of NRQM which is internally consistent and consistent with subsequent extension by information arguments.

1.1 Conventions for Order of Mathematical Operations and Labeling

For definiteness, absent further parenthesization, operations in mathematical expressions and within parenthetical expressions and exponents are performed in the following order: exponentiation; multiplication; division; addition or subtraction. For example, $a/bc = a/(bc)$, not $(a/b)c$.

Formal statements herein are in context of all logically preceding formal statements herein. Except as noted, every formal reference to a numbered condition (C(*integer*)) refers to the last logically preceding condition so numbered.

Prefix 'A-' designates definitions, lemma, theorems, and corollaries appearing in the axiomatic treatment in Section 3.

Decimal points in numeric or alphameric character strings as labels of subsections, lemmas, corollaries, and links are treated as separating numeric strings treated as separate integers, ordered lexicographically with precedence from left to right. For example, Corollary 6.2 precedes Corollary 8.1, Corollary 8.9 precedes Corollary 8.16.

Quotation marks are used herein as symbol (string) delimiters variously for referencing symbol strings (as in naming and/or defining), to designate special use(s) and/or meaning(s), and/or to express what might be approximately stated in words as 'so called'. All quotation marks herein, whether single (' ') or double (" ") are used to 'raise' the enclosed symbol(s) one such level relative to surrounding text. For this reason, most quotation marks herein are single; double marks are used occasionally, for example, for convenience in distinguishing nested ranges, and/or reducing likelihood of confusion with apostrophe(s) and/or prime(s) in and/or at end(s) of range(s).

Square brackets '[ ]' are used herein to enclose portions, of text and/or expressions, relating to spectra other than $E_j = jE_1$ derived in Section 3 hereof.

'It', 'its', 'who', and/or 'which' may be used generically to refer to (an) observer(s).

1.2 Compatability With Prior Knowledge (Section 2) (2.1–2.4)

Information exists only in so far as it is represented. In NRQM, a finite closed system (in a 'box' B) has a number of (discrete) states. Therefore, specifying any state represents an amount of information, I. The amount of information in a closed finite system is a conserved quantity (Theorem 1 and Cor 1.1). Noting that without using another 'code book' one cannot distinguish between complementary information representations, that is, that there is not any information represented by distinction therebetween, from information conservation and such symmetry we find that two such complementary information representations represent equal amounts of information (Cor 1.2).

Consider information I represented by photons in B, with total energy approximately E. Such representation of I corresponds to a set of numbers giving the number of photons in each state, without phase (Theorem 2). Next, suppose there were a smaller box B_1 in larger box B_2. In B_1 is placed internal information I_1, represented by photons in photon states in B_1 (Def 3). B_1 is in some NRQM state in B_2. B_2 contains total energy E_2 and information I_2. If B_1 were suddenly opened in B_2, I_2 would still be conserved (Theorem 3). Here is a place where sloppy thinking in terms of entropy might lead to confusion. Thus, I represented by photons in box B

depends only on energy E in B and B's geometry, not on details of photon arrangements. One can think of $I_2 - I_1$ as information represented by B_1 in B_2, which I called external information I_1' of B_1 in B_2 (Def 4).

This sets the stage for the first deep result of this approach. The question is, can one extend these information 'bookkeeping' concepts to include mass particles? Here you, the reader, get to read a result in *much* less time than it took me to do it.

The key here was to realize that I should construct one configuration of photons which represented the same information as a mass particle; Theorems 1–3 showed that one would be as good as any other. I did this as presented in Lemma 4.1, Theorem 4, and Corollaries 4.1–4.3, in the following general way:

Let B_0 denote a box which itself has no mass, containing when substantially at rest two photons with equal and opposite momenta in its lowest state, with total rest contained photon energy E_0, with nonzero nonpropagating field outside B_0 (Def 5 and 6). Put B_0 in much larger box B_2. Require that B_0's two photons be in photon states of B_2. With B_0 moving with velocity v in B_2, B_0's electromagnetic energy may be decomposed into forward and backward components, (Doppler) frequency shifted up and down respectively. For both up- and down-shifted components to satisfy electromagnetic boundary conditions at B_2's walls, both must be able to be in phase there. This is equivalent to requiring that the spatial period λ_b of their difference (beat) frequency fit into B_2 an integer number of times.

Calculating λ_b in the nonrelativistic approximation ('NRA'), $\lambda_b = c^2h/E_0v$, where 'c' denotes the speed of light and 'h' denotes Planck's constant (Lemma 4.1). Note the $1/v$ dependence. This $1/v$ dependence matches the $1/v$ dependence of the de Broglie wavelength λ_P of a mass particle P, with mass m and velocity v, $\lambda_P = h/mv$ (NRA). This means that for one value of E_0, electromagnetic theory ('EMT') allowed states of B_0 in B_2 can be put in one to one correspondence with, and interchanged with, the allowed states of P in B_2. In this way, putting P in one of its allowed states in B_2 is just as effective for representing information as putting B_0 in one of its allowed states in B_2. Thus, B_0 and P represent the same information in B_2. From Theorems 1–3, B_0's photon energy E_0 would then be the same as would be required for any representation of that information by photons in B_2. Thus, I found that information is conserved when photon and mass particle representations are interchanged if and only if ('iff') mass m is interchanged with total electromagnetic energy mc^2 (Theorem 4, Cor 4.1–4.3). Note especially that this result is obtained nonrelativistically, by an information argument. This begins to give a glimmer of why special relativity ('SR') works. I obtain a foundation for all of SR as an example later.

With this, a nonrelativistic photon particle representation ('NRPPR') can be used to replace photon energy with mass, or vice versa, conserving information (Def 7). Thus, the information bookkeeping is extended to include mass particles. The information content of a box then depends on just its geometry and total contained energy, counting total mass m as mc^2 (Theorem 5). This result is then extended by enlarging the box to be our universe (Cor 5.2, 5.3).

For this link to mass particles, I resorted to proof by construction, which depends on finding one way. Deduction is often much easier, and most of this argument is deductive. I would like to emphasize that constructions are often relatively less interesting without a strong context, such as uniqueness proofs. For example, Lemma 4.1 would be much less interesting without Theorems 1–4 and Corollaries 4.1–4.3. Usually, I have not constructed detailed descriptions of information representations except where necessary to proceed.

Next, I showed that information and energy are substantially proportional (exact, in 'infinite enclosure', and in axiomatic form), by a deductive argument: given two total energies, each respectively may be divided into n_1 or n_2 times a smaller energy which divides both. This smaller energy can be used as a B_0 or particle P, to represent information I_P. Each larger energy is then n_1 or n_2 substantially noninteracting Ps. From NRQM, each large system's total state is then a product state of its separate Ps. The log of the number of product states equals the sum of the logs of the numbers of states in the factors, which are all identical. So, from the logarithmic measure of information, $I_1 = n_1 I_P$ and $I_2 = n_2 I_P$. Thus, $I_1/I_2 = E_1/E_2$, and information is proportional to energy (Theorem 6). One consequence is that conservation of energy and conservation of information are equivalent (Cor 6.1–6.3); another is that amplitudes A_j of co-possible modes j of representing information, where a representation in mode j represents minimum information I_j, are proportional to $1/I_j^{1/2}$ (Cor 6.4).

The next question I treated was extension of information bookkeeping to include moving systems and changes in system size (Theorem 7–Cor 8.4). The key to this is to note that, in NRQM, a small perturbation induces transitions between states at a rate proportional to the *square* of the perturbation. In reaching velocity v in time t, a system would experience a perturbing acceleration v/t, inducing transitions at a rate proportional to $1/t^2$, for a total transition probability proportional to $1/t$. This result also holds for sums over finitely many intermediate states. I am treating a finite system. (This is discussed in more detail in Comment 4.1, pp. 298–303.) Thus, I found that in an infinite enclosure a finite box containing a system could gently be given a velocity v ('gentle acceleration', 'GA', Def 10) without affecting its internal state. Thus, any information represented by states of that system would be preserved as to both amount and form. Similarly, the size of a box can be changed slowly, gently, and proportionately ('gentle scaling', 'GS', Def 11) without affecting its internal state or internal information. Note that this is not at all like classical mechanical behavior: this result depends on 'graininess' of finite quantum mechanical systems.

By shifting to a slowly moving system in NRA, we can examine a somewhat faster moving system, and so to higher velocity, and show that in an infinite enclosure, for any $v < c$, the accumulated changes in state go to zero for long acceleration times. However, the only way information is represented is as an internal or external state of a system. Thus, two important results are obtained:

1. Information is proportional to energy (exact in 'infinite enclosure' and in axiomatic form).
2. For any closed system containing photons and/or mass particles and moving uniformly with velocity $< c$ in an infinite enclosure, internal and external information are separately conserved and have independent representations. I called this the *principle of detailed information conservation* (*PODIC*).

To really see that all this works, I divided this part of the argument into many small steps, and examined each approximation and limit. I do not claim that the result is an elegant derivation, but it seems right (Theorem 7, Cor 7.1, Lemmas 8.1–8.11, Theorem 8, Cor 8.1–8.4).

1.3 Special Relativity and a Limitation on its Domain of Applicability (2.4.1–2.4.2)

Why does special relativity work? Consider a closed system B in uniform motion with velocity v in an infinite enclosure, containing photons and/or mass particles. Its velocity is a state descriptor in an external system, and thus is, or represents, external information. From PODIC, internal information is separate from external information. Thus, an observer in B does not have access to external information represented by v. Therefore, experimental results in B do not depend on $v < c$, including measurements of the speed of light, for all experiments using photons and/or mass particles. Thus, all laws and constants of physics pertaining thereto, including the speed of light, are independent of uniform motion of B with velocity $v < c$. Thus, taking NRQM as a starting theory, one may regard special relativity as working because of the graininess of quantum mechanics.

Quite generally, a finite observer can have its external state varied by a process without affecting its internal state if and only if all laws and constants of physics as seen by it are not changed by that process (Principle 3). For photons and/or B_0s (with photon particle representation, mass particles), this follows from PODIC, and includes translational invariance and isotropy.

As an exercise, one can show special relativistically that B_0's states correspond to quantum mechanical ('QM') mass particle states (Theorem 4'). Some results for B_0s are collected in Corollaries 4.4 and 4.5. This theory is special relativistically covariant within SR's domain of applicability, and includes SR as a special case.

Incorporating PODIC by using computational methods of SR, by examining properties of moving B_0s, I found a limitation on the domain of applicability of PODIC (Cor 8.5–8.16):

As v increases, B_0's associated down-shifted photon goes into lower states in B. When B_0's down-shifted photon is in a (compared to B_0's 'rest frame' energy E_0) relatively low state in B, I call B_0 a 'fast B_0' (Def 13).

I solved for E_0 of B_0 moving with velocity v, with down-shifted photon having energy E_{Bj}, the jth photon energy level in B, as $E_0(v) = 2E_{Bj}(1+v/c)/\sqrt{1-v^2/c^2}$ (Cor. 8.6). *If this formal method be applicable*, the fastest a B_0 can go is determined by the lowest energy for which its down-shifted photon can satisfy electromagnetic boundary conditions in B, E_{B1} (Cor 8.7). *(See main text for cautionary note.)* So, I called a B_0 a 'fastest B_0' or 'fB_0' if its associated down-shifted photon has the lowest allowed energy in B (Def 14).

I solved for energy E transported by B_0 as seen by an observer substantially at rest in B, and found $E(v) = 2E_{Bj}(1/(1-v/c))$ (Cor 8.8). This I then solved for B_0's velocity as a function of energy transported, and found $v(E) = c(1-2E_{Bj}/E)$ (Cor 8.9). A fast B_0 moves with velocity close to c, and in any frame of reference moving with velocity $v < c(1-\epsilon)$, $\epsilon \gg 2E_{Bj}/E$, a fast B_0 would appear to have velocity $\approx c$, and its momentum would appear to transform substantially as would a photon's (Cor 8.10).

For a fast B_0, rest frame energy, transported energy, and lower photon energy E_{Bj} are approximately related by $E_0 \approx 2\sqrt{EE_{Bj}}$ (Cor 8.11).

Fast B_0s have such large time dilations that even if a fast B_0 only lasted for one period of its associated photons in its rest frame, its lifetime $\tau \approx h/2E_{Bj}$. For fastest B_0s, $\tau \approx h/2E_{B1} \sim \pi R_B/c$ where R_B denotes a 'radius' for 'box' B (Cor 8.12). In our universe, that is not a very short time. An fB_0 goes so fast that in such lifetime it would lag behind a photon by only about one wavelength of a photon carrying the same energy (Cor 8.13).

Next, I examined a fast B_0's allowed energies (momenta). When it is being formed, a fast B_0 can be given any energy which can be expressed as a sum of suitable photon energy levels in B, so, for $E \gg E_{B1}$, the difference between two adjacent allowed energy levels of B_0 is a very small fraction of either level (Cor 8.14).

However, changing the energy of an already formed fast B_0 without changing its rest frame energy E_0 involves shifting B_0's associated down-shifted photon, with energy E_{Bj}, among relatively low photon energy levels j in B. These are relatively coarsely spaced. The coarsest case is for restriction to a single direction, for which $E_{Bj} = jE_{B1}$.

In the axiomatic form, I found (years later) formally that for photon state energy levels in (a) universe U $E_j = jE_1$, j finite integer >0, where E_1 is least photon energy >0 possible in U (A-Theorems 3 and 4, A-Def 22). For explication, I have retained, in square brackets, steps and parts of expressions relating to spectra other than $E_j = jE_1$.

The finest case that I found for this from any prior discrete theory appeared to be that found by taking the counting of photon states used in quantum mechanics ('QM') to calculate 'black body' radiation, and assuming that every state has different energy. Thus, the fractional energy 'graininess' of B_0's successive allowed energy levels E and $E' < E$, is approximately given [bounded] by

$$\left[\frac{1}{\dfrac{4}{3\pi}\left(\dfrac{E_{Bj}}{E_{B1}}\right)^3 + 1} \lesssim \right] \delta \underset{[\lesssim]}{\approx} \frac{1}{\dfrac{E_{Bj}}{E_{B1}} + 1} = \frac{1}{j+1} \qquad \text{(Cor 8.15).}$$

Thus, an already formed fast B_0 has much grainier allowed energies than are available for formation of a fast B_0 with $E \gg E_{B1}$.

If a moving system B_3 contained a B_0 which was moving so fast that its energy states in B were (observably) coarser than the lowest photon energy in B_3, an observer in B_3 might see peculiar effects. Thus, separability of internal and external information, and therefore PODIC, holds only for B_3's velocity $v < \approx c(1-(R_3/R_B)^2)/(1+(R_3/R_B)^2)$ and PODIC fails by loss of separability for $v \geqslant \approx c(1-(R_3/R_B)^2)/(1+(R_3/R_B)^2)$ (Cor 8.16). SR depends on separability of internal and external information, and so fails when PODIC fails. This defines a limitation on the domain of applicability of special relativity. This failure would occur at a velocity so great that an observer moving with B_3 much smaller than B would see B (e.g., 'a' universe) as shrunk parallel to B_3's velocity to about twice the length of B_3.

With these tools, I examined several interesting questions, which had not seemed ('locally', to me) accessible by prior theory.

1.4 Inertia (2.5–2.11)

What is inertia? This question might be divisible into four parts: position; momentum; equivalence of inertial and gravitational mass; and discreteness of mass, charge, and statistics.

1.5 Momentum and Position (2.5)

Why does a moving object transport energy? Consider a box B_2 containing information I_2 represented by photons. Detecting each photon and unambiguously assigning it to its correct photon state (unambiguous detection and assignment, 'UDA') would extract I_2. One might do this by placing a perfect QM detector D contiguous with B_2, and waiting forever. Then, all photons in B_2 would be detected, with no energy uncertainty. To detect I_2 in finite time, D and B_2 must pass through each other, so that all of B_2 is swept clear of electromagnetic energy. However, finite detection time means that D has nonzero energy uncertainty. This entire system of D and B_2 may be regarded as contained in a large closed box B, within which information is conserved. For D and B_2 to pass through each other, they must have nonzero relative velocity v. D sees B_2 approach with velocity v, and so sees all photon states in B_2 as raised and spread apart in energy. Information is not lost, so this spreading must be at least enough to allow UDA. If the spreading were more than enough for UDA, the time for B_2 to pass through D could be divided into shorter times without energy ambiguity, allowing both energy and phase to be detected. But, that would create information. So, from information conservation (in B), photon energy levels in B_2 are raised and spread apart by relative velocity v as seen by D exactly enough for unambiguous detection of I_2 (Theorem 9). However, we can replace the photon representation of I_2 with a mass particle representation in B_2. Thus, the energy carried by any moving system may be fully attributed to its transport of information in a 'quantum mechanical' universe (Cor 9.1).

Note that this is a special relativistically covariant theory. The three spatial components proportional to information flow have the form $m_{\text{rel}}c^2\vec{v}$, and the fourth component $(d/dt)(Ect) = Ec$; representing, respectively, spatial and temporal flow of energy. Note that the 4-vector formed is c^2 times the 4-momentum. In particular, rest mass energy of an object represents propagation of information from the past to the future, for example, by presence of a localized object at one place out of all the places and/or combinations of places in which it might have been and/or among which it might have been divided. This can be thought of as defining the concept of *position*.

In this picture, kinetic energy of inertia appears as a 'consequence' of what is usually thought of as a local phenomenon, the 'graininess' which is approximately partially described by Heisenberg's noncommutativity (uncertainty) relations in QM.

1.6 Detection and Transcription (2.6)

When information is detected, it must be transcribed into some storage system. Two kinds of transcription seemed historically interesting. One of these involves recoding information in highly redundant form. For example, a small detector might have a large transcription box: a photon in a small box entering the detector could then be transcribed as a large number of lower energy photons in the corresponding photon state of the large box. The contents of this large box could then be examined many times with approximately 'negligible' disturbance. This redundant transcription appears to correspond to the classical concept of detection, and I called it 'classical transcription' or 'CT' (Def 15).

Another type of transcription involves creating a box of the same size as the detected box, emitted from the transcriber as the detected box enters the detector, into which entering information is transcribed with the same energy and redundancy of coding. I called this 'quantum mechanical transcription' or 'QMT' (Def 16). In particular, any 3-surface in 4-space may be thought of as such a system of detector and transcriber, with traversal of the surface regarded as a QMT process.

1.7 Position, 'Red Shift', Nearness, and Gravity (2.7–2.8)

What is gravity, and why are gravitational and inertial mass equivalent? A gross feature of gravity is that bringing two masses closer together releases energy. This gross feature appears understandable in terms of information. Each object may be regarded as a relatively localized time-like flow of information in a position representation, that is, in which information is represented by the object being in one of all the distinguishable places and/or combinations of places in which it might have been and/or among which it might have been divided. Two independent objects represent a total amount of information equal to the sum of their separate amounts of information. However, if two objects are required to be within some distance of each other, the number of possible configurations is reduced, so their being in any one such configuration represents less information. Some portion of their total information has been subsumed by nearness to each other. Energy is proportional to information, so a portion of their total mass energy is released in bringing them closer together.

How can one understand quantitatively this effect? If box B_1 initially fills box B_2, of linear size L_2, and is then gently and proportionately shrunk to linear size L_1 by factor L_1/L_2, from PODIC, amount I_1 of B_1's internal information is unchanged. However, if I_1 were represented by photons in B_1, each photon would have L_2/L_1 times as much energy as before, so B_2 would contain amount $(L_2/L_1)I_1$ of information, of which amount I_1' of B_1's external information in B_2 would be $I_1' = I_1(L_2/L_1 - 1)$ (Theorem 10). Thus, localization can be defined just in terms of information, as the ratio of total information to internal information (Def 17). In particular, detecting external information of box B without detecting its internal information reduces B's localization (Cor 10.1).

By examining conversion of representation of information from position to momentum representation, I found that in an enclosure of linear size R_B, light would appear red shifted with distance d of uninterrupted ('free') propagation, expressable, for $d \ll R_B$, in terms of a fictitious velocity of recession $v \approx Hd$, where $H \approx c/R_B$ (Cor 10.2, 10.3). Such relation appears, to the extent tested as presented herein before Correspondence Corollary 10.4, allowed by observation prior to the time of this writing, with H corresponding to local Hubble's constant (Correspondence Cor 10.4).

From PODIC, gently scaling (GS, Def 11) B's size by factor q does not affect its internal information. Thus, all laws and constants of physics seem unchanged to an observer in B, independent of B's localization (Theorem 11, Cor 11.1). This requires that units of measurement in B change. In particular, units of mass M, length L, and time T used by an observer in B appear to an unchanged observer outside B as M/q, Lq, and Tq (Cor 11.2).

A principle of multiobserver physics is that, if an observer can not have internal history that tells it its unit of length is changed, we honor its report(s) of length measurement(s) (Principle 4). In some sense, length comparison is different from time comparison: length is measured locally, and calibration involves transporting a standard. Time standards can be compared by transmitting pulses, and the steady state number arriving equals the number transmitted, per unit time.

With this in mind, I looked at properties of quantum mechanical transcription (QMT) among regions S_i which had experienced GS by different factors q_i relative to a common initial state. This process corresponds to propagation of a test object. In QMT, a test box B_1 is transcribed into a new box of the same size, defined locally. Consider B_1 containing I_1 in a photon representation. In moving slowly from S_1 with q_1 to S_2 with q_2, B_1 is transcribed into a box of linear size q_2/q_1 times its original size. Each corresponding photon energy level is changed. Work is done, in an amount $W_1 = E_1(q_1/q_2 - 1)$, where W_1 and B_1's internal energy E_1 are measured by an observer O_1 in S_1 (Cor 11.3). If B_1 moves freely (energy conserved) and has enough energy to enter S_2, then its new velocity turns out to be just such that its new Lorentz contracted length (including q effects) exactly equals its previous Lorentz contracted length, both as seen by an undistorted (outside) observer O_0 (Cor 11.4, 11.5). In an assembly of contiguous regions with slowly spatially varying q, O_0 would see freely moving B_1 as varying in width as $1/q$, with constant length parallel to its velocity. Local observers would report B_1 as having constant width, with its length parallel to its velocity varying as q. O_0 would see B_1 describe a trajectory, with varying velocity.

If the contiguous regions have suitable qs, relations among neighbors can be put together to construct a differential manifold, with metric distortions. Such an assembly of regions might be said to contain a gravitational field.

So, next I looked at the question of how much external information is represented by nearness to a mass M, and from this approximately calculated local metric effects of M. In the weak field case, the resulting metric near M seems to be substantially the same as in Einstein's general theory of relativity. However, the gravitational constant appears as a derived quantity, which does not disagree with the measured value. The strong field case seems qualitatively different.

Here is how I proceeded: Consider a large box B with linear size R_B, containing total information I_B, including a substantially isolated time like information flow I_M in a position representation, which serves to provide the 'gravitational field', and a much smaller box B_1 in B, with linear size $R_1 \ll R_B$, internal information I_1, external information I_1' in a position representation in B, and total information $I_T = I_1 + I_1'$. How much information would B_1 give up by being within distance r, $R_B \gg r \gg R_1$, of I_M? Suppose B_1 were in a box B_3 with linear size r in B. If we were not able to know where B_3 was in B, we would not have reduced the freedom of B_1, so we would not have detected any of its information. The position of B_3 is specified by position information, and is known only that well. Next, suppose B were free in much larger box B'. B_1 has position information only relative to B in B'. I_M contributes a fraction I_M/I_B of the position information of B. Returning to B_3, if we had all of I_B available to locate B_3, we could detect all of B_1's external information in B that was outside

B_3. If we only have I_M to locate B_3, we detect I_M/I_B of that part of B_1's external information outside B_3. This detected information is very nearly $\Delta I \approx I_T R_B I_M / I_B r$, neglecting I_T/I_M in comparison to 1 (Theorem 12).

Energy, mass, and information are proportional. So, in a box B with linear size R_B, containing total mass energy M_B, including isolated mass $M \ll M_B$, we find $\Delta I \approx (R_B M / M_B r) I_T$ (Cor 12.1). All such detected information would come from B_1's external information, without changing B_1's internal information. Thus, near M, B_1's localization would be changed as by GS by factor $q(r) \approx 1 + R_B M / M_B r$ compared to r very large (Cor 12.2).

With such approximate description for $q(r)$ about M, we can explore effects of M on a small length standard. From the discussion of QMT, B_1 moving freely radially toward M would be seen by successive observers as having its nominal width (Lemma 13.1). In the weak field case, B_1 would be seen by local observers as shortened by factor $1/q(r) \approx 1 - R_B M / M_B r$ (Lemma 13.2). Also in the weak field case, clocks near M would appear to a remote observer to run slower by factor $q(r) \approx 1 + R_B M / M_B r$ (Lemma 13.3). And, in the weak field case, test mass m_1 slowly displaced toward M would release energy

$$\Delta E \approx \left(\frac{m_1 c^2 R_B M}{M_B} \right) \left(\frac{1}{r_2} - \frac{1}{r_1} \right)$$

in approaching M from distance r_1 to r_2 (Lemma 13.4). Collecting these results, we see that, in the weak field case, M would appear to have a gravitational field, and in the weak field limit the 'gravitational constant' $G \approx c^2 R_B / M_B$ (Theorem 13).

The curvature of space about a massive object thus appears as a consequence of subsuming position information by nearness, not as a separate postulate. This also shows any form of mass or energy as being by its position information equally effective as a gravitational field source, and affected equally by gravity, from the proportionality of mass, energy, and information.

To examine the strong field case, consider two equal masses M and $M' \ll M_B$, separated by distance r as observed by a hypothetical observer located at, but not affected by, M. Solving for $q(r)$, I found

$$q(r) \approx \frac{1 \pm \left(1 + 4 \dfrac{R_B M}{M_B r} \right)^{1/2}}{2} \qquad \text{(Cor 13.1).}$$

Thus, if M and M' started far apart ($r \sim R_B$), the total mass energy released, as observed by a remote observer, in proximating M and M' to separation r would be approximately describable by

$$M_r(r) \approx 2M\left(1 - \frac{2}{1+\left(1+4\dfrac{R_B M}{M_B r}\right)^{1/2}}\right) \qquad \text{(Cor 13.2).}$$

Note that there is not any case in which gravity would prevent photons from escaping. Thus, gravity does not produce any region(s) which cannot exchange information with the rest of B: information is conserved everywhere in B (Cor 13.3). Thus, there is not any 'black hole'. With this qualitative behavior in the strong field case, one might imagine stable, massive, luminous objects, with a substantial part of their red shifts due to gravitational ($q(r)$) effects.

Theorem 13 $q(r)$ effect due to M at time t if observable as locally propagated to distance $\approx r$ from M is there so observable at $t' \approx t + r/c$ (Cor 13.4).

Note that the 'gravitational constant' in Theorem 13 is fixed by R_B, M_B, and c. This offers a way to compare this picture with observation.

With Corollary 10.3 and Theorem 13, the mean density ρ of mass (energy) in our universe can be approximately expressed in terms of the 'gravitational constant' G and Hubble's constant H, by

$$\rho \approx \frac{3H^2}{4\pi G}.$$

A. R. Sandage reported[2] H as $\sim 15 < H < 40$ km/sec megalightyear, or $H \sim (1.587 \text{ to } 4.232) \times 10^{-18}$/sec. This range for H yields an approximate calculated density range of 9.012×10^{-27} to 6.408×10^{-26} kg/m^3. The lower value corresponds to about 5.055×10^9 eV per cubic meter. This range appears roughly one to two orders of magnitude higher than various previously reported values, of which I am at the time of this writing aware, for the density of luminous matter in our universe, and substantially allowed by such knowledge prior to the time of this writing[3] (as in Correspondence Cor 13.5).

1.8 Discreteness of Mass, Charge, and Statistics (2.9–2.11)

B_0 and 'Elementary Particles' (2.9)

Generally, I approached this question by examining the properties of a B_0 in which the two photons were gravitationally coupled. More specifically, I looked first at the nature of B_0 with any form of coupling: for two photons to be observably coupled in B, their interaction must release at least enough energy to represent 1 bit of information in B, which is just the energy E_{B1} required to put one photon in the lowest photon mode in B, for example, in our (finite) universe. So, for every form of photon interaction, there is a lower bound to the energy which can participate in forming a coupled B_0, just enough to release E_{B1} by that interaction (Theorem 14).

Viewing B_0 as two coupled harmonic oscillators, E_{B1} also equals the normal mode splitting. Thus, B_0's electromagnetic energy must be decomposable into two photons separated by energy $\geqslant E_{B1}$, as another representation of the same amount of information. B_0's minimum velocity thus appears to correspond to one difference frequency wavelength fitting into B: B_0's lowest state has $v>0$, like a corresponding QM mass particle (Cor 14.1).

Suppose that a few (N) bits of information can be represented by different relations among B_0's coupled photons. From effects of phase coherence and superposition for this alternative representation of information represented by such stable entity, the energy required to represent these few bits is approximately that energy required to put one photon in each of the lowest N photon states in B.

To form such a B_0, the interaction must give up at least this much energy (Cor 14.2). Let $Y(N)=\Sigma_{i=1}^{N}E_{Bi}/E_{B1}$, where the E_{Bi} are the lowest N photon states' energies in B, of which E_{B1} is the lowest (Def 18). (Note that more than one photon state might have the same energy.)

The only energy releasing interaction which I had studied thus far in my analysis was the $q(r)$ effect in Theorem 13, which appears to correspond to gravity. This photon interaction appears able to assume two forms. One is an extended interaction, in which two photons sweep together through all of B, interacting 'parametrically' by the effects of their mutual $q(r)$ on their mutual propagation. Because both photons travel together with their $q(r)$ effects, an observer at the center of these converging 'waves' would be looking out at earlier times, and thus see the arriving electromagnetic energy as a ('historical') 'solid' as the wave arrived. For this (approximate) calculation, the reader is referred to the main text. The result is that in a box B with linear size R_B, containing total mass M_B, an upper bound for the least mass m_0 for a B_0 containing N bits of information, coupled by $q(r)$ interaction (Theorem 13, gravity) is approximately

$$m_0\left(32\pi^2(\ln(R_B 2m_0 c/\hbar))^2\right)^{1/3} \approx \left(\frac{\hbar^2 M_B Y(N)}{c^2 R_B^2}\right)^{1/3} \qquad \text{(Cor 14.3).}$$

The 'parametric' nature of this interaction allows two types of B_0, distinguished by 1/4 cycle relative phase lead or lag between the photons in this extended interaction state and those in the other, less coupled, normal mode in which the B_0's two coupled photons may exist. These are the two 'normal modes' of the 'weakly coupled oscillators' for B_0 (Cor 14.4). Including effects of 1 bit coded in terms of such two possible relative phases leads to a term $1+1/N^2$ in the denominator on the right side of the Corollary 14.3 expression for m_0,

$$m_0\left(32\pi^2(\ln(R_B 2m_0c/\hbar))^2\right)^{1/3} \approx \left(\frac{\hbar^2 M_B Y(N)}{c^2R_B^2\left(1+\frac{1}{N^2}\right)}\right)^{1/3} \qquad \text{(Cor 14.5).}$$

Also, this extended interaction allows exactly two different relative polarizations of B_0's two photons (Cor 14.6). Thus, 3 bits of ('internal') information are required for formation of this type of B_0: existence of coupling, relative phase (1 out of 2 states), and relative polarization (1 out of 2 states) (Cor 14.7).

The other form which this type of photon interaction may have is for the two photons to be localized as closely as their wavelength allows, and be as close together as their localization can define. I calculated the minimum mass required for this localized interaction to release the same energy as the extended coupling. More mass is required. The extended interaction appears more effective for releasing energy (information) by a factor

$$F \approx \left(32\pi^2(\ln(R_B 2m_0c/\hbar))^2\right)^{1/3} \qquad \text{(Cor 14.8).}$$

Please see main text for details.

Next, I explored interactions among B_0s. For equal rest mass B_0s, like types (lead or lag) repel, and unlike types attract (Cor 14.9); an assembly of equal rest mass like B_0s at low energy acts like an assembly of 'spin $\frac{1}{2}$' particles (fermions) with approximate 'Pauli' exclusion (Cor 14.10); and unlike equal mass B_0s can mutually annihilate (Cor 14.11). These effects all arise from photon superposition. Attraction and repulsion are due to interaction between the localized forms of two B_0s, by superposition of their localized electromagnetic fields.

From Corollaries 14.10 and 14.11 one notes that like B_0s are unable, and unlike B_0s are able, to represent information with complementary codings (Cor 14.12).

With Corollary 6.4, I approximately calculated the interaction between two equal mass B_0s, and found, for $r \gg \hbar/m_0c$, an energy change $\Delta E \approx \pm\hbar c/Fr$, where '+' denotes repulsion of like B_0s, and '−' denotes attraction of unlike B_0s (Cor 14.13, 14.14). (See also Section 6 question 80.)

In Corollary 14.15, I collected these results for $q(r)$ (gravity) coupled photon B_0s. Remember that this is specifically calculated to be the least massive such particle which is stable when substantially at rest. Writing (from Cor 10.3) R_B as c/H, where H is local Hubble's constant, I solved for m_0 and F. The agreement with the electron mass m_e and charge was remarkable: mass to about +18% to +64.5%, charge to about −0.24% to −0.72%. The ranges are due to range of H assumed.

At the time of this writing, H appears only approximately known (to me) directly observationally, so, to test this correspondence more sharply, I combined the expressions for m_0 and F to eliminate H, and set $m_0 = m_e$ and $F = \alpha^{-1}$, where 'α' denotes the 'fine structure constant' ($\alpha = \{\mu_0 c^2/4\pi\} e^2/\hbar c$, where the term in curly brackets is used with Système Internationale (SI) units, and 'e' denotes the electron charge), thereby obtaining a relation among G, c, m_e, $\hbar$, and α, all of which appear, at the time of this writing, much better so known observationally than H:

$$\alpha^{-3}\exp\left(\frac{\alpha^{-3}}{32\pi^2}\right)^{1/2} \approx \frac{7.2\hbar c}{m_e^2 G}.$$

Substituting for G, c, $\hbar$, and m_e values recommended by Cohen and Taylor[4] ('CT', 1973), and solving by successive approximation for α^{-1}, I found $\alpha^{-1} \approx 137.0387173$, which, although differing from CT's recommended value for α^{-1} by about +19.683 parts per million (ppm), and from that recommended by Cohen and Dumond[6] (1963) by about −0.6035 ppm, appears, on consideration of observational data tabulated by Taylor, Parker, and Langenberg[5] (1969) and reported[4] (preliminary) interpretation of positronium measurements, to be allowed by knowledge prior to the time of this writing. In the main text appear considerations relating to theory underlying CT's recommended value for α. I then solved for H, obtaining the value $H \approx 9.7191 \times 10^{-19}$/sec, which appears not reasonably excludable by prior reports of observation, and checked by solving numerically for m_0 and F to see if the apparent compatability with observation of the tested expression for α might have come about by cancellation of any errors in these separate expressions on combining them to eliminate H, and found $m_0 \approx m_e$ and $F \approx \alpha^{-1}$, from which it would seem that compatability with observation of the expression relating G, c, $\hbar$, m_e, and α was likely not in this way spurious.

Together with all the other features of wavelength, momentum, kinetic energy, statistics, particle and antiparticle, annihilation, and attraction and repulsion, these close numerical correspondences strongly suggest that, in our universe, these B_0s correspond to electrons and positrons (Correspondence Cor 14.16).

Using F so obtained, I calculated an approximate value for e. Using H so obtained, I calculated approximate values for the 'radius' of our 'box', energy per bit in our 'box', total mass in our 'box', total amount of information in our 'box', average density, and mass in a 1 m^2 column to the 'wall' of our box. These various values are tabulated in Table 2, p. 155, together with expressions for recalculating approximate values for such quantities with different sets of input values. Please look at Table 2 before continuing with this informal introduction.

With the closeness of correspondence of such B_0s with the electron and positron, likely electrons and positrons are (with Theorem 13) affected in the same way by, and equally effective as sources of, gravity; and complementarity of coding of internal information appears to offer some insight into conservation of the properties which appear to correspond to coding of such B_0's 3 bits of internal information: 'charge' sign, 'spin', and 'lepton number'.

I called such B_0's 'photonic particles' (Def 19).

1.8.1 Fast B_0's (2.10). For fast B_0s (Subsection 2.10, starting on p. 156), we find breakdown of the 'principle of equivalence', that an observer in a small laboratory in free fall not too close to a mass M not see any effects due to M within that laboratory. The physical insight here is that position information associated with a B_0 cannot be localized more closely than the smallest enclosure within which both B_0's associated photons can satisfy boundary conditions, and it is B_0's position information which participates in Theorem 13 (gravity) interaction. Thus, B_0's down-shifted photon must fit into a region small enough for the principle of equivalence to be a good approximation in a region that big. For lab size r much less than distance R to source M, the principle of equivalence would be a good approximation for $E_{Bj} \geqslant \hbar c/r$ and $E-2E_{Bj} \geqslant \hbar c/r$; for fast B_0, $E_0 \gg \hbar c/r$ and $E \leqslant E_0^2 r/4\hbar c$ (Theorem 15). The principle of equivalence would be substantially irrelevant for B_0s whose position information were spread much further than the distance to M, or $E_{Bj} \ll \hbar c/R$; for fast B_0, $E \gg E_0^2 R/4\hbar c$ (Cor 15.1). The faster a B_0, the more its information is in a momentum representation and the less in a position representation, and a fastest B_0 fills B and has no position information in B (Cor 15.2). Thus, an observer would see B_0 and M as substantially unattracted by a Theorem 13 (gravity) interaction with each other if $E_{Bj} \ll \hbar c/R$, for fast B_0 if $E \gg E_0^2 R/4\hbar c$ (Cor 15.3). As the extreme case, a fastest B_0 and mass M do not attract each other, because interaction with a fastest B_0's position information does not give any information about M other than that it be in B, which was already known (Cor 15.4).

Using Corollary 6.4, a portion $1/F_f$ of B_0's rest frame energy E_0 appears in localized form, with

$$F_f \approx \frac{1}{E_0}\left(\frac{M_B Y(N)\hbar^2 c^4}{R_B^2}\right)^{1/3} \qquad \text{(Cor 15.5).}$$

This can be expressed in terms of E and E_{Bj}, as

$$F_f \approx \frac{1}{2\sqrt{EE_{Bj}}}\left(\frac{M_B Y(N)\hbar^2 c^4}{R_B^2}\right)^{1/3} \qquad \text{(Cor 15.6).}$$

The interaction strength of B_0's localized form can be expressed in terms of a 'charge magnitude' $|a| = (\hbar c/F)^{1/2}$ (Def 20). This 'charge magnitude' $|a_f|$ for a fast unbound B_0 can be expressed in terms of that of the least massive bound B_0 of Theorems 13 and 14 and their corollaries, $|a_f|$, as

$$|a_f| \approx |a_b|\left(\frac{F_b}{F_f}\right)^{1/2} \approx |a_b|\sqrt{2}\, E^{1/4}\frac{E_{Bj}^{1/4}}{m_b^{1/2}c}\left(\frac{0.9\,Y(3)}{Y(N)}\right)^{1/6},$$

where N is the number of bits of internal information in B_0 (Cor 15.7).

In the unbound state, a fast B_0 carries at least 2 bits of internal information, so $N \geqslant 2$ (Cor 15.8); for fB_0, $N=2$ (Cor 15.9).

Using the relation between 'charge magnitude', E_{Bj}, and E in Theorem 15, and the graininess appearing in Corollary 8.15, I found an upper limit for B_0's next allowed energy $E' < E$ (see Subsection 1.1 for explanation of special use herein of square brackets),

$$\frac{E}{1+\dfrac{4E_{B1}E}{m_b^2c^4}\left|\dfrac{a_b}{a_f}\right|^4\left(\dfrac{0.9\,Y(3)}{Y(N)}\right)^{2/3}} \approx E' \; [\lesssim]$$

$$\left[\lesssim \frac{E}{1+\dfrac{48\pi E_{B1}^3E^3}{m_b^6c^{12}}\left|\dfrac{a_b}{a_f}\right|^{12}\left(\dfrac{0.9\,Y(3)}{Y(N)}\right)^{2}}\right] \qquad \text{(Cor 15.10).}$$

By looking at coding of internal information, I found that, for a process which involves making a 3 internal bit B_0, 'B_{03}', as in Corollary

14.15, and supplies or complements coding for only one bit of B_{03}'s internal information: said 1 bit is coded as relative phase of B_{03}'s associated two photons; such process involves making a fast B_0, 'B_{02}', representing internally 2 bits of information in coding complementary to B_{03}'s other 2 bits of internal information; B_{02} does not represent any internal information by relative phase of its associated two photons; B_{02} represents 1 bit of internal information by relative polarization of its associated two photons; and such process involves transcription of 5 bits of information to be represented internally by B_{02} and B_{03} (Cor 15.11).

Using various of these approximate relations, I (approximately) calculated a set of approximate numerical results for fast B_0s in our universe. ($E_{B1} \approx 6.397 \times 10^{-34}$ eV.) From Corollary 8.10, a 1 MeV fastest B_0's momentum would appear to transform substantially like a photon's in frames of reference moving with velocity $v < c(1-\epsilon)$, $\epsilon \gg 2E_{B1}/E \approx 1.279 \times 10^{-39}$. From Corollary 8.11, a fastest B_0 carrying 1 MeV energy as seen by us would have rest frame energy $E_0 \approx 5.059 \times 10^{-14}$ eV.

From Corollaries 15.1–15.3, a fastest B_0 near the earth likely would not be attracted by the earth.

Using the correspondence between the least massive bound B_0 and the electron (Cor 14.16), and Corollary 15.7, I compared interaction strength $|a_f|$ of a fastest B_0 with energy E to the electron's interaction strength e, as $|a_f| \approx |e| \times 1.097 \times 10^{-11}\ E^{1/4}$, with E in eV. For $E = 1$ MeV, $|a_f| \approx 3.47 \times 10^{-10}|e|$ in B_0's rest frame. Please see main text for cautionary note on applicability of formal method in the case of fastest B_0s.

Note that these calculations do not treat effects of source and/or detector motion on fast B_0s, if any. Note also that calculations of scattering cross sections might be affected by momentum graininess.

At the time of this writing, these various aspects of fast B_0s appear, to the extent tested as presented herein before Correspondence Corollary 15.2, allowed by prior knowledge, with correspondence between fast B_0s and neutrinos (Correspondence Cor 15.12).

1.8.2 Information and Inertia (2.11) and Compatibility with Prior Knowledge (2.12). By the derivation outlined, a physical picture based on information appears obtainable from a consistent part of prior physical theory, and, therefore, compatible with all prior knowledge described, in their respective domains of applicability, by the various prior theories with which formal results of such picture have been compared in Section 2, although differing from such theories in ways compatible with prior knowledge as noted in Section 2. As an example, used as a common 'thread' through various parts of the material presented, collecting results on momentum, gravity, mass, charge, and 'statistics', a physical picture of inertia in terms of information appears, to the extent tested as presented herein before Correspondence Theorem 16, allowed by knowledge prior to the time of this writing (Correspondence Theorem 16).

1.9 Axiomatization (Section 3)

Having found that a physical picture based on information appeared compatible with prior knowledge, I sought to isolate the premises in such picture, in terms of information. It seemed to me that, if I could find a set of postulates in terms of information from which all of the formal structure could be derived, then by such axiomatization the nature of the physical picture might be starkly delineated.

1.10 Looking for Postulates (3.1)

In Subsection 3.1, Looking for Postulates, I described relatively informally some of the process by which, starting with material presented in Section 2, I attempted to find sufficient postulates in terms of information. Such description, (pp. 174–179) is informal enough so that you might read it as part of this informal introduction, and then continue here. This is another instance in which you, the reader, get to read a result in *much* less time than it took me to do it.

1.11 Postulates (3.2)

I then list the postulate set which I found, together with a formal definition of information. Because they form a sequence independent of derivation appearing in Section 2, definitions, lemma, theorems, and corollaries appearing in the axiomatic form are designated by a prefix 'A-'. Postulates appear in only the axiomatic form, so such a prefix is not needed for them.

In A-Definition 1, I defined amount in bits of information as $\log_2$ of the number of (distinguishable) possibilities from which one is designated.

In terms of such information, the postulates which I chose may be stated as

1. Information is conserved.
2. Information is communicable.
3. Information is finitely accessible.

or, as condensed in a subsequent comment, 'Information is conserved, communicable, and finitely accessible.', which form was not used in the test of sufficiency, separate statement facilitating reference.

I do not claim that this set of postulates is unique, nor have I shown formally that every part of such postulate set is necessary (see also Comment 4.13, p. 314), and you may wish to find and test other set(s) of postulates. In some ways, one interesting aspect of a physical picture or theory is the equivalencies it induces on postulate sets, and comparisons of such equivalent postulate sets and corresponding formalisms seems helpful in elucidating relations between physical content and formalism (e.g., comparison of wave and matrix formalisms for QM). One reason for presenting the material in Section 2 in such detail, instead of presenting an axiomatic form as a 'fait accomplis', is to provide you with a basis from which you might proceed differently from the way I have.

However, this postulate set appears sufficient, testing of which is described in Subsection 3.3.

1.12 Testing Postulates (3.3)

I defined 'adequate' and/or 'adequately', with respect to description by mathematical function, to mean 'with any and all error less than 1 bit of information for the description and/or any observable dependence thereon' (A-Def 2). Usefulness of such criterion is discussed formally in A-Corollary 4.1. For convenience, 'information representation adequately describable by' may be written 'irad' (A-Def 2).

1.13 Describing Communication (3.3.1)

I then proceeded to examine basis functions spanning a mathematical function space in terms of which communication could be adequately described locally, in the following general way (A-Theorem 1):

Communication requires that a source and a receiver exist, existence of each of which requires 1 bit of information. Therefore, mathematical functions describing communication are functions of four independent variables, which, from finite accessibility of information, are adequately describable for small enough region (locally) in terms of an orthonormal set of four linear variables, x_0, x_1, x_2, and x_3. (I chose 0–3, rather than 1–4, to correspond to the four 2-bit binary numbers 00, 01, 10, and 11.)

That encoding of information for transmission exist requires 1 bit, so such function space must be expressable as a product of 2-spaces spanned by the two independent variables corresponding to that existence of encoding bit. By an argument involving local linearity; the need, from information conservation, to be able to describe a 'wave' which does not leave information behind, with positive- and negative-going functions; and various other features; I found that such function space is spanned by a finite set of solutions $w_{z,k}$ ('waves') of the coupled equations

$$\left.\begin{aligned} &\nabla \times w_{z,k} + (-1)^k \frac{1}{c}\frac{\partial}{\partial x_0} w_{z,1-k} = 0 \\ &\nabla \cdot w_{z,k} = 0 \end{aligned}\right\} \quad k = 0, 1$$

writing '$\nabla \times$ ' and '$\nabla \cdot$' for 'curl' and 'divergence' with respect to x_1, x_2, x_3, where z is a label of a finite set (A-Theorem 1).

The constant c arises solely from the requirement, from finite accessibility of information, of local linearity, and may be regarded as a 'linear coefficient of local propagative communication'. Please see main text for details.

For convenience in reference, such 'wave equations' in A-Theorem 1 are designated 'WE', and such solutions thereof are designated variously 'w', 'w_z', 'w_j', '$w_{j,k}$', '$w_{z,k}$' and/or '$w_{j,z}$' (A-Def 3).

1.14 3-Space and Apparently Unidirectional Time (3.3.2)

In Subsection 3.3.2, I defined 'observer' (A-Def 4) and 'observation' (A-Def 5). Noting that observation requires 2 existence bits, one each for event and observer, I found that an observer can adequately describe observed events by mathematical functions of x_0, x_1, x_2, x_3 (A-Cor 1.1), locally a flat 4-space, and defined an observer O's 'universe' 'U' to be all that about which O can have information (A-Def 6).

Then, by detailed examination of information possessable by an observer, I found that an observer can adequately describe x_0 as homogeneous, unidirectional, and distinct from the 3-space spanned by x_1, x_2, x_3, which 3-space would appear isotropic and homogeneous (A-Cor 1.2). The derivation is relatively long and detailed, and the reader is referred to the main text for such derivation. This singles out x_0 in a way corresponding to the usual notion of time, and I defined 'space' as said 3-space, 'time' as x_0, and 'spacetime' as the 4-space spanned by x_0, x_1, x_2, x_3 (A-Def 7), and restated A-Corollary 1.2 in terms of space and time as A-Corollary 1.3.

Nonlinear 'coordinates' would represent information. Because much of the formal development of this physical picture is concerned with describing amounts and representations of information, I chose to use for U ('global') coordinates which would themselves not represent information, and, from formal considerations, found that this corresponds to choosing globally flat coordinates as in A-Definition 7, and chose such (A-Def 8). Such choice has the effect of making information represented in physical systems appear in descriptive functions over flat coordinates, although one can, as in Theorems 12 and 13, choose to describe phenomena in terms of nonflat coordinates.

Isotropy of 3-space, finite accessibility of information, and choice of globally flat coordinates allow 3-space sizes in a universe U to be scaled by assigning a linear size R_U to U (A-Cor 1.4), as is done in A-Definition 9, which may be thought of as calibrating length measuring units. Linear size R may be written in terms of 'scale' $s = R/R_U$ (A-Def 10).

From Postulate 3, A-Theorem 1, A-Cor 1.3, and A-Definitions 3, 6, and 8, in an observer O_U's universe U, there is a finite maximum spatial period possible for any solution(s) of Theorem 1 'wave equations', and/or linear combination of such solutions, having amplitude >0 (A-Cor 1.5), which maximum spatial period I called 'λ_1' (A-Def 11).

1.15 Separability of Internal and External Information (3.3.3)

From finite accessibility of information, each $w_{z,k}$ possible in U has finitely many possible amplitudes >0, and, therefore, a least such, $A_{z,k}>0$, describing representation of least amount I_z of information, independent of k. By considering encoding and conserved information in irad $w_{z,k}$, I found $A'_{z,k}(lI_z) = l^{1/2}A_z(I_z)$, where l is a finite integer $\geqslant 0$ (A-Lemma 2.1). This appears to be the 'basis' of dependence on squares of amplitudes in both electromagnetic theory ('EMT') and QM.

Next, I defined 'B_0' in observer O_U's universe U in terms of least amplitude>0 w's propagating in opposite directions, and defined, for $f_+ > f_-$, 'w_+', 'w_-', 'λ_+', 'λ_-', 'f_+', 'f_-'; velocity v for observer O' who sees w_+ and w_- with f_+ and f_- as both having frequency f_0 and wavelength λ_0; operations on B_0 (not creating or destroying B_0) as those for which O' would be able to adequately describe f_0 and/or λ_0 as unchanged; and 'scale' s_0 of B_0 by $s_0 = (\lambda_0$ of B_0 substantially at rest as seen by $O_U)/\lambda_1$ (A-Def 12); and transitions among allowed states of B_0 (A-Def 13). This facilitated guiding axiomatic formal development by insights obtained in the development appearing in Section 2.

I next defined 'gentle acceleration 1' ('GA1', A-Def 14) and 'gently scaling 1' ('GS1', A-Def 15) for a system containing an irad one w with least amplitude >0 or one B_0, defining one 'step' of GA1 or GS1 as in effect just enough GA1 or GS1 to change the state in U of such contained irad w or B_0, consistent with finite accessibility of information, and defined 'internal information' and 'external information' of a system in a box, in a larger box (A-Def 16).

By looking at cumulative performance by GA1 and/or GS1 of transitions of system S irad w with least amplitude >0 and longest wavelength possible in S, or two ws with different k values such that S would in effect be as a B_0, among possible states thereof in S with scale s starting substantially at rest as seen by O_U in U, I found that, for N_1 steps of GA1 and/or N_2 steps of GS1, if

$$\sum_{i=1}^{N_1+N_2} s_i^2 < 1,$$

taking s_i before each step of GA1 or GS1, both amount and representation of S's internal and external information are adequately describable as separable (A-Theorem 2); which result I called 'adseparability' (A-Def 17). Because information conservation is a postulate in this axiomatic form, adseparability has a role similar to PODIC in Section 2.

In this formal context, within its domain of applicability, adseparability implies Einstein's SR, and may be formally included in calculations by using computational methods of SR. For convenience in using adseparability in further consideration of *ws* and B_0s, I collected results of some calculations, relating to B_0s, using computational methods of SR to formally include effects of adseparability, and defined 'n_+', 'n_-', and 'n_0', as being the number of times the shorter (n_+) or longer (n_-) wavelength w, or hypothetical 'rest frame' (n_0) w, would fit into a distance L (A-Cor 2.1; see A-Cor 2.1 for definitions of n_+, n_-, and n_0). Part of the reason I collected so many results in A-Corollary 2.1 was to help make it apparent subsequently that, with certain constructs, the mechanics of B_0s includes a relativistic quantum mechanics.

Partly by counting steps of GA1, I found that, in U, adseparability applies to information represented by irad one w with least amplitude >0 and longest wavelength possible in B_1 in U with allowed velocities as a B_0 with two such *ws* in U, or by irad one B_0, treated as having *ws* with different k values for purposes of A-Theorem 2, in any possible external state in U with velocity v less than its maximum possible velocity v_m; and that A-Corollary 2.1 adequately describes such B_0 in any possible state in U with $v < v_m$: roughly stated, shrinking B_1 by GS1 would perform transitions of *ws* in B_1 to states in which their wavelengths would fit into B_1 a smaller number of times, and the longest wavelength w in B_1 cannot be so changed; and irad such B_0 could get to one step below its highest possible velocity by a number of steps of GA1 too small to perform transitions 'inside' it. A B_0 may be regarded in substantially its rest frame as two *ws* with least amplitude >0 and longest possible wavelength possible in a box (of scale s_0), so both arguments can be used to adequately describe such *ws* and B_0s (A-Cor 2.2). Not any transitions implies not any change in information represented, hence adseparability, on which A-Corollary 2.1 depended, so A-Corollary 2.1 is also adequate for such B_0 in any state possible in U with $v < v_m$.

Therefore, the results collected in A-Corollary 2.1 may be used to adequately describe such B_0s in considering the domain of applicability of adseparability and SR, without the description of each such B_0 becoming inadequate.

1.16 Spectrum of *w* States in *U* (3.3.4)

How does one find the possible solutions of the A-Theorem 1 'wave equations' (WE) in a universe *U* without boundary conditions?

I proceeded in the following general way (A-Theorem 3): From finite accessibility of information, I had found that there must be a longest wavelength, λ_1, for any solution of WE possible with amplitude >0 in *U*. Call $w_{j,z}$ with $\lambda_j = \lambda_1$ $w_{1,z}$. From isotropy, homogeneity, flatness, and absence of boundary conditions, for any two possible wavelengths λ_{j_1} and λ_{j_2} (j not initially assumed to be an integer) in *U* it is possible to find $w_{j_1,z}$ and $w_{j_2,z}$, with wavelengths λ_{j_1} and λ_{j_2} respectively, propagating in opposite directions, which can be regarded as a B_0. However, such B_0, being a linear combination of solutions, is itself a solution of WE in *U*, and, from A-Corollary 1.5, has wavelength $\lambda_b \leqslant \lambda_1$. Using this 'spacing apart' of possible λ_js, I 'pieced together' the spectrum, and found that, in universe *U*, all possible $w_{j,z}$ solutions of WE have respectively wavelengths $\lambda_j = \lambda_1/j$, where j is a finite integer >0.

1.17 Describing Information Representation by *ws* (3.3.5)

From A-Lemma 2.1, we already have that there is a least amount I_j of information which can be exchanged with representation by irad $w_{j,z}$: what is the dependence of I_j on j? By considering a B_0 with w_j and w_{j+1}, I found that observer O' moving substantially with B_0 would see both *ws* as having the same frequency and exactly two possible relative phases, for a total amount of information 1 bit more than twice the amount representation of which would be adequately described by only one of the *ws*, which, together with $I_1 = 1$ bit (an 'existence of encoding' bit), leads to $I_j = j$ bits of information (A-Theorem 4).

Therefore, information is always exchanged in integer numbers of bits (A-Lemma 2.1 and A-Theorem 4), so error less than 1 bit is not representable, communicable, or observable; this provides a formal basis for adequate (A-Def 2) description never being observably different from exact description (A-Cor 4.1), and assures that the formalism does not require exact descriptions which might involve infinite information. 'adequately describable as equal to' does not appear to be transitive. In view of A-Corollary 4.1, it appears formally consistent to write correspondences with observation using adequate descriptions as the strongest statements ever necessary; and to define '*x*', where '*x*' is '$=$', '$>$', '$<$', '$\geqslant$', or '$\leqslant$', as the (prior) mathematical '*x*' with respect to all calculational manipulations, and as 'adequately describable as x' with respect to satisfaction of such expression in (a) (finite) universe U (with discrete quantities).

Although possible λ_js are discrete, this does not present a problem in finding B_0 states, because O' does not see changes less than ± 1 in $n_0 = \sqrt{n_+ n_-}$ (equation from A-Cor 2.1), which allows ranges of n_+ and n_- sufficient for n_+ and n_- to both be integers (but does not eliminate the graininess discussed in relation to breakdown of PODIC in Section 2). Thus, the formalism is consistent with respect to B_0s (A-Cor 4.2).

With λ_1 as in A-Theorem 3, A-Theorem 4 implies $I(f) = (\lambda_1/c)f$ (A-Cor 4.3). Generalizing this to real f would lead one to reject w spectra of all solutions of WE with 'hard' boundary conditions in a flat space cavity with single connectivity as a model for U, generally on the ground that such conditions would lead to a spectrum containing incommensurable frequencies, and, with finitely accessible information, information would not be conservable in exchanges involving incommensurable 'lumps'. Partly from such considerations, and for other reasons stated in the main text, R_U is taken to be $= \lambda_1/2\pi$ (A-Def 18).

Considering amplitudes A_y of co-possible modes y, where presence of information in irad mode y would involve representation therein of amount I_y of information, I found A_y proportional to $1/I_y^{1/2}$ (A-Cor 4.4).

For convenience, I defined an integer 'occupation number' 'l' for information representations in which information is representable in integer multiples only of a least amount >0 of information (A-Def 19), and with l discussed information representation variously by irads ws, $I = \sum_j l_j j$ (A-Cor 4.5); externally by irads B_0s, $I = \sum_j l_j (n_{+,j} + n_{-,j})$ (A-Cor 4.6); by irad a 'wave' of $w_{j,z}$'s, where z represents any other indices,

$$W = \sum_{j,z} l_{j,z}^{1/2} j^{1/2} w_{j,z}, \qquad I = |W|^2 = \sum_{j,z} l_{j,z} j \qquad \text{(A-Cor 4.7);}$$

and by irad a B_0 'wave packet'

$$\psi = \sum_z l_z^{1/2} (n_{+,z}^{1/2} + i n_{-,z}^{1/2}) w_{n_+,z} w_{n_-,z},$$

$$I = |\psi|^2 = \sum_z l_z (n_{+,z} + n_{-,z}),$$

where $i = (-1)^{1/2}$ (A-Cor 4.8).

1.18 Adseparability for Systems (3.3.6)

Next, I considered decomposition of information represented by irad system S into internal and external representations, and found that, where internal information I_0 and external information I_1 of irad system S

representing in U information $I = I_0 + I_1$ are adseparable, irad S can be adequately described by one element of a set with cardinality 2^I; by a pair of elements one each from two sets with cardinality 2^{I_0} and 2^{I_1} respectively; by mathematical functions of 2^{I_j} identical variables, each of which variables is a mathematical function adequately describing representation of amount I_{1-j} of information, $j = 0, 1$; by two mathematical functions, each of which adequately describes representation of amount I_0 or I_1 of information, respectively; or by one mathematical function adequately describing representation of amount I of information (A-Theorem 5).

In particular, irad B_0 representing internally amount I_0 of information, in U, can be adequately described by a mathematical function of 2^{I_0} identical variables, each of which is a ψ as in A-Corollary 4.8 (A-Cor 5.1). This appears, with Theorem 14–Corollary 14.16, to offer insight into such prior formalisms as Dirac's 'spinors', and you might wish to look at various examples appearing in the main text following A-Cor 5.1 (pp. 218–219).

By considering effects of GA and GS on irad system S of B_0s and/or ws in a box B_1 with scale s and velocity v in U, I found that information represented by irad S in B_1 is adseparable iff v is adequately describable as $v < c(1 - s^2)/(1 + s^2)$ (A-Theorem 6). Please see main text for details. This includes, as the 'breakdown' of adseparability for $v \ll c$, $s \approx 1$, and for $s \ll 1$, $v \approx c(1 - 2s^2)$, at which velocity the universe would look, to an observer moving substantially with B_1, as about twice as long in the direction of v as B_1. Thus, both the moving and substantially 'stationary' observers would regard B_1 as able to be in U.

That this result, obtained from consideration of possibility of presence in an observer's internal history of information indicating occurrence of acceleration, is consistent with the result obtained in Corollary 8.16, from consideration of effects observable by an observer moving uniformly, appears to illustrate in a relatively direct way consistency of formal structure presented herein with the concept that an observer's 'physics' depends only on information accessible to such observer.

Observer O in adseparable system S in U would not be in S if S did not exist; therefore, O cannot observe the alternative to S's existence, and so ($\log_2 1 = 0$) does not have access in adseparable S to S's existence bit in U (A-Cor 6.1).

1.19 Coding (3.3.7)

As in Section 2 Definition 2, I defined 'complementary' with respect to information representations and codings (A-Def 20), and found that complementary information representations represent equal amounts of information (A-Theorem 7).

From this, iff an amount of accessible information is an even number of bits, that information can be coded in complementary representations (A-Cor 7.1), and fully transcribed into some other representation without any other information ('code book') (A-Cor 7.2); which total even number of bits internally accessible in an adseparable system S would correspond to S's internal information appearing to an observer outside S as an odd number of bits (A-Cor 7.3).

1.20 Quantized Electromagnetic Theory and Relativistic Quantum Mechanics (3.3.8)

By choosing a set of constructs, with I in bits, 'energy' $E=(\hbar c/R_U)I$, 'mass' $m=(\hbar/R_U c)I$, and 'momentum' $p=(\hbar/R_U c)vI$, $(v \leqslant c)$ (A-Def 21); or, letting $E_1=\hbar c/R_U$, $E=E_1 I$, $m=(E_1/c^2)I$, and $p=(E_1/c^2)vI$, $(v \leqslant c)$ (A-Def 22); one can express various formal results appearing in Section 3 in terms recognizable from various prior formalisms. By expressing various quantities in terms of n_+ and n_-; and writing a 'wave function' or set of 'wave functions' ψ, for example, as in A-Corollary 4.8, A-Theorem 5, and A-Corollary 5.1, and defining operators $\mathbf{n}_+$ and $\mathbf{n}_-$ such that $\mathbf{n}_\pm\psi=n_\pm\psi$; one can obtain (e.g., from A-Cor 2.1), substantially by inspection, both number and operator formalisms which appear to substantially correspond to various prior formalisms in their various domains of applicability. Please see main text for details. With these correspondences, formal structure appearing in Section 3 appears to subsume (a) quantized electromagnetic theory and (a) relativistic quantum mechanics (consistent with respect to finite accessibility of information and movability of material particles).

1.21 Sufficiency of Postulates (3.3.9)

With these constructs, it appears that Postulates 1, 2, and 3 are sufficient for all formal structure presented in Section 2 and in that part of Section 3 preceding statement in Section 3 of A-Theorem 8 (A-Theorem 8). A-Theorem 8 includes lists of substituted referents for merging Sections 2 and 3.

1.22 Axiomatic Form (3.4)

The axiomatic treatment through A-Theorem 8 in Section 3 is, following A-Theorem 8, merged with the treatment appearing in Section 2 after definition of PODIC, providing an axiomatic form for the theory, with three postulates as stated, and three numerical quantities: total amount I_U

of information in universe U ($\approx 3.643\times 10^{122}$ bits), $R_U \approx 3.0846\times 10^{26}$ m, and $c\approx 299792458$ m/sec; to which set may be added, for convenience in using instruments calibrated in terms of energy, $E_1 \approx 1.025\times 10^{-52}$ J, $\approx 6.397\times 10^{-34}$ eV, 'energy' per bit of information represented in U. (Please see Subsection 3.4 for discussion of significance of I_U, R_U, c, and E_1.) Other sets of numerical quantities may be used, such as G, H, c, and h; or G, c, h, and m_e; or G, h, m_e, and α. Thus, concepts in this physical picture appear to be expressed in such axiomatic form formally and without any arbitrary parameter.

1.23 Partial Test of Axiomatic Form (3.5)

One aspect of the theory presented herein which seems, in comparison with other theories of which I am at the time of this writing aware, peculiarly within the domain of this physical picture and formal structure, is description of 'particle' properties in terms of representation internally of a few bits of information, or only about 3.756×10^{-39} (for an electron) of total information represented by such entity in U. Also, the concept of a finite 'size' of U appears quantitatively in such calculation, for example, as in calculating an approximate value for $F\approx\alpha^{-1}$.

In view of this, as a partial test of the axiomatic form, with decay schemes of 'particles', I traced, as presented in a 'chain' of reasoning, Chain 17 and its links ('L17. '), likely internal information, and obtained expressions for likely approximate 'rest mass', of photons (γ, L 17.2), electron neutrino and antineutrino (L 17.3), electron and positron (L 17.4), four muons $\mu^{\pm}_{L,H}$ (L 17.5), muon neutrino and antineutrino (L 17.6), four charged pions $\pi^{\pm}_{L,H}$ (L 17.7) and one neutral pion π^0 (L 17.8, 17.9).

Various of these results are collected and compared with reports of prior observations in Table 3 (p. 268), and you may wish to turn to Table 3 before continuing in this informal introduction.

To briefly mention a few aspects of such reasoning, muons appear to be approximately describable as substantially localized two photon entities, somewhat like the relatively localized form of the electron and positron, and pions appear approximately describable as substantially localized three photon entities. For convenience in referring to various classes of entities, I defined a set of notations, appearing in Definition 21, p. 254, to which you are referred. Approximate expressions for amounts of 'rest frame information' represented by entities of various classes are also collected into Definition 21. Because such definitions depend on various internal coding properties of such entities, such formulas are not repeated here. Table 3 provides comparisons of approximate values obtained with such expressions and various reports of observations prior to the time of this writing. I also defined a comparison quantity δ with postscripts, for convenience in comparing two and three photon forms without some of their internal coding details (Def 22).

With proximity of one entity, x, to another, substantially more massive entity, y, it appears that, under some circumstances, Theorem 13 interaction between x and y might subsume x's internal information, and that x would thus 'lose' its separate 'identity' (L 17.10, 'L 17.10 interaction'), and least 'rest frame information' I_y'' for y for L 17.10 interaction with $e^{\pm}$, $\mu^{\pm}_{L,H}$, π^0, and $\pi^{\pm}_{L,H}$ was approximately calculated (L 17.11).

Your attention is specifically directed to entries in column 10 of Table 3, including fractional discrepancies between approximate calculated likely values and previously reported values for various particle masses. These discrepancies appeared sufficiently small to warrant interpretation in more detail, for testing various specific aspects of the physical picture and formal structure presented herein.

As noted more fully following Table 3, approximate correspondence between pion properties and the approximate description and picture for pions, and apparent absence of a pionlike entity less massive than π^0, appears to substantially test the spectrum $Y(N)$ for $N \leqslant 13$ (E_j, $j=1,2,3$) to less than $\sim$0.552%, which appears sufficient to substantially exclude various other spectra, such as square root of sums of squares of integers as used in NRQM calculations relating to radiation spectral distribution in cavities with interactive walls.

Internal coding in $e^{\pm}$, $\mu^{\pm}_{L,H}$, and $\pi^{\pm}_{L,H}$ appears to offer insight into $\pi^{\pm}_{H}$ decaying more to $\mu^{\pm}_{H}$ than 'directly' to $e^{\pm}$.

The physical picture, and formal structure based thereon, presented herein before Correspondence Link ('CL') 17.12, appear, at the time of this writing, to the extent tested as presented herein before CL 17.12, substantially able to subsume approximate description of all neutrinos and antineutrinos reported prior to the time of this writing as observed, and of phenomena apparently regarded in prior theory(s) in terms of such theory(s)'s 'weak interaction(s)' (CL 17.12).

With L 17.10 interaction, the physical picture, and formal structure based thereon, presented herein before CL 17.13, appear, at the time of this writing, to the extent tested as presented herein before CL 17.13, substantially able to subsume approximate description of phenomena apparently regarded in prior theory(s) in terms of such theory(s)'s 'strong interaction' (CL 17.13).

For other features of the derivations and calculations in Chain 17–CL 17.13, various of which are summarized in Table 3, and other aspects tested by correspondences tabulated in Table 3, please see the main text.

In view of all material presented herein before CL 17.14, the physical picture, and formalism based thereon, presented herein before CL 17.14

appear, at the time of this writing, to the extent tested as presented herein before CL 17.14, to be substantially allowed by knowledge prior to the time of this writing (CL 17.14).

1.24 Comments (Section 4)

In 'Comments' (Section 4) appear various thoughts which occurred to me in relation to the physical picture and formal structure presented herein, and some explanatory material, some of which did not seem to belong with only one specific portion of the formal treatment.

1.25 Some Predictions and Experiments Which Might Be Interesting (Section 5)

In Subsection 5.1 (predictions) appear various results which might be testable, some of which appear to differ, in domains in which there appears at the time of this writing to not be ('locally') sufficient observational data, from prior theory(s). In Subsection 5.2 (experiments) appear in tabular form various experiments which, in the context of material presented herein, might be interesting, various of which might also be interesting for other reasons.

1.26 Questions (Section 6), and Appendix 1

It seemed to me appropriate, in breaking the development of the physical picture presented herein to make part of it available to others, to mention some open questions, and some such questions appear in 'Questions' (Section 6).

With respect to one question relating to fast B_0s in gravity, there appeared to be various conjecturable approaches, and I collected discussion relating thereto into Appendix 1. In view of the extent to which various results presented in Sections 2 and 3 of the main text appear compatible with knowledge prior to the time of this writing, and of the importance of the consequences if one of the conjectured alternatives proves substantially correct, experimental resolution of questions raised in Appendix 1 appears warranted.

1.27 Lists (Appendix 2) and Map (Appendix 3)

For convenience in following in detail derivations presented herein, I have included reference lists with page numbers and 'reminders' of the various postulates, definitions, lemmas, theorems, corollaries, chain, links, principles, and conjectures appearing herein. I made such kinds of lists, except for page numbers, for my own use in making these derivations; having found them helpful, I have included such lists, with page numbers added, for your use, and have added, for convenience in interrupted and/or incomplete reading, a list of abbreviations.

Part of Section 2 appears in both nonaxiomatic and axiomatic forms presented herein. For convenience in following logical precedence in such forms, I have included a 'map' of the sequences of both (Appendix 3).

By way of ending this informal introduction, your attention is directed to the warning in the Preface: this is a report of work in progress and not yet complete, and I do not yet know whether, in what ways, or to what extent material presented herein might be changed or discarded.

So far, this appears to be an interesting path of inquiry.

So, on to Section 2.

Section Two

Compatibility with Prior Knowledge

I wish to present in this section consideration of whether a physical picture in terms of information is allowed by knowledge prior to the time of this writing, rather than whether it is necessary. If it is allowed, then the reader may be guided by an esthetic sense regarding insights presented. *This Section* 2 *is under the interpretation stated on p.* 2.

2.1 INFORMATION REPRESENTATIONS

If we start from nonrelativistic quantum mechanics ('NRQM'), then, in our universe, quantum mechanical effects in all detectors have the effect of introducing fluctuations into all measurements, or of constraining them to assume a set of discrete values. Either of these descriptions of quantum mechanical effects on detectors has essentially the same consequences for reception of information by such detectors: to make disjoint the admissible representations by means of which information may be unambiguously encoded for reception by such a detector.

Here, information is used in its conventional sense of 'reduction' of 'ambiguity', and is measured by the number of possible choices from which one is specified. A conventional way of measuring this is to give the number of binary digits, or bits, required to specify one out of that many. For additivity of measure, we use hereinafter number of bits as our measure of information.

In a classical description of a detector, noise affects measurement of continuously defined quantities, such as electric potential. In the absence of such noise, specification of the magnitude of such a quantity could in such theory be used to represent an infinite amount of information.

Were there not any noise in such a classical system, one could in theory distinguish infinitely many different states of the information channel. Thus, specification of one such state would represent an infinite amount of information.

In a classical mechanical detector, noise introduces ambiguity in observing the state of an information channel. To allow reception in the presence of such noise, each usable element of a code for representing information must be regarded as surrounded by a finite nonzero neighborhood. Any state of the channel found within that neighborhood is assigned, during the reception process, to the code element belonging to that neighborhood. Usually, such neighborhoods are defined as continuous regions; the use of several disjoint regions to represent the same code element is usually described as a form of redundancy in the code, although classically any region might be partitioned with arbitrary fineness and then regrouped by coding redundancy to form the neighborhoods described above. The measure of the neighborhood or collection of neighborhoods assigned to a code element, divided into the measure of the signal channel, gives the number of choices among which reception of a code element specifies one. In electrical systems, this is the nature of the $\log_2(1+S/N)$ term in the modified Hartley theorem of channel capacity ($C = W\log_2(1+S/N)$, where C is channel capacity in bits per second; W is channel bandwidth in cycles per second; and S and N are signal and noise, respectively, measured in power units). In the presence of noise, specification of a state in a continuously variable channel would represent finite information. One way to think of this classical case is to note that presence of finite nonzero noise restricts admissible codes to those with disjoint elements, whose neighborhoods form a covering of the signal channel with finite nonzero mesh determined by the noise.

When a quantum mechanical detector is to be used for measuring the state of an information channel carrying a nonzero flow of information, it can be allowed only finite time for each such measurement. If the channel is described by a continuous variable, finite measurement time has as a consequence nonzero uncertainty. Usable code elements form a disjoint set, whose neighborhoods form a covering of the signal channel with finite mesh determined by the uncertainty. A signal channel with disjoint allowed quantum states has explicitly the property that usable code elements form a disjoint set, corresponding to allowed states. Note that in quantum mechanics code element neighborhood size cannot be reduced below that resulting from such uncertainty, regardless of detector design. Note, also, that the effect of uncertainty for a quantum mechanical detector is directly analogous to the effect of noise for a classical (mechanical) detector.

A classical mechanical transmitter is presumed able to precisely determine the state of a channel at signal injection. Thus, it could, in theory, inject an infinite amount of information. However, in the classical case, presence of noise does not necessarily lead to loss of information; rather, total system noise, from transmission through detection, restricts choice of codes for which no information will be obscured to those with large enough neighborhoods.

In quantum mechanics, a transmitter is also limited by the uncertainty relations, and in finite time ultimately determines the state of a channel at signal injection no better than can a quantum mechanical detector measure it. Unlike the classical case, it is not necessary to assume infinite information at transmission. Both transmitter and detector are limited by the same uncertainty relations; thus, codes allowed by transmitter limitations are exactly acceptable by detector limitations. One way to think of this is to note that there is not any controlled measurement, that is, experiment, in which either transmitter or detector is omitted. This may seem trivial, but it has the important consequence that there is not any experiment which can be performed which would allow one to show that, in a quantum mechanically perfect system of transmitter and detector, the presence of irreducible quantum mechanical noise had obscured information. That is, information coded more finely than to allow complete and unambiguous detection by a detector with irreducible quantum detector noise cannot be injected into and/or detected in an information channel, and thus does not have any operational consequences. (Occasionally herein '(has) no x' may be used to mean '(has) not any x' or '(is) devoid of x', and 'there is no x' may be used to mean 'there does not exist any x'.)

Definition 1. *'Information' is specification of one from $j > 1$ alternatives, and is measured in 'bits' by $\log_2 j$; restricted hereinafter in this Section 2 to 'that information which can be encoded for complete and unambiguous transmission and detection in a perfect quantum mechanical system'.*

'Information' may occasionally be used hereinafter to mean 'amount of information'.

To exist, such information must be represented in some suitable way. Such information representation is what is usually meant by a 'signal'. When such signal is received in a communication process, the process called 'detection' consists of comparing such signal with other information, suitably represented, so as to 'decode' the signal and re-encode the information represented thereby in some suitable ('new') representation in accordance with a code for such new representation, which code is itself information suitably represented.

Although the process of detection is usually thought of as including both the steps of decoding and re-encoding, it appears interesting to divide such process conceptually into decoding and re-encoding. In such case, the signal and the 'code book' for such signal are each information which to exist must be suitably represented in some code. In particular, if both the signal and the 'code book' contain suitable amounts of information, one might choose codings for the signal and its 'code book' such that either could be treated as a signal and the other as its 'code book.' With such coding, such 'complementary' representations would be able to completely 'decode' each other, so that there would not be any information 'left' represented thereby. Together, two such complementary representations may be thought of as 'pure' information, fully available for representation by any other signal and 'code book' combination.

Definition 2. *Two information representations which can together decode each other so that all information in both is available for representation in any other way are said to be 'complementary', and the codes therein are said to be 'complementary'.*

Transmission and detection of information at nonzero rate in the presence of noise or quantum mechanical 'uncertainty' requires that the flow of information be accompanied by a flow of energy. Classically, this can be seen in the modified Hartley theorem. Quantum mechanically, this is a consequence of the effect of the uncertainty relation between time and energy on the neighborhoods which must be defined about acceptable code elements to allow their associated states of an information channel to be distinguished by measurements made in finite time, using a detector with irreducible quantum mechanical 'noise'. All known evidence is that our universe is to this extent describable as 'quantum mechanical'. Generalizing from this, I assume for this Section 2 that all detectors in our universe have 'quantum noise'. Recalling the definition of information (Def 1), we are thus free to say that any process which ('locally') transports information must transport energy. (See also text near after A-Theorem 1, Section 3.)

2.2 MOTION OF A MATERIAL PARTICLE

Consider motion of a material particle. We can assign motion with a velocity in one range to one element of a code, and with velocity in another, disjoint, range to another code element. For example, an electron with a velocity in the range of 0.4 to 0.6 c could represent a '1', and in the range 0.1 to 0.3 c a '0'. By detecting the velocity of the particle, we can detect this information. Detectors exist by which this might be implemented, for example, a germanium diode detector one centimeter cube, used with a measuring time of one second, with suitable electronics. Thus, motion of a material particle can be used to transport information. From the preceding, motion of a material particle must therefore also transport energy. That is, at least part of the energy of motion of a material particle must be attributable to the requirement that it transport energy while transporting information. I propose to show, in a 4-dimensionally covariant way, that we are free to attribute all energy transported by a material particle to its transport of information, and thus construct a conceptual picture in which this kinetic energy (momentum) aspect of inertia appears in terms of quantum mechanical 'noise', that is, in terms of quantum mechanical noncommutativity of observables. In the axiomatic form presented hereinafter, beginning in Section 3, this is expressed in terms of information, without using quantum mechanics as a starting theory.

2.3 INFORMATION CONSERVATION

I proceed here by a sequence of information 'bookkeeping' theorems:

Theorem 1. *With NRQM, in an isolated finite quantum mechanical system, information is conserved.*

1. From NRQM, in such a system, all eigenstates are discrete. (Classically, one could think of a phase space for the system, with allowed states forming a set of points about which can be defined disjoint nonzero neighborhoods.)
2. From NRQM, and 1, all eigenstates are distinguishable. (That is, a quantum mechanically perfect detector can unambiguously detect which of all eigenstates the system is in. Classically, the disjoint neighborhoods would be greater than or equal to the noise, or h^n, for n degrees of freedom, where 'h' denotes Planck's constant.)
3. From NRQM, density of states is preserved by time development.

(Classically, the volume of each disjoint neighborhood remains the same, although the neighborhood may be distorted.)

4. From 3, all previously distinguishable states are not rendered indistinguishable by time development, and previously indistinguishable states do not become distinguishable. (Classically, detector noise remains the same, the neighborhood volumes remain the same, and by time developing the detector one could cause it to relate all points at which the system might be found in the new, distorted, neighborhoods to their appropriate contained time-developed allowed states.)
5. From 2 and 4, eigenstates qualify for use as code elements in a system using a detector and having only quantum noise.
6. From 5, all eigenstates are usable code elements.
7. From 4 and 6, all possible codes based on quantum states conserve information.
8. From 7, all information represented in such system at any time is conserved.

●

I regard conservation of information as more basic than the Hermiticity of the Hamiltonian used in steps 1 and 2 or the unitarity of the time development operator used in step 3. With a postulate set including information conservation, unitarity and hermiticity appear derivable. (See Section 3.)

By isolation of a box, we mean specifically that energy is not exchanged between the box and its environment. From the preceding discussion, this implies that information is not exchanged. Thus, we obtain

Corollary 1.1. *With NRQM, in a system energy of which is not exchanged with an environment, information is conserved.*

●

In transcription of information into or from representation by a pair of complementary information representations without using any other 'code book', there is not any information represented by distinction between signal and 'code book'. Therefore, from information conservation, by symmetry,

Corollary 1.2. *With NRQM, in a system energy of which is not exchanged with an environment, complementary information representations represent equal amounts of information.*

●

Theorem 2. *Box B containing only electromagnetic energy; all information represented in B is representable by photon state occupation numbers, without phase.*

1. Place a perfect quantum mechanically limited detector contiguous with B.
2. Unambiguous detection and assignment of each photon to its correct photon state ('UDA') requires unambiguous energy measurement.
3. From NRQM noncommutativity of energy and time, and 2, UDA requires infinite time.
4. From electromagnetic theory ('EMT') and NRQM, in case as in step 1, with infinite time uncertainty, phase is not observed.
5. From EMT and NRQM, in case as in step 1, in infinite time, with any nonzero coupling to the detector, the expectation value of electromagnetic energy remaining in B goes to zero.
6. From 5, after infinite time, there is not any portion of the signal present in B and not received by the detector.
7. From Def 1, and 2 and 6, all information in B is extracted by UDA.
8. From 3 and 4, the result of UDA is an unambiguous set of photon state occupation numbers without phase.
9. From 7 and 8, all information in B is representable by this set of photon state occupation numbers, without phase.

●

Consider box B_1, containing photons with total energy E_1 in B_1's rest frame. B_1 is free to move in larger box B_2, with some set of allowed states and energy E_1' in B_2, measured in B_2 excluding internal energy E_1. Total energy in B_2 is $E_2 = E_1 + E_1'$. To enumerate all the ways B_2 can contain total energy E_2, one must include variation of B_1, division of E_2 into E_1 and E_1', state constructed in B_1, and state of B_1 in B_2. There is not any reason to assume complete freedom in choosing B_1 and E_1 for given B_2

and E_2: B_1's freedom to move with remaining energy E_1' in the (remaining) volume of B_2 presumably must satisfy boundary conditions in a way related to its energy content. I discuss this in more detail for Lemma 4.1 and Theorem 4.

For the following analysis, I found it useful to distinguish internal and external information for B_1 in the above system:

Definition 3. *Information coded as photon state occupation number in B_1 is said to be 'internal information' of B_1 with internal energy E_1 (substantially at rest).*

Definition 4. *Information coded in B_2 by size relationship of B_1 to B_2, decomposition of E_2 into E_1 and E_1', and state of B_1 in B_2 is said to be 'external information' of B_1 with E_1 in B_2 with E_2.*

As used herein, 'state of B_1 in B_2' refers to external states of B_1, not to internal states in B_1.

In this portion of the analysis, I have not found it necessary to discuss the size of B_1 as measured internally thereto, nor do I have reason to expect such an internal measurement to agree in general with an external measurement.

Because of the intricacies of labeling states in B_2 when B_2 contains boxes B_1 of photons, the following theorem appears useful:

Theorem 3. *Amount of information represented in isolated box B_2 does not depend on decomposition into internal and external information of a contained box B_1.*

1. With Def 4, write total energy in B_2 as $E_2 = E_1 + E_1'$.
2. Opening B_1 and freeing its contained photons into B_2 does not change E_2. (B_2 remains isolated.)
3. From Cor 1.1, and 1 and 2, amount of information in B_2 is not changed by process as in step 2.
4. From Def 3 and 4, and 3, amount of information represented in isolated box B_2 does not depend on decomposition into internal and external information of B_1 in B_2.

●

Note that statistical physics uses the contrivance of at least weak coupling to an external system to prove entropy increase theorems. B_2 is an isolated system, and the second law of thermodynamics does not describe processes occurring in the entire box. It might be used to approximately describe coupled subsystems within B_2 the internal states of at least one of which were ignored, but that would seem to obscure the underlying physics.

We might have constructed a box B_2 containing total energy E_2, in the form of combinations of boxes of electromagnetic energy and their states in B_2, together with photons in the remaining volume of B_2, in any combination. In the same way as for Theorem 3, we obtain

Corollary 3.1. *The information content of a box containing photons within or without contained boxes depends only on its energy content and geometry.*

Next, consider a small box containing two photons. That these photons be contained requires only that they not transport energy away from the box. Classically, this does not require that the field strength be zero outside the box: an example of this is the nonpropagating field present outside a dielectric when light is reflected within by total internal reflection. I chose two photons, so as to be able to give them equal and opposite momenta; this allows a rest state to be defined for a box which itself has no mass. As an example out of sequence, such a 'box' might be a region of space with higher index of refraction, such as a tiny gravitational pit.

Definition 5. *A soft box is one which contains photons while allowing nonpropagating nonzero field outside.*

Definition 6. *Let B_0 denote a soft box which itself has no mass, appearing to contain when seen as substantially at rest two photons with equal and opposite momenta in its lowest state, with total rest contained photon energy E_0.*

('standing wave' 'photons' are *not* used in defining B_0. Representation by 'standing waves' is discussed hereinafter in Section 3, in relation to A-Cor 5.1.)

Note that Theorem 3 applies to soft boxes B_0 in B_2.

Were B_0 in a pure quantum state in larger box B_2, it could be thought of as nonlocalized and 'filling' B_2. Were B_0 to be 'opened', remarkably little would happen: energy would be conserved, information would be conserved, and the two photons would 'fill' B_2. There does not appear to be any natural break in the range from a totally soft, or nonexistent, box to a fully containing, or 'hard', box. Therefore, the allowed states of motion for any softness of B_0 could be expected to satisfy the 'softest' condition, that the two photons simultaneously satisfy electromagnetic boundary conditions in larger box B_2.

Next, I extend the representation of information as photon state occupation numbers to include information represented by mass particles in the nonrelativistic limit $v \ll c$. I wish to use the following analysis to explore a conceptual foundation for special relativity. Therefore, for this derivation, I use nonrelativistic quantum mechanics (NRQM) and ignore terms of order $(v/c)^2$ and higher (nonrelativistic approximation, 'NRA'). Hereinafter the derivation is tested fully relativistically, but to use a relativistic derivation here might be logically improper.

Generally, the derivation proceeds by exploring the states of motion allowed by electromagnetic boundary conditions for B_0 containing energy E_0 in a larger box B_2. It appears that there is one condition which allows these states to in all 'containers' be in one to one correspondence with NRQM allowed states of a mass particle P having (rest) mass m_p.

Lemma 4.1. *In NRA, for states of B_0 moving with velocity $v \ll c$ in B_2 allowed in EMT, v is such that $\lambda_b = c^2h/E_0v$ fits into B_2 an integer number of times.*

1. From EMT and Def 6, for B_0 in an allowed state (of motion) in B_2, each of B_0's associated two photons must be in an electromagnetic eigenstate in B_2.
2. From EMT, NRQM, and Def 6, for B_0 substantially at rest, its contained energy E_0 is substantially equally divided between its associated two photons.
3. From NRQM, and 2, each such photon has frequency $f_0 \approx E_0/2h$.
4. For B_0 moving in B_2 with $v \ll c$, there exists a preferred direction, the line of motion. Decomposing electromagnetic energy in B_0 into two photons, one traveling forward and one backward, we obtain (nonrelativistically) two shifted frequencies,

$$f_+ \approx f_0\left(\frac{1}{1-v/c}\right) \qquad \text{forward,}$$

$$f_- \approx f_0\left(\frac{1}{1+v/c}\right) \qquad \text{backward.}$$

5. From EMT, for both f_+ and f_- to simultaneously satisfy electromagnetic boundary conditions in B_2, they must be able to be spatially in phase at B_2's boundaries.

6. From EMT, NRQM, and 4, spatial periodicity λ_b with which photons with frequencies f_+ and f_- are in phase is equal to the speed of light divided by their difference (beat) frequency f_d,

$$\begin{aligned}\lambda_b &= c/f_d \\ &= c/(f_+ - f_-) \\ &\approx c\Big/\left(f_0\left(\frac{1}{1-v/c} - \frac{1}{1+v/c}\right)\right) \\ &\approx c\Big/\left(f_0\left(\frac{2v/c}{1-v^2/c^2}\right)\right) \\ &\approx \frac{c^2}{2f_0v}(1-v^2/c^2)\end{aligned}$$

which, neglecting terms of order v^2/c^2 in comparison to 1 (NRA), is

$$\lambda_b \approx \frac{c^2}{2f_0v}.$$

Substituting from step 3 $E_0/h = 2f_0$,

$$\lambda_b \approx \frac{c^2h}{E_0v}.$$

7. From 5 and 6, for states of B_0 in B allowed in EMT, v is such that such λ_b fits into B_2 an integer number of times.

●

Theorem 4. *With EMT and NRQM, in NRA, allowed states of B_0 in B_2 are in one to one correspondence with allowed states of a mass particle P with rest mass m_P if and only if ('iff') $E_0 \approx m_Pc^2$, for $v \neq 0$.*

1. In NRQM, a mass particle has characteristic wavelength $\lambda_P \approx h/m_Pv$, and its 'allowed states' in B_2 are defined to be those for which λ_P fits into B_2 an integer number of times, with $v \neq 0$.
2. From Lemma 4.1, in NRA, $\lambda_b \approx c^2h/E_0v$.
3. From 1 and 2, both λ_b and λ_P are linear in $1/v$.

4. From Lemma 4.1, and 1 and 3, in NRA, we can construct B_0 with suitable energy E_0, to have exactly the same spectrum as P of allowed states in B_2, by setting $\lambda_b = \lambda_P$.
5. From 1, 2, and 4, in NRA, B_0 and P would have the same spectrum of states in B_2 if $c^2h/E_0v \approx h/m_Pv$.
6. From 5, in NRA, B_0 and P would have the same spectrum of states in B_2 if $E_0 \approx m_Pc^2$.

   (Note that for $v \ll c$, kinetic energy is of order $(v/c)^2$, and thus in NRA does not appear.)
7. From 6, in NRA, setting $E_0 \approx m_Pc^2$ puts the allowed states of B_0 in B_2 into one to one correspondence with the allowed states of P in B_2, for every allowed velocity.
8. From Lemma 4.1, and 1, boundary conditions restrict allowed wavelengths to the same integer fits for both B_0 and P in B_2.
9. From EMT, NRQM, and 8, solution to the boundary value problem is unique, giving a set of allowed wavelengths $\{\lambda_j\}$.
10. From 1, 2, and 9, requiring that allowed states in NRA of B_0 in B_2 be in one to one correspondence with allowed states of P in B_2 for all sizes and shapes of B_2 implies $\lambda_b = \lambda_P$.
11. From 6 and 10, in NRA, allowed states of B_0 in B_2 are in one to one correspondence with allowed states of P in B_2 iff $E_0 \approx m_Pc^2$, for $v \neq 0$.

●

Exclusion of $v=0$ as a state of B_0 is discussed hereinafter (see discussion following Cor 14.1 and A-Cor 2.2).

Corollary 4.1. *In NRA, with $E_0 = m_Pc^2$, the external information of B_0 in B_2 is the same as the external information of P in B_2.*

1. From Theorem 4, in NRA, allowed states of B_0 in B_2 and P in B_2 are in one to one correspondence iff $E_0 \approx m_Pc^2$.
2. From 1, for any code constructed from allowed states of B_0 can be made an identical code constructed from allowed states of P, and vice versa.
3. From Def 1, only codes constructed from allowed states can represent information.
4. From 1–3, with $E_0 = m_Pc^2$, any information representable externally by B_0 in B_2 is so representable by P in B_2, and vice versa.

●

Note that, for B_2 much larger than B_0, nearly all information represented by B_0 in B_2 is external to B_0. Only a few bits of internal information might be contained in B_0, for example, represented by projections of photon spins. A mass particle can also carry a few bits, for example, corresponding to 'charge' sign, spin, particle or antiparticle, descriptors. If more descriptors are desired for P, a 'larger' 'box' with more accessible photon states might be constructed, and/or more photons used. Without departing from the spirit of the argument, I assume for this part of this derivation that the internal information of the box of photons can be set equal to the internal information of the mass particle. This is discussed formally herinafter, and this part of the derivation is replaced by axiomatic treatment appearing in Section 3. With this assumption, we obtain:

Corollary 4.2. *In NRA, the same amount of information is represented by a mass particle P in* B_2 *as is represented by* B_0 *with* $E_0 \approx m_P c^2$ *in* B_2.

1. From Cor 4.1, P and B_0 with $E_0 \approx m_P c^2$ have the same amount of external information in B_2.
2. By assumption, P and such B_0 represent the same amount of internal information.

●

Corollary 4.3. *In NRA, information carried by mass particle P with mass* m_P *in box* B_2 *can be represented by any configuration of photons in* B_2 *if and only if their total energy* $E_2 \approx m_P c^2$.

1. From Cor 4.2, B_0 in B_2 represents the same amount of information as P in B_2 with mass m_P iff $E_0 \approx m_P c^2$.
2. From Cor 3.1, any configuration of photons and/or boxes of photons with energy E_0 in B_2 represents the same amount of information.
3. From 1 and 2, photons in B_2 represent the same amount of information as P in B_2 iff $E_2 \approx m_P c^2$.

●

We see in Theorem 4 and its corollaries a remarkable result. Conservation of information was derived from NRQM and energy conservation. In this nonrelativistic approximation, any photon configuration with energy E can be used to represent the same amount of information as a mass particle P if and only if $E \approx m_P c^2$. Thus, *if any physical process can interconvert mass and energy, information conservation and energy conservation are both satisfied if and only if the energy and mass satisfy the relation* $E \approx mc^2$.

Definition 7. *Call a photon representation for the information represented by a mass particle, using total energy $m_P c^2$, a 'nonrelativistic photon particle representation', or 'NRPPR'.*

Note that this derivation so far has not used Einstein's special theory of relativity.

Conversion of mass to energy can be observed without use of moving frames of reference, for example, in nuclear power plant operation or radioisotope decay. By observation, $E \approx mc^2$. Therefore, information and energy are both conserved in the conversion of mass to energy. We can include this observation in itself, or by making information conservation a law in the theory: Corollary 4.3 shows that these two approaches are equivalent. With NRQM, expressing the result of Corollary 4.3 and the observation that $E \approx mc^2$ as a law of information conservation allows Corollary 1.1 to be extended to systems in which mass and energy are interconverted. I find it more intuitively satisfying to state this as a law of conservation of information. Rather than drop reference to this interconversion, I refer to it explicitly in several of the following theorems. In some of the following theorems, I use a photon representation for information represented by mass particles.

Photon representations (or other energy propagating at the speed of light) are especially pleasant to use, because photons leave no awkward rest mass after reception by a detector. This appears related to the special role of the speed of light. Note that this comment does not itself make use of special relativity. More on this later.

Theorem 5. *With NRPPR, information content of box B_2 depends on its energy content, independent of partitioning into photons, boxes of photons, mass particles, or combinations thereof.*

1. From Cor 4.2, information and energy contained in B_2 are unchanged if all mass particles P are replaced by boxes B_0 with energy $E_0 \approx m_P c^2$.
2. From Cor 3.1, and 1, information content of B_2 then depends on E_2, without details of boxes, particles, or photons in B_2.

●

Note that Theorem 5 allows us to choose any convenient representation for information in subsequent analyses.

Following customary practice, we can attribute energy and information as follows:

Definition 8. *All energy in box B_2 containing only one box B_1 or particle P is attributed to B_1 or P, respectively.*

Definition 9. *All information in box B_2 containing only one box B_1 or particle P is attributed to B_1 or P in B_2, respectively.*

With these definitions, Theorem 5 yields directly

Corollary 5.1. *With NRPPR, information represented by box B_1 or particle P in box B_2 depends only on its energy and the geometry of B_2, counting mass m as energy mc^2.*

●

Note that we can choose B_2 (finite) as large as we wish. Thus,

Corollary 5.2. *With NRPPR, in a finite universe, information represented by any configuration composed entirely of electromagnetic energy and/or mass depends only on its total energy, counting mass m as energy mc^2.*

●

Note that most of the information represented by any energy configuration much smaller than the universe can be external information, as is discussed more fully hereinafter in relation to Theorem 10 and Definition 17. Most of this external information would not appear in operation of a transmitter and receiver for information transfer: only those states (in the universe) of the transmitted box which allow it to pass through a receiver while the receiver is sensitive (on) are useable for encoding information for transmission and reception. With this interpretation of accessibility of external information, there seem to be no observational difficulties in generalizing Corollary 5.2 to

Corollary 5.3. *With NRPPR, information represented by any electromagnetic energy and/or mass configuration depends only on its total energy, counting total mass m as energy mc^2.*

Finiteness of universe is formally treated hereinafter.

Theorem 6. *With NRPPR, total information contained within isolated box B_2 is substantially proportional to total energy E_2 contained therein.*

1. Choose for reference amount E_R of energy, which in B_2 can represent amount I_R of information.
2. For any energy E_2 in B_2, choose an approximate common divisor E' of E_R and E_2, $E_2 \approx n_2 E'$, $E_R \approx n_R E'$.
3. From Theorem 5, the information content of B_2 can be approximately preserved by replacing the configuration having energy E_2 with a set of n_2 noninteracting boxes B_0, each with energy E'.
4. Express the states of the system in B_2 as product states of the n_2 B_0 states.
5. From Def 1, and 4, noting logarithmic definition of amount of information, amount I_2 of information represented by specifying one product state out of all is the sum of amounts I' of information represented by specifying one state of the possible single box B_0 states in B_2 for each B_0.
6. From 2, E' is the same for each single B_0 configuration in B_2.
7. From Cor 5.1, and 6, I' is the same for each single B_0 configuration.
8. From 3 and 7, $I_2 = n_2 I'$.
9. From 1, 2, and 7, $I_R = n_R I'$.
10. From 2, 8, and 9, $I_2 \approx (E_2/E_R) I_R$.
11. From 1–10, any pair of energies and associated information contents in box B_2 can be related in this way to a reference or to each other.

In the axiomatic form of this theory, this result appears formally exact, which exactness appears related to commensurability of all frequencies possible in an observer's universe. This is treated more fully in

Section 3 hereinafter, for example, in A-Theorems 3, and 4 and its corollaries, and discussion relating thereto.

For an 'infinite enclosure', the common divisor may be regarded as exact, and this result regarded as exact.

By inspection, from Corollary 5.3 and Theorem 6,

Corollary 6.1. *With NRPPR , amount of information represented by any electromagnetic energy and/or mass configuration is substantially proportional to its total energy, counting total mass m as energy mc^2.*

●

(This result is also exact in the axiomatic form.)

From the substantial (exact in axiomatic form) proportionality of information to energy in Theorem 6, we obtain directly

Corollary 6.2. *With NRPPR, in any box B_2, conservation of information implies substantially conservation of energy.*

●

(This result is also exact in the axiomatic form.)

From Corollaries 1.1 and 6.2, we obtain:

Corollary 6.3. *With NRQM and NRPPR, conservation of information is substantially equivalent to conservation of energy.*

●

(This result is also exact in the axiomatic form.)

From conservation of information, substantial (exact in the axiomatic form) proportionality of information to energy, and proportionality of energy to square of amplitude in EMT and NRQM, we obtain a result helpful for computation:

Corollary 6.4. *With NRQM and NRPPR, in isolated box B, amplitudes A_j of different co-possible modes j, each if transcribably present representing at*

least amount I_j of information, by which information I might be encoded, are substantially proportional to $1/I_j^{1/2}$.

1. From Theorem 6, information is substantially proportional to energy.
2. From EMT and NRQM, with NRPPR, energy is proportional to square of amplitude.
3. From 2, 'occupation number' of representation of mode j is proportional to A_j^2. ('occupation number' as in NRQM)
4. Representation in mode j if transcribably present represents at least amount I_j of information. ('occupation number' = 1)
5. From co-possibility of modes, substantially all information I may be obtained by detection of exclusively any one mode j of representation. (a definition of 'co-possibility'; note that in NRQM an amplitude corresponding to 'occupation number' < 1 might seem definable)
6. From Theorem 1, I is conserved in B.
7. From 4–6, detection of substantially all of I as represented in mode j involves detection of I/I_j 'occupation' in mode j.
8. From 1, 3, and 7, amplitude A_j for detecting representation in mode j is substantially proportional to $1/I_j^{1/2}$.

●

(This result appears exact in axiomatic form.)

Note that 'co-possibility' is a kind of 'coherence' condition.

This result might offer insight into 'equipartition', NRQM 'phase space' arguments, parallel impedances, Fermi's 'Golden Rule', etc.

2.4 SEPARABILITY OF INTERNAL AND EXTERNAL INFORMATION

In this Section 2.4, parts of NRQM and EMT are used as a starting place for derivation of behavior of moving systems. Such starting theoretical basis is not formally relativistically covariant, and is not 'global' in the sense that the formalism is defined only in local flat spacetimes without intrinsic restriction formally incorporating effects due to (a) 'size' of universe. This is formally similar to regarding the enclosure as infinite.

For this reason, effects of finite 'size' of universe are treated by first extending NRQM and EMT on their own terms, as if in an infinite enclosure, to relativistic and global form (Theorem 7 through Cor 8.4, and

Theorem 4′), and then treating effects of finiteness of 'enclosure' as limitations on domain of applicability of, and/or corrections to, results so obtained (Cor 8.5–8.16, and Theorems 12, 13, and 15 and their lemmas and corollaries). Such a procedure, although in some respects awkward and requiring care to assure that each result is not used before it is derived, respects the limitations of nonrelativistic and nonglobal form of the starting theories, and allows an information picture to be derived in the context, and thus shown to be allowed by an internally consistent part, of such starting theories.

In particular, Definitions 10 ('gentle acceleration') and 11 ('gentle scaling') are stated in the context of the formally nonrelativistic and nonglobal nature of the starting theories, and presume therewith an infinite enclosure.

After consideration in this Section 2 of whether an information picture is allowed by prior knowledge, in the axiomatic form presented beginning in Section 3 hereinafter, the starting place is a set of postulates for such information picture, and the formal derivations proceed more directly, with 'finiteness' of 'enclosure' appearing in a more integral way.

With the correspondence between B_0 and P constructed for Theorem 4 (NRPPR), we could, for example, choose to describe any system in a box B_1 containing mass particles and/or electromagnetic energy by a set of photon state occupation numbers.

Suppose that B_1 is moving with velocity $v \ll c$ in a box B_2:

***Theorem** 7. In NRA, NRPPR applies to particles and/or photons in B_1 moving at velocity $v \ll c$ in B_2.*

1. From EMT, NRQM, and Lemma 4.1, seen by an observer in B_2, the wavelength of B_0 still agrees with that of P if and only if $E_0 \approx m_P c^2$, for the same v. (Kinetic energy in NRPPR is of order $(v/c)^2$ and does not appear in NRA.)
2. From EMT and NRQM, boundary conditions at the walls of B_1 restrict P and B_0 to the same integer fit.
3. From Def 1, and 1 and 2, if and only if $E_0 \approx m_P c^2$, the allowed states of B_0 and P in B_1 will be in one to one correspondence, and B_0 and P thus represent the same amount of information in B_1.
4. From EMT, for each angle between the direction of propagation and v, the Doppler shift is linear in frequency.
5. From 4, frequencies of photon states transform in proportion for each direction of propagation considered separately. (e.g., forward, backward, perpendicular to v)

6. From EMT and NRQM, for $v \ll c$, in NRA, the nonrelativistic Doppler shift is linear in v.
7. From 5 and 6, for B_0 moving with velocity $v' \ll c$ relative to B_1, in NRA, B_0's two photons bear the same relation to allowed photon states in B_1 as when B_1 is substantially at rest in B_2 and B_0 has velocity v' in B_1 in B_2.
8. From Lemma 4.1, Theorem 4, and 3 and 7, Lemma 4.1 and Theorem 4 and its corollaries apply in B_1 moving with $v \ll c$ in B_2.
9. From Def 7, and 8, NRPPR applies in B_1 moving with $v \ll c$ in B_2.

●

In changing speed, the box would experience acceleration. To avoid analyzing effects of acceleration of a containing box on boundary conditions for a particle P and a photon representation B_0, I treat such velocity changes in the limit of slow acceleration, through Corollary 8.4. Such acceleration would not begin or end abruptly, but rather with large time uncertainties, which time uncertainties would be negligibly small compared to the time for acceleration. Effects of acceleration are treated explicitly in Section 3 hereinafter.

Definition 10. *Call the long time limit of slow acceleration through a fixed velocity change, with large time uncertainty for beginning and ending acceleration, 'gentle acceleration' or 'GA'.*

Note that in finite enclosure B, with linear size R_B, allowed states of motion of contained box B_1 with linear size $R_1 < R_B$ are discrete, so, for R_B finite, GA becomes the making substantially one by one of smallest distinguishable velocity increases substantially in the desired direction by transitions among B_1's allowed states in B. This is discussed more fully in Section 3 hereinafter.

Corollary 7.1. *With NRPPR and NRQM, in NRA, during GA of box B_1, boundary conditions for mass particle P in B_1 are in effect the same as for a corresponding B_0 in B_1.*

1. In EMT and NRQM, 'allowed' electromagnetic and/or mass particle states depend continuously and continuously differentiably on geometry.
2. From 1, for small acceleration, these allowed states depend continuously and continuously differentiably on the acceleration. (One might

think of this as dependence on how far the walls of the box are displaced from the position they would have reached at constant velocity, by the acceleration acting during the time for a photon, B_0, or P to 'traverse' the box.)

3. From Def 10, and 1 and 2, in NRA, GA of B_1 does not change these boundary conditions differently with respect to contained P and B_0 from those which apply for constant velocity.
4. From Theorem 7, for any uniform velocity $v \ll c$, NRQM boundary conditions on P in B_1 are in effect the same as EMT boundary conditions on a corresponding B_0.
5. From 3 and 4, in NRA, during GA of B_1, boundary conditions on P in B_1 are in effect the same as for a corresponding B_0 in B_1.

●

One might think of GA of B_1 as if in an infinite enclosure as a 'quasi-static' process, in the sense used in thermodynamics.

To simplify the following argument, I have divided it into eleven lemmas, a theorem, and sixteen corollaries. *The following analysis is derived for systems containing electromagnetic energy and/or mass particles only.*

Lemma 8.1. *In NRQM, with NRPPR, in NRA, neglecting terms of order* $(R_1/R_B)^2$ *in comparison to* 1, *in box B with linear size* R_B, *gently accelerating by* $v \ll c$ *contained box* B_1 *with linear size* $R_1 \ll R_B$ *containing information* I_1 *leaves both amount and representation of* I_1 *unchanged.*

1. With NRPPR, for every mass particle P construct its corresponding B_0.
2. From Theorem 7, Cor 7.1, and Def 10, Theorem 7 and Corollary 7.1 apply for all intermediate velocities during gentle acceleration of B_1.
3. From 1 and 2, each mass particle P is always in one to one correspondence with its original B_0.
4. From EMT and Def 6, allowed states of each B_0 correspond to occupation by its associated two photons of photon states in B_1.
5. From NRQM and EMT, using $E = hf$, and summing the terms in Lemma 4.1 step 4,

$$< E(v) > = E_0\left(\frac{1}{1 - v^2/c^2}\right) = E_0(1 + v^2/c^2 + \cdots).$$

6. From EMT, and 5, for $R_1/R_B \ll v/c$, total probability that acceleration to $v \ll c$ induce transitions between photon states is at most of order $(v/c)^2$.
7. From NRQM, EMT, and Def 1 and 10, least observable GA in B of B_1 containing B_0 or photon is such that, were the B_0 or photon to be released in B after GA in B_1, its wavelength would fit into B along the direction of GA one more time than before GA.
8. From EMT and NRPPR, photon and/or B_0 states in B_1 have wavelength spacings $\sim R_B/R_1$ times as great as corresponding photon states in B.
9. From EMT and NRPPR, and 7 and 8, least observable GA would have probability of order $(R_1/R_B)^2$ of inducing transitions of photon or B_0 among allowed states in B_1.
10. From 6 and 9, neglecting terms of order $(v/c)^2$ (NRA) and $(R_1/R_B)^2$, gentle acceleration to $v \ll c$ does not induce transitions among photon states in B_1.
11. From Def 6, and 10, to that order, similarly transitions are not induced among allowed states of any B_0 in B_1.
12. From Theorems 4 and 7, Cor 7.1, and 3 and 11, with NRPPR, to that order, similarly transitions are not induced among allowed states of mass particles in B_1.
13. From 12, with NRPPR, in NRA, to that order, the state of motion of B_1 can by GA be changed by $v \ll c$ without changing its internal state.
14. From 13, with NRPPR, in NRA, to that order, any representation of information by states in B_1 is unchanged.
15. From Def 1, and 14, with NRPPR, in NRA, to that order, information contained in B_1 is unchanged as to both amount and representation.

●

Note that criterion of least observable GA in step 7 treats B_1 as containing only one photon or one B_0, giving the coarsest spacings of velocity of B_1 in B; thus, in NRA, $(R_1/R_B)^2$ serves as an estimate of the largest probability that least GA induce state transition of one photon or B_0 in B_1. GA of B_1 containing more than one B_0 and/or photon is treated formally in A-Theorem 6.

For R_B finite, n successive transitions to higher velocity allowed states of B_1 in B cumulate probability of inducing transition(s) of B_0 or photon among states in B_1, which, for $n \ll (R_B/R_1)^2$, cumulates as $\approx n(R_1/R_B)^2$.

Lemma 8.1 is fundamental to formal derivation in this Section 2 of special relativity (SR), and cumulation of probability of transition turns out to give a size and velocity relation as a limitation on SR's domain of applicability *consistent with* such limitation derived from constraint to satisfaction of EMT boundary conditions on B_0 by fitting an integer number of its down shifted photon's wavelength(s) into B. Although one can begin to see indications of these results in Lemma 8.1, the formal basis in the starting theories is nonrelativistic, and, therefore, such results are taken only after extension to relativistic form. Neglecting terms of order $(R_1/R_B)^2$ in comparison to 1 is, in this context, a formal way of respecting the formally nonrelativistic and nonglobal nature of the starting theories, as noted hereinbefore, and seems similar to neglect of B's boundary conditions conventionally presumed, and thus a formal statement of the usually unstated presumption of NRQM and EMT that the enclosure be infinitely large. After derivation of a relativistic form under this formal presumption, such derived form is used to consider effects of finiteness of R_B. This is the awkwardness noted hereinbefore, and the statement of Lemma 8.1 conforms to the requirement of not using a concept or result before it is properly obtained.

Lemma 8.2. *With NRPPR, energy conservation implies a sum rule on photon occupation numbers in ('stationary') box B, $\Sigma_i n_i E_{Bi} = constant$.*

1. Start with B containing any combination of photons and/or mass particles.
2. From Theorems 4 and 5, and 1, using NRPPR, for every mass particle construct its corresponding B_0.
3. From Theorem 2–Cor 6.3, and 1 and 2, form a representation for the contents of B, consisting of photon occupation numbers $\{n_i\}$, including therein the two photons associated with each B_0.
4. From Theorem 5, and 1 and 3, with NRPPR, these photons have the same total energy as was originally in B.
5. From Theorem 2, and 1–4, with NRPPR, every state of the system in B has corresponding to it a representation as in steps 2 and 3.
6. From 5, changes (for example, time development) of the system in B can be represented by changes in this photon occupation number representation (e.g., as action of photon creation and annihilation operators).
7. From EMT and NRQM, to each photon state i in B corresponds an energy E_{Bi}.
8. From 1, 2, 3, and 7, $\Sigma_i n_i E_{Bi} = E$ total in B.

9. From 8, conservation of energy during changes in B implies that the photon occupation number representation always satisfies the sum rule $\Sigma_i n_i E_{Bi}$ = constant. (the constant is E)

●

('stationary', applying NRQM in 'infinite enclosure')

Lemma 8.3. *With NRPPR, in NRA, energy conservation implies the same sum rule in box B moving with velocity $v \ll c$ as in B 'stationary'.*

1. Start with B containing any combination of photons and mass particles.
2. Form a photon occupation number representation as in steps 2 and 3 of Lemma 8.2.
3. As in Lemma 8.1, gently accelerate B to velocity v, without changing its internal state.
4. From Theorems 4 and 7, application of NRPPR allows construction of a box B for each mass particle in B, whose two photons occupy the same photon states in B moving as they would if B were 'stationary'.
5. From 3, photon state occupation numbers in B moving are unchanged from B 'stationary'.
6. From 4 and 5, all of the state of the system in B moving (mass particle(s) and/or photon(s)) has corresponding to it the same photon occupation number representation as in B 'stationary'.
7. From EMT, in the long time limit, time averaged electromagnetic energy in a photon state is shifted by velocity as the sum of the shifted forward and backward components.
8. From EMT, the largest such shifts occur for propagation parallel and antiparallel to velocity v.
9. From EMT and NRQM, using $E = hf$, and summing terms in Lemma 4.1 step 4,

$$\langle E(v) \rangle = E_0 \left(\frac{1}{1 - v^2/c^2} \right) = E_0 (1 + v^2/c^2 + \cdots).$$

10. From 9, in NRA (neglecting terms of order $(v/c)^2$ and higher), $E(v) = E$.

11. From 8 and 10, in B moving with $v \ll c$, in NRA, the same energies E_{Bi} appear for each photon state i as in B 'stationary'. (This is a key step, because it is isotropic in B moving.)
12. From 6 and 11, with NRPPR, in NRA, total energy in B moving is $\Sigma_i n_i E_{Bi}$ with the same n_i and E_{Bi} as for B 'stationary'.
13. From 1–12, step 12 applies for any state of the system in B moving.
14. From 12 and 13, in NRA with NRPPR, energy conservation in B moving implies $\Sigma_i n_i E_{Bi}$ = constant as in B 'stationary'. (constant is E)

●

Lemma 8.4. *With NRPPR, $\Sigma_i n_i E_{Bi}$ = constant implies information conservation in box B.*

1. From Theorems 4 and 5, with NRPPR, a photon representation can be constructed equal in energy and information content to any representation using photons and/or mass particles.
2. From EMT and NRQM, and 1, for such photon representation, $\Sigma_i n_i E_{Bi}$ is its total energy.
3. From Theorems 1 and 4, with NRPPR, conservation of total energy in B implies conservation of information in B.
4. From 1–3, with NRPPR, $\Sigma_i n_i E_{Bi}$ = constant implies information conservation in B.

●

Lemma 8.5. *With NRPPR, $\Sigma_i n_i E_{Bi}$ = constant for any photon representation of a system in B 'stationary' implies substantially energy conservation in B, for any system in B formed of combinations of photons and/or mass particles.*

1. From Lemma 8.4, $\Sigma_i n_i E_{Bi}$ = constant implies information conservation in B.
2. From Cor 6.2, with NRPPR, conservation of information implies substantially conservation of energy in B for any representation of that conserved information. ('substantially' from Theorem 6)

●

Lemma 8.6. *With NRPPR, in NRA, energy conservation in B moving with velocity $v \ll c$ is substantially equivalent to information conservation in B so moving.*

1. From Lemma 8.3, with NRPPR, in NRA, energy conservation in B moving implies $\sum_i n_i E_{Bi} = \text{constant}$.
2. As in Lemma 8.3 steps 7–11, and 1 in NRA, the E_{Bi} are the same in B moving or 'stationary'.
3. From Lemma 8.4, and 1 and 2, $\sum_i n_i E_{Bi} = \text{constant}$ implies information conservation in B, moving or 'stationary'.
4. From Lemma 8.1, in NRA, with any contained system, B can be brought to rest from $v \ll c$ without changing the amount or representation of its contained information.
5. From Cor 6.3, and 4, in NRA, applying Corollary 6.3 to the resulting rest system, information conservation in B moving with $v \ll c$ implies substantially energy conservation in the resulting rest system. ('substantially' from Theorem 6)
6. From Lemma 8.2, and 5, in NRA, information conservation in B moving with $v \ll c$ implies $\sum_i n_i E_{Bi} \approx \text{constant}$.
7. From EMT and NRQM, in NRA, photon energy states E_{Bi} are the same in B moving with $v \ll c$ as in B substantially at rest.
8. From 6 and 7, information conservation in B moving with $v \ll c$ implies energy in photon representation is substantially conserved in B so moving.
9. From Theorems 4 and 7, NRPPR applies in NRA in B moving with $v \ll c$, and photon representation is equivalent in energy to any representation of the same information with photons and/or particles.
10. From 8 and 9, in B moving with $v \ll c$, in NRA, with NRPPR, information conservation implies substantially energy conservation for all representations.
11. From 3 and 10, with NRPPR, in NRA, energy conservation in B moving with $v \ll c$ is substantially equivalent to information conservation in B so moving.

●

(The 'substantially' is from Theorem 6, and would not appear in the axiomatic form; but this entire portion of the derivation does not appear in the axiomatic form.)

Lemma 8.7. *With NRQM and NRPPR, in NRA, all physical processes contained in B moving with velocity $v \ll c$ conserve information.*

1. In NRQM, all allowed physical processes in B moving conserve energy.
2. In NRA, with NRPPR, energy and mass interconversion is specified to be $E \approx mc^2$.
3. From Lemma 8.6 (step 3), and 1 and 2, in NRA, all physical processes contained in B moving with $v \ll c$ conserve information. ('substantially' in Lemma 8.6 was in other direction (step 5): see Lemma 8.6 step 3 for this direction.)

●

Definition 11. *Call changing the size of a box proportionately, with large time uncertainty for beginning and ending that change, in the limit of long time for such change, 'gently scaling' or 'GS'. (used as noun, verb, or adjective).*

In finite enclosure B with linear size R_B containing B_1 with linear size $R_1 \ll R_B$, GS becomes the making of least transitions substantially one by one for all transitions among any discrete states affected by changing the size of B_1 in B.

For several of the following lemmas to Theorem 8, the following condition ('C 1') is used:

Condition 1. *For box B with linear size R_B containing box B_1 with linear size $R_1 \ll R_B$, neglecting terms of order $(R_1/R_B)^2$ in comparison to 1, for GS such that R_1 always $\ll R_B$, B_1 with velocity $v \ll c$ in B*

(This conforms to formally neglecting finiteness of enclosure.)

Lemma 8.8. *C 1; gently scaling B_1 does not change amount or representation of its contained information I_1.*

1. From EMT, NRQM, and Def 1 and 11, least observable GS of B_1 containing photon, B_0, or mass particle is such that after GS wavelength of such photon, B_0, or mass particle would fit into B one time more or less than before such GS.
2. From EMT and NRQM, least allowed wavelength change of photon, B_0, or mass particle in B_1 is R_B/R_1 times as great as least such change in B.

3. From EMT and NRQM, and 1 and 2, least observable GS has probability of order $(R_1/R_B)^2$ of inducing change of state of photon, B_0, or mass particle in B_1.
4. From 1 and 2, for R_1 before GS and R_1' after GS both $\ll R_B$, least number n of least steps of GS necessary for GS from R_1 to R_1' is $n \ll R_B/R_1$.
5. From 3 and 4, cumulative probability of inducing transition of photon, B_0, or mass particle among states in B_1 during GS from R_1 to R_1' is $\sim n(R_1/R_B)^2 \ll R_1/R_B$, of order $(R_1/R_B)^2$.
6. From 1, 4, and 5, neglecting terms of order $(R_1/R_B)^2$ in comparison to 1, GS as in steps 1 and 4 preserves all states in B.
7. From 6, to that order, any coding of information by states in B is preserved by such GS.
8. From Def 1, and 7, to that order, such GS preserves both amount and representation of I_1.

●

Lemma 8.9. *For GS, NRPPR applies in a box B′ with changed size the same as in B.*

1. From EMT, NRQM, and Def 6, boundary conditions at the walls of B are in effect the same for a mass particle P and the two photons of its corresponding B_0.
2. From EMT, NRQM, and Def 11, these boundary conditions are not affected by GS other than to be applied at the walls in their new locations.
3. From 1 and 2, GS of B at all times keeps P in correspondence with the same B_0 *with the same photon state occupation in B.*
4. From 3, for GS, NRPPR applies in box B' with changed size the same as in B.

●

Intuitively, changing to any other photon states during GS would represent a discrete 'error', and, therefore, would compete only to an extent $\sim(R_1/R_B)^2$ (square of matrix element) with following GS in the same photon states.

Note that at any time during GS of B, we could apply the argument that both information and energy are conserved iff P were replaced by the B_0 photons which had 'tracked' with it during deformation. This seems to have interesting consequences.

With this interpretation and Lemma 8.9, we can discuss both energy and information in B deformed by GS.

Lemma 8.10. *C*1; *with NRPPR, all physical processes in B deformed by GS conserve energy and information.*

1. From EMT and NRQM, all physical processes in B deformed which do not interconvert mass particles and photons conserve energy.
2. From Cor 1.1, and 1, all processes in B deformed not interconverting mass particles and photons conserve information in B.
3. From Lemma 8.9, NRPPR under GS in B deformed by GS allows both energy and information to be conserved during interconversion of mass particles and photons.
4. From 2 and 3, with NRPPR, under GS, all physical processes in B deformed by GS conserve energy and information in B deformed.

●

Note that Lemmas 8.7 and 8.10 can be applied in any sequence. Thus, putting together Lemmas 8.7 and 8.10, we obtain the useful result:

Lemma 8.11:. *C*1; *with NRPPR, all physical processes in* B_1 *deformed and/or moving with* $v \ll c$ *conserve both energy and information.*

●

Lemmas 8.1 and 8.8 offer a process for changing the state of a box without affecting its information content:

Definition 12. *Call allowing a box to 'expand' by GS, obtaining energy thereby, then returning this energy to the box as energy of motion, or vice versa, 'gentle conversion' or 'GC'.*

Theorem 8. *C*1; *with NRPPR, in NRA, for a moving box* B_1 *initially containing energy* E_1 *for* B_1 *substantially at rest, contained alone within* *box B containing energy E_2; the amount and representation of information I_1 represented internally in B_1 is independent of the amount and representation of information I_1' represented externally by B_1 in B; the amount and representation of I_1' is independent of the representation of I_1 for all physical processes in B_1; and I_1 and I_1' are separately conserved.*

1. Starting with B_1 substantially at rest in B, place in B_1 information I_1 represented by any combination of photons and/or mass particles.
2. Gently change B_1's state of motion to an allowed velocity v of B_1 in B representing any desired amount of information I_1' thereby, subject to $v \ll c$. Then isolate B.
3. From Lemma 8.1, neglecting terms of order $(R_1/R_B)^2$ in comparison to 1, step 2 does not change either amount or representation of I_1.
4. Using GC, change to any desired representation for I_1'.
5. From Lemmas 8.1 and 8.8, Def 12, and 4, neglecting terms of order $(R_1/R_B)^2$ in comparison to 1, step 4 does not change either amount or representation of I_1.
6. From Lemma 8.11, and 4, all subsequent physical processes in B_1 conserve I_1.
7. From Lemma 8.11, Def 12, and 5 and 6, to that order, energy E_2 is conserved in B. (Note that, for infinite enclosure, common divisor in Theorem 6 may be regarded as an exact common divisor, and Theorem 6 in such case may be regarded as exact.)
8. From EMT and NRQM, specifying a state of B_1 with total energy E_2 in B consists of specifying
   - 8.1 division of E_2 between initial internal energy E_1 in B_1 substantially at rest and energy E_1' associated with state of B_1 in B, $E_2 = E_1 + E_1'$; and initial size relationship between B_1 and B, with B_1 'substantially' at rest: these together are equivalent to specifying ratio of I_1 to I_1';
   - 8.2 state in B_1 (for 8.1 fixed, this is a code element in the representation for I_1); and
   - 8.3 state of B_1 in B (for 8.1 fixed, this is a code element in the representation for I_1').
9. From Lemma 8.11, energy is conserved in B_1.
10. From 7–9, neglecting terms of order $(R_1/R_B)^2$ in comparison to 1, E_1, E_2, and, therefore, E_1', are constant.
11. From EMT and NRQM, initial size relation between B_1 and B is a fixed initial statement, independent of subsequent changes by GC.
12. From 8, 10, and 11, I_1 and I_1' specify all remaining information in B. ('8' as including its substeps)

13. From Theorem 1, and 7, to that order, total amount I_2 of information in B is conserved.
14. From Theorem 5, and 12 and 13, $I_1 + I_1' = \text{constant}$.
15. From 6 and 14, all physical processes in B_1 conserve I_1'.
16. From Def 1, a code element in a representation for I_1 bits of information is an element of a set containing 2^{I_1} elements, each of which corresponds to a distinct state S_i in B_1; for I_1' there are $2^{I_1'}$ distinct states S_j' of B_1 in B; and the total, $I_1 + I_1'$, corresponds to specification of one code element from a set of $2^{I_1 + I_1'}$ elements.
17. From 16, the set of states representing $I_1 + I_1'$ can be written $\{P_{ij}\} = \{S_i\} \otimes \{S_j'\}$, with elements $P_{ij} = S_i S_j'$.
18. From 16 and 17, with states representing code elements for $I_1 + I_1'$ written as product states, any change in notation for either set $\{S_i\}$ or $\{S_j'\}$ which preserves its cardinality leaves the notation for the other set unchanged.
19. From Def 1, all changes in representation for I_1 or I_1' which conserve I_1 or I_1' preserve the cardinality of their respective sets $\{S_i\}$ or $\{S_j'\}$.
20. From 6, 18 and 19, all physical processes in B_1 do not affect representation of I_1'.
21. From 15 and 20, amount and representation of I_1' are unaffected by physical processes in B_1.
22. From Lemmas 8.1 and 8.8, and 18 and 19, any *change* in representation for I_1' which conserves I_1' leaves both amount and representation of I_1 unchanged.

●

By letting B become the universe, as in Corollaries 5.3 and 6.1, attributing all energy and information in B to B_1, we obtain directly

Corollary 8.1. C1; *with NRPPR, in NRA, for closed system B_1 moving with uniform velocity, amounts of internal and external information are separately conserved for all physical processes in B_1, and B_1's internal and external information representations are independent of each other.*

●

Theorem 8 can be extended, subject to the caveat of limitations to be calculated after such extension, to, in an infinite 'enclosure', all velocities less than the speed of light, as follows in Corollaries 8.2–8.4:

Corollary 8.2. *C*1, *in NRA, with NRPPR, as seen by an observer in* B_1, *all laws of physics have the same form in* B_1, *containing mass particles and/or photons, moving with uniform velocity* $v \ll c$, *as in* B_1 *at rest, and the velocity of light in* B_1 *is independent of* v.

1. Choose a momentum representation for external information I' of B_1.
2. From 1, with such representation, velocity v of B_1 becomes a code element descriptor for I'.
3. From Cor 8.1, neglecting terms of order $(R_1/R_B)^2$ in comparison to 1, internal information I of B is separately conserved and has the same representation, independent of I'.
4. From 3, to that order, every experiment or physical process contained in B_1 does not have access to any of I'.
5. From 2 and 4, to that order, every experiment or physical process contained in B_1 does not depend on v.
6. From 5, to that order, as seen by an observer in B_1,
   6.1 All physical laws have the same form in B_1; and
   6.2 The velocity of light in B_1 is independent of v.

●

However , these laws of physics and velocity of light were all that were used in Theorems 1 through here; thus, we obtain a 'bootstrap' corollary:

Corollary 8.3. *C*1; *in NRA, with NRPPR, Theorem* 1 *through Corollary* 8.3 *apply in closed system B moving with uniform velocity* $v \ll c$.

●

With Corollary 8.3, we can apply the same argument in B moving to a moving system with slightly higher velocity. For any velocity $v < c$, the number of steps of (nonrelativistic) change of frame of reference with relative velocity dv required to reach v is of order $n = v/dv$, not $(v/dv)^2$. Note that criteria NRA and NRPPR also shift their frames of reference with successive steps. Thus, noting squaring of 'matrix element', cumulated effect in shifting to velocity v in an 'infinitely large' 'enclosure' is expressable by a factor of form

$$\left(1+\frac{1}{n^2}(v/c)^2\right)^n = 1+\frac{1}{n}(v/c)^2+\cdots,$$

which goes to 1 in limit of large n. For B_1 in an 'infinite enclosure', as n increases, cumulative error becomes small compared to least change representable in any finite system (finite contained information), and might then be regarded as exactly zero. Thus, we can drop limitation to nonrelativistic cases (NRA, NR of NRPPR), and thus proceed in extending this formalism to treat effects due to cumulative acceleration of B_1 in finite 'enclosure', by looking at effects of velocity (so accumulated, or otherwise acquired).

In the approximation used, finite 'size' of 'enclosure' has been explicitly formally neglected by neglecting terms of order $(R_1/R_B)^2$ in comparison to 1. To calculate effects of finite 'size' of 'enclosure' by the two-step procedure outlined hereinbefore, I have taken this approach so as to formally respect the nonglobal nature of the starting theories, by explicitly neglecting terms which would be zero if R_B were infinite. Also, as noted following Theorem 6, for an infinite enclosure, Theorem 6 may be regarded as exact even in this approximate derivation (the Theorem 6 result is exact in the axiomatic form).

Thus, the formalism which results can be stated as if exact in an 'infinite enclosure', formally dropping restriction to nonrelativistic cases, as

Corollary 8.4. *In 'infinite enclosure', with PPR, all laws of physics, the speed of light, and Theorem* 1 *through Corollary* 8.4 *are the same in all frames of reference moving with uniform velocity* $v<c$.

●

References hereinafter in derivation to any of Theorem 1 *through Corollary* 8.3 *are, unless otherwise noted, construed as references also to Corollary* 8.4.

An axiomatic form is used in Section 3, which takes the place of this part of this line of derivation, and explicitly includes treatment of limitations on information separability. The axiomatic form in Section 3 joins into, and provides an axiomatic basis for subsequent parts of, this derivation.

Before proceeding with formal discussion, in Corollary 8.5 through Corollary 8.16, of effects of finite 'size' of 'enclosure', I would like to discuss here some conceptual results of part of the formal structure already presented.

In Corollary 4.3, I showed that information and energy are both conserved during interconversion of electromagnetic energy and mass iff $E \approx mc^2$. From there through Corollary 8.4, I used NRPPR or PPR, referencing this result. Within the domain of applicability of NRQM, measurements could be made to show that $E \approx mc^2$, for example, fuel mass consumed versus energy released in a nuclear power plant. Therefore, with Corollary 4.3, *information is conserved.* Thus, we can remove the restriction to NRPPR or PPR, subsuming this observation by principle 2 hereinafter.

We can now compactly state two principles of information for all systems containing only electromagnetic energy and/or mass as if in an 'infinite enclosure':

Principle 1. *Information is proportional to* (*mass*) *energy*.

Principle 2. *For any closed system containing photons and/or mass particles and moving uniformly as if in an 'infinite enclosure' with velocity less than velocity of light, internal and external information are separately conserved and have independent representations.* (*called herein* '(*the*) *principle of detailed information conservation*' ('*PODIC*'); *domain of applicability specified more fully hereinafter in Cor* 8.16)

ENTRY OF AXIOMATIC FORM

This is where the axiomatic form beginning in Section 3 enters.

In conformity with, and based on, the axiomatic treatment formally logically preceding this entry and appearing in Section 3 hereof, energy (mass) is treated as exactly proportional to information, and results stated in Theorem 6 through Corollary 6.3 and Principle 1 are construed as restated without the word 'substantially' and without restriction to 'infinite' 'enclosure'.

The axiomatic form incorporates substantial portions of material appearing hereinafter in this Section 2, and various steps in such material are stated in such way as to be consistent with such use thereof. Because the axiomatic treatment from A-Definition 1 through A-Theorem 8 formally logically precedes this entry, reference hereafter in this Section 2 to any formal statement or result in such part of the axiomatic form in Section 3 is not in itself logically improper. (See map in Appendix 3.)

PODIC appears to be a profound conceptual principle. Consider some of its implications:

1. An observer outside closed box B could choose a momentum representation for external information I'. Then, PODIC implies that for any physical process contained within B, for any state of motion of B, I' is conserved. This yields the law of *conservation of momentum*, including as the time component *conservation of energy with* $E = mc^2$.
2. An observer outside B could choose a position representation for I'. Then, PODIC implies that for any physical process contained within B, for any coding of I' by means of position of B, I' is conserved. Representation of information by position depends on specifying one position out of a set of distinguishable positions and combinations of positions. Thus, PODIC implies that the number of distinguishable positions and combinations of positions is conserved, or a law of *conservation of localizability of position* or, more simply, *conservation of position*.
3. For an observer inside B, PODIC implies no access to any information externally represented by B. We could, for example, imagine a linear uniform momentum representation for I'. Then, the velocity of B would be a state descriptor in I'. Thus, PODIC implies that every process occurring within B is not in any way affected by uniform motion of B. This applies to all processes involving photons and mass particles. Thus, PODIC implies
   3.1 *all laws and constants of physics pertaining to mass particles and photons, including the velocity of light, in a closed system moving within an 'infinite enclosure' are independent of the state of uniform translational motion of that system with velocity less than the speed of light.*

The closed system could (in an 'infinite enclosure') be any size, and so becomes merely a frame of reference. This is a foundation from quantum mechanics for Einstein's special theory of relativity. I used a nonrelativistic derivation for Theorem 4 to preserve the purity of this result. Hereinafter, I use this approach to discuss a limitation on domain of applicability of special relativity in a finite enclosure. (See also axiomatic form, Section 3, for discussion of a limitation on domain of applicability of special relativity, and Comment 4.1, p. 298.)

4. With an observer inside B, we could imagine a position representation for I'. In that case, PODIC implies that an observer watching a physical process occurring in a closed system will not see any dependence upon

position of B, or *relativity of position*. We might also gain some insight into how it feels to spread quantum mechanically: *quantum mechanical spreading of B is not observable in B*. This corresponds to independence in B of amount of external information in a position representation, or localization.

(If one wished, one could choose more intricate internal and/or external representations for information; examples 1–4 correspond to relatively simple pure position and momentum representations. Treatment of 'cosmological red shift' and of 'gravity' hereinafter involves further discussion of position and momentum representations together.)

(example 5 is not included in axiomatic form: it would be redundant to formal treatment in A-Cor 1.2)

5. With PODIC, events appear treatable as nodes through which conserved information passes. If 'time' is taken as designating that variable as function(s) of which availability of information is described, then this appears to offer insight into apparent unidirectionality of time, in the following way:

   For observation by observer O (an 'it', used generically) at about time t_0 of event E, O does not have information of (occurrence of) E for time $t' < \sim t_0$, and for $t' > \sim t_0$ has such information; for 'reversed observation', O has such information for $t' < \sim t_0$, and for $t' > \sim t_0$ does not have such information. Therefore, at any time t, O does not have any information of occurrence with it of reversed observation at $t' < t$, and, therefore, O does not have any information of in its past ever having had prior observational information about its (then) future which has subsequently (before time t) been 'confirmed' by observation.

   With respect to $t' > t$, O does not have information at time t of observation at $t' > t$, and O describes reversed observation for $t' > t$, noting that information from O enters the node called E, as causation; it seems that the only way O can have information of occurrence of E for times before *and* after the time $\sim t'$ of occurrence of E is for O to make observations for $t'' < \sim t'$ such that observation of E would not represent any information for O, that is, there would not be any alternative ($\log_2 1 = 0$), a condition consistent with interpretation of reversed observation as causation.

   Thus, O can interpret its own time as unidirectional, and regard itself as transporting information from its past to its future.

   Suppose there were two observers, O_1 and O_2, carrying information in opposite directions in time: could either obtain from the other any information which would show that its notion of the direction of time was not unique?

Suppose O_2 carried information I from $\sim t' > \sim t$ to time $\sim t$, communicated I to O_1 at about t, and O_1 carried I from $\sim t$ to $\sim t' > \sim t$: in that case, both observers would regard I as existing only for times $> \sim t$, and both would describe information as not conserved by such communication (not constant for all t).

Similarly, suppose in addition both, and only, O_1 and O_2 interact with event E at about time $t' > \sim t$: from PODIC, information would be conserved in E, therefore, E may be regarded as a node in communication between O_1 and O_2, forming a loop in which information would be carried by O_1 to about t' and by O_2 to about t; information so carried would exist only in the range of the loop (between $\sim t$ and $\sim t'$), and both O_1 and O_2 would describe information as not conserved by such loop communication. From Definition 1, there is not any virtual information, so making the range $(\sim |t' - t|)$ of the loop smaller seems to not avoid this.

Suppose O_1 and O_2 communicated at about t, with each transferring to the other exactly the same amount of information: both would describe information as conserved; and each appears free to describe such communication entirely within its own sense of direction of time, as transformation of representation of conserved information. For such conserved information to indicate that one observer had time direction opposite to the other's, both would have to communicate through another node representing unidirectional communication between them, that is, have in common an 'event' in one's 'future' and the other's 'past', which appears precluded by PODIC, as noted hereinbefore.

Thus, PODIC appears to have as another consequence that, for all observations and communications which can occur, an observer may regard time as always unidirectional. (This is treated more formally in Section 3, A-Cor 1.2; but see also Comment 4.5.)

2.4.1 Special Relativistic Covariance of Theory

We have reached a conceptual foundation for Einstein's special theory of relativity (SR), and the theory and notation of SR can now be introduced without compromising the purity of this foundation.

In reaching this conceptual foundation, we have followed a path of reasoning which might be expressed as a principle:

Principle 3. *A finite observer can have its external state varied by a process without affecting its internal state if and only if all laws and constants of physics as seen by it are not changed by the process.*

This appears to be a fundamental principle underlying theories of physics which allow more than one observer, and appears to correspond to the concept that observers be treated equivalently. This is discussed in more detail in Comment 4.1, pp. 298.

For example, with this foundation, Theorems 1–8 and their lemmas and corollaries can be applied in many different frames of reference.

It appears logically acceptable to use SR at this point to check Lemma 4.1 and Theorem 4 in an 'infinite enclosure' to all orders in v/c, for all velocities $v<c$ (Theorem 4′, and Cor 4.4, 4.5, and 8.5, are formally redundant to, and not included in, the axiomatic form):

Theorem 4′. *(Relativistically in an 'infinite enclosure') In box B_2 containing B_0, the states of B_0 with total rest energy E_0 and total energy E allowed by electromagnetic boundary conditions in B_2 are in one to one correspondence with the quantum mechanically allowed states of a mass particle P with rest mass E_0/c^2 and total energy E, $v\neq 0$.*

1. From EMT, quantum mechanics (QM), and SR, each of the two photons in B_0 has energy $E_0/2$ measured at rest relative to B_0.
2. From QM, and 1, each has frequency $f_0=E_0/2h$ measured at rest relative to B_0.
3. From EMT and SR, for B_0 moving with velocity v relative to an observer at rest with respect to B_2, there is one preferred direction in B_0, parallel to B_0's motion. Decomposing B_0's contained electromagnetic energy into one photon moving forward and one backward along this direction, the photon moving forward appears to such observer as shifted to frequency

$$f_+=f_0\sqrt{1-v^2/c^2}\left(\frac{1}{1-v/c}\right)$$

and the photon moving backward

$$f_-=f_0\sqrt{1-v^2/c^2}\left(\frac{1}{1+v/c}\right).$$

4. From Def 5 and 6, B_0 is a soft box, with nonzero field outside it in B_2.
5. From EMT, and 4, both B_0's photons, with f_+ and f_-, must satisfy electromagnetic boundary conditions in B_2.
6. From EMT, outside B_0, f_+ and f_- are in phase at locations with a spatial periodicity equal to the wavelength λ_b of their difference (beat) frequency, $\lambda_b=c/(f_+-f_-)$.
7. From 5 and 6, both B_0's photons will satisfy EMT boundary conditions in B_2 if λ_b fits into B_2 an integer number of times.

8. From 2, 3, and 6,

$$\lambda_b = c(f_+ - f_-)^{-1} = c\left((E_0/2h)\sqrt{1-v^2/c^2}\left(\frac{1}{1-v/c} - \frac{1}{1+v/c}\right)\right)^{-1}$$

$$= c\left(\frac{E_0\sqrt{1-v^2/c^2}}{2h}\left(\frac{1+v/c-1+v/c}{1-v^2/c^2}\right)\right)^{-1}$$

$$= c\left(\frac{E_0 v}{ch\sqrt{1-v^2/c^2}}\right)^{-1} = \frac{h}{\left(\dfrac{E_0}{c^2\sqrt{1-v^2/c^2}}\right)v}.$$

9. From SR and QM, for a mass particle with relativistic mass m_{rel}

$$\lambda_p = \frac{h}{m_{\text{rel}}v}, \qquad v \neq 0.$$

10. From SR, and 9, setting the particle's rest mass $m_0 = E_0/c^2$,

$$m_{\text{rel}} = \frac{E_0}{c^2\sqrt{1-v^2/c^2}}.$$

11. From 8–10, $\lambda_b = \lambda_p$.
12. From QM, allowed states of motion of a relativistic mass particle are those for which λ_p fits into B_2 an integer number of times.
13. From 7, 11, and 12, for every allowed state of B_0 in B_2 there is a state of P in B_2 with the same wave number, and vice versa.
14. From 2 and 3, sum frequency f_s for B_0's two photons is

$$f_s = f_+ + f_- = \frac{1}{h}\left(\frac{E_0}{\sqrt{1-v^2/c^2}}\right).$$

15. From SR, QM, and 14, relativistic mass of B_0 is thus

$$m_{\text{rel}\,B_0} = \frac{1}{c^2}\left(\frac{E_0}{\sqrt{1-v^2/c^2}}\right).$$

16. From 10 and 15, the relativistic masses and energies of B_0 and P agree for all allowed $v \neq 0$.

(Note also agreement between f_s of B_0 and $m_{rel}c^2/h$ of mass particle P.)

17. From 13 and 16, for every allowed state of B_0 in B_2 with $v \neq 0$, there is a state of P in B_2 with the same wave number and energy, and vice versa.

●

(Restriction to $v \neq 0$ is discussed hereinafter formally.)

This result is a necessary consequence of PODIC, which included Einstein's special theory of relativity as a special case. Thus, in a fundamental sense, it does not tell us anything new. However, it is interesting to display this proof, because it confirms explicitly the deduction that the form of Theorems 4–8 is preserved in the relativistic case in 'infinite enclosure', and provides a computational method embodying effects of separability of internal and external information, exact for the case of complete separability.

Thus, we drop restriction, in 'infinite enclosure', to NRA from Theorems 1–8 and their corollaries, without otherwise changing their form.

This leaves a theory of information conservation which, within the domain of applicability of special relativity, is fully special relativistically covariant.

For convenience, I have collected several simple algebraic consequences of steps in Theorem 4′:

Corollary 4.4. *For B_0 moving uniformly in infinite enclosure with velocity $v < c$, with energy E, rest frame energy E_0, associated up-shifted photon energy E_+, down-shifted photon energy E_-, and wavelength λ:*

$$(a)\ E_{\pm} = \frac{E_0}{2}\frac{\sqrt{1-v^2/c^2}}{1\mp v/c},$$

$$(b)\ E = E_+ + E_- = \frac{E_0}{\sqrt{1-v^2/c^2}},$$

$$(c)\ \frac{hc}{\lambda} = E_+ - E_-,$$

$$(d)\ E_0^2 = E^2 - \left(\frac{hc}{\lambda}\right)^2 = 4E_+E_-,$$

$$(e)\ E_{\pm} = \frac{E \pm \sqrt{E^2 - E_0^2}}{2},$$

$$(f)\ \frac{v}{c} = \frac{E_+ - E_-}{E_+ + E_-} = \frac{E_+ - E_-}{E},$$

$$(g)\ \lambda = \frac{hc^2}{Ev} = \frac{hc^2\sqrt{1-v^2/c^2}}{E_0 v}.$$

1. From Theorem 4′ steps 1–3, multiplying both sides by h, recalling that $f_0 = E_0/2h$,

$$E_{\pm} = \frac{E_0}{2}\frac{\sqrt{1-v^2/c^2}}{1 \mp v/c}.$$

2. By definition, $E = E_+ + E_-$.
3. From 1 and 2 (or from SR and Theorem 4′ step 15),

$$E = E_+ + E_- = \frac{E_0}{2}\left(\frac{\sqrt{1-v^2/c^2}}{1-v/c} + \frac{\sqrt{1-v^2/c^2}}{1+v/c}\right)$$

$$= \frac{E_0}{2}\left(\frac{(1+v/c)\sqrt{1-v^2/c^2} + (1-v/c)\sqrt{1-v^2/c^2}}{1-v^2/c^2}\right)$$

$$= \frac{E_0}{2}\left(\frac{(1+v/c+1-v/c)\sqrt{1-v^2/c^2}}{1-v^2/c^2}\right)$$

$$= \frac{E_0}{2}\left(\frac{2}{\sqrt{1-v^2/c^2}}\right)$$

$$E = \frac{E_0}{\sqrt{1-v^2/c^2}}.$$

4. From Theorem 4′ step 8, inverting and multiplying both sides by hc, $hc/\lambda = E_+ - E_-$.
5. From 2 and 4, squaring both and subtracting, $E^2 - (hc/\lambda)^2 = 4E_+E_-$.
6. Substituting from 1 in 5, $E_0^2 = E^2 - (hc/\lambda)^2 = 4E_+E_-$.
7. Solving 6 for $E_{\pm}$, $E_{\pm} = E_0^2/4E_{\mp}$.
8. Substituting 7 in 2, and solving the resulting quadratic equation,

$$E_{\pm} = \frac{E \pm \sqrt{E^2 - E_0^2}}{2}.$$

9. From 1,

$$\frac{E_+ - E_-}{E_+ + E_-} = \frac{\dfrac{E_0}{2}\dfrac{\sqrt{1-v^2/c^2}}{1-v/c} - \dfrac{E_0}{2}\dfrac{\sqrt{1-v^2/c^2}}{1+v/c}}{\dfrac{E_0}{2}\dfrac{\sqrt{1-v^2/c^2}}{1-v/c} + \dfrac{E_0}{2}\dfrac{\sqrt{1-v^2/c^2}}{1+v/c}}$$

cancelling,

$$= \frac{\dfrac{1}{1-v/c} - \dfrac{1}{1+v/c}}{\dfrac{1}{1-v/c} + \dfrac{1}{1+v/c}}$$

$$= \frac{\left(\dfrac{1+v/c-1+v/c}{1-v^2/c^2}\right)}{\left(\dfrac{1+v/c+1-v/c}{1-v^2/c^2}\right)}$$

$$= \frac{2v/c}{2}$$

$$= \frac{v}{c}.$$

10. From 4, solving for λ,

$$\lambda = \frac{hc}{E_+ - E_-}.$$

11. From 2 and 9,

$$\frac{v}{c} = \frac{E_+ - E_-}{E}.$$

12. From 11, inverting, multiplying by hc/E, and cancelling,

$$\frac{hc^2}{Ev} = \frac{hc}{E}\left(\frac{E}{E_+ - E_-}\right) = \frac{hc}{E_+ - E_-}.$$

13. From 10 and 12,

$$\lambda = \frac{hc^2}{Ev}.$$

14. From 3 and 13 (or Theorem 4′ step 8),

$$\lambda = \frac{hc^2\sqrt{1-v^2/c^2}}{E_0 v}.$$

●

This set of results appears compatible with identifying momentum p with h/λ, as was used before. Thus, (d) becomes $E_0^2 = E^2 - (pc)^2$ of Einstein's special theory of relativity. This appears to offer insight into the nature of wave behavior of matter, in a relativistically covariant way.

With caution as to domain of applicability of PODIC and (computational method of) special relativity in finite enclosure, for B_0 in a finite enclosure, various results collected in Corollary 4.4 can be expressed in terms of the numbers of times the various wavelengths of the B_0 and its associated photons fit into the enclosure:

Corollary 4.5. *In box B with linear size R_B, containing B_0 moving with velocity v, number of times that wavelength λ_b of B_0, n_b; associated up-shifted photon, n_+; and down-shifted photon, n_-; fit into B:*

(a) $n_b = n_+ - n_-$,

(b) $\dfrac{v}{c} = \dfrac{n_+ - n_-}{n_+ + n_-}$,

(c) $n_+ n_- = n_0^2$, *(construct n_0^2 defined herein).*

1. From Theorem 4′ step 6, λ_b is defined as difference (beat) frequency of B_0's associated photons.
2. From 1, and definition of n_+, n_-, and n_b, λ_b fits into B $n_b = n_+ - n_-$ times.
3. From Cor 4.4 (f), $v/c = (E_+ - E_-)/(E_+ + E_-)$.
4. From EMT and QM, in finite isotropic B, $E_\pm = n_\pm C_1/R_B$, where C_1 is a suitable constant.
5. From 3 and 4, substituting and cancelling, $v/c = (n_+ - n_-)/(n_+ + n_-)$.
6. From Cor 4.4 (d), $4E_+E_- = E_0^2$.
7. From 4 and 6, as in 4, defining n_0 as a construct for the (approximate) number(s) of times each of B_0's two associated photons would fit into B

if B_0 were substantially at rest and both such photons were free in B, $n_0 = E_0/2C_1$. (This is treated without approximation in A-Cor 2.1, and may be taken as an exact result therefrom.)

8. From 4, 6, and 7, $n_+n_- = n_0^2$.

●

Corollary 4.5 provides a convenient notation for considering effect, if any, of GA in finite enclosure on internal information of a B_0 or, with PPR, a mass particle, and thereby examining applicability of PODIC and computational methods of special relativity to description of B_0s and their associated photons and, with PPR, corresponding mass particles.

Corollary 8.5. *With PPR, in isolated box B with linear size R_B containing B_0 with associated up- and down-shifted photons fitting into B n_+ and n_- times respectively, or box B_1 with linear size $R_1 < R_B$ containing two photons in lowest photon mode in B_1, with the two photons of B_0 or in B_1 having perpendicular polarizations: PODIC applies to such B_0, or such two photons in lowest photon mode in B_1, with any allowed velocity less than maximum velocity v_m possible for such B_0 or B_1 in B, or to any corresponding mass particle P with any allowed velocity $< v_m$ in B.*

1. From Def 6, Lemma 8.1, and Cor 4.5, for B_0 as B_1 in Lemma 8.1, R_1 of such B_1 in Lemma 8.1 has the same proportional relationship to the wavelength of each of B_0's two associated photons as seen by an observer at rest with respect to B_0 substantially at rest in B as in Corollary 4.5.
2. From Cor 4.5 step 7, and 1, noting that 'linear size' is defined in the same way for B and B_1, $R_B/R_1 = n_0$.
3. From EMT, QM, Lemma 8.1 step 9, and 2, each step of least observable GA of B_0 in B would perform transition of photon among allowed states in B_0 as $\sim 1/n_0^2$, without changing polarization.
4. From Def 1, and 3, such GA would separately for each polarization perform changing of internal information in B_0 as $\sim 1/n_0^2$.
5. From Theorem 1, information is conserved in B.
6. From 4 and 5, such GA would separately for each polarization transcribe information between internal and external information of B_0 in B as $\sim 1/n_0^2$.
7. From Cor 8.3 and 8.4, noting that formal neglect of finiteness of R_B was by neglect of terms of order $(R_1/R_B)^2$ in comparison to 1 in

Lemma 8.1, each such step of GA would perform photon transition among allowed states in such B_0 as described in 6.

8. From definition of PODIC (Principle 2), PODIC applies to B_0 in B for all cases in which transfer of information between internal and external information of B_0 in B does not occur.
9. From Def 6, B_0 consists of two (single) photons.
10. From Def 1, information is never virtual.
11. From 6–10, PODIC applies to such B_0 in B for a number N of such steps of GA of such B_0 in B such that $N(1/n_0^2)<1$.
12. From Lemma 8.1, Cor 4.5, and Def 1 and 10, each step of least observable GA of B_0 in B changes $n_+ - n_-$ by 1.
13. From Cor 4.5(c), $n_+ = n_0^2/n_-$.
14. From EMT, for B_0's down-shifted photon to satisfy boundary conditions in B, $n_- \geqslant 1$.
15. From 13 and 14, $n_+ \leqslant n_0^2$.
16. From 12 and 15, for GA to $v<v_m$, $n_+ \leqslant n_0^2 - 1$.
17. From EMT, QM, and Def 10 and 11, for B_0 substantially at rest to exist for long enough time for substantial GA and/or GS, B_0's associated photons must be 'bound'.
18. From Def 1 and 6, binding of B_0 involves association of two photons.
19. From Def 1, and 18, for two photons to be associated in bound B_0, detectable coupling interaction must occur between them.
20. From Def 1, and 18 and 19, for binding of B_0, 1 bit of information representing occurence of such interaction must be represented in B.
21. From EMT and QM, least amount of photon energy which can represent 1 bit in B is that energy required to create a photon in lowest energy photon state in B.
22. From Theorem 6, with PPR, least amount of energy required to represent 1 bit of information in B is independent of whether electromagnetic energy and/or mass particle(s) is (are) used for such representation, and/or choice of representation.
23. From 20–22, least energy which can represent occurrence of coupling between two photons in B is energy of lowest energy photon in B.
24. From EMT and QM, for bound two photon B_0 nearly at rest in B, its associated two photons have nearly equal energy.
25. From EMT, QM, and 19 and 23, the photons in states of B coupled by the interaction in step 19 can be decomposed into two modes as seen by an observer substantially at rest in B, 'split' by such coupling. (an alternative way of describing representation of the same information)

26. From EMT, QM, and Theorem 4′, and 19, 23, 24, and 25, least such frequency split which can be defined in B for these modes is frequency of lowest frequency photon in B.
27. From 25, electromagnetic energy associated with B_0 can be decomposed into one photon in each of these two modes.
28. From 26 and 27, associated with such bound B_0 are two photons with different frequencies, whose frequencies differ by at least the frequency of the lowest frequency photon in B.
29. From QM, Cor 4.5(b), and 28, lowest velocity state of bound B_0 in B has velocity $v \neq 0$.
30. From EMT, QM, Theorem 4′, and 28, lowest velocity state of bound B_0 in B has wavelength equal to the wavelength of the lowest frequency photon possible in B.
31. From EMT, QM, and 30, direction of bound B_0's least $v>0$ is not observable.
32. From Def 1, and 31, such direction does not represent any information in B.
33. From 28, 29, and 32, bound B_0 substantially at rest in B could have $n_+ - n_- = 1$ associated with $v>0$ in direction substantially opposite to direction of GA.

    (In the axiomatic form, steps 17–33 might be replaced by reference to A-Cor 2.2 steps 11–25.)
34. From 12, 16, and 33, any allowed direction and velocity $v < v_m$ of B_0, having with velocity v n_+ and n_-, in B, can be reached by N steps of GA, $N \leqslant n_+ - n_- + 2 - 1 \leqslant n_+ - 1 + 1 < n_0^2$. ('2' from possibility of starting in direction with n_+ of least velocity as n_- with respect to desired direction for velocity; first '-1' from number of steps between states being 1 less than number of states; and second '-1' from least $n_- = 1$)
35. From 17 and 34, any allowed velocity $< v_m$ of B_0, susceptible to substantial GA, in B, can be reached by N steps of GA, $N(1/n_0^2) < 1$.
36. From EMT and QM, with PPR, allowed states in B of B_1 containing total electromagnetic energy E_0 when substantially at rest are the same as for a B_0 containing total energy E_0 in such B_0's rest frame.
37. From Def 6, and 11, 35, and 36, GA to $v < v_m$ of B_1 containing such two photons in lowest photon state in B_1 would not perform transition of such photons among allowed photon states in B_1.
38. From Def 11, during GS to smaller linear size of B_1 containing one photon, any transitions of photon among allowed photon states in B_1 would be to states in which the photon satisfied boundary conditions in B_1 by fitting into B_1 fewer times.

39. From EMT and QM, there is a smallest number >0 of times that a photon in B_1 can fit into B_1 and satisfy electromagnetic boundary conditions in B_1 (the 'lowest' state in B_1).
40. From 38 and 39, during GS of B_1 to smaller linear size, such photons in lowest state in B_1 would not undergo transition to any other photon state in B_1.
41. From Def 6, and 40, GS of such B_0 to smaller linear size would not induce transitions of such B_0's associated photons among photon states in such B_0.
42. From Def 1, and 41, GS of such B_0 to smaller linear size would not change such B_0's internal information.
43. From EMT and QM, any two linear sizes can be related by process involving GS only decreasing linear size.
44. From 35, 37, 40, 42, and 43, PODIC applies to such B_0, and/or such two photons contained in and in lowest photon state in B_1 with any allowed size, with velocity $< v_m$ in B.
45. With PPR, from 44, PODIC applies to mass particle P, corresponding to such B_0, with any allowed velocity $< v_m$ of P in B.

●

Thus, it appears that, for all velocities less than the maximum which can be reached thereby, PODIC applies to such B_0s and, with PPR, corresponding mass particles in a finite enclosure, and, therefore, in such domain, computational methods of special relativity (Theorem 4′, Cor 4.4 and 4.5) can be used in description of B_0s and their associated photons in considering limitations on separability of internal and external information, and associated limitations on special relativity, for an extended system larger than any contained B_0(s) and moving in a finite enclosure, without the computational methods used for such analyses themselves 'breaking down'. Various aspects of the concept of a maximum velocity for a B_0 are discussed in various of the following corollaries of Theorem 8.

2.4.2 A Limitation on Applicability of PODIC

As the second step of the aforementioned two-step approach to separability of internal and external information, in the next eleven corollaries appears a detailed study of properties of motion of B_0s, by which is obtained what appears at the time of this writing to be a velocity and size relation describing a limitation on the domain of applicability of PODIC.

Consider B_0 which in its rest frame contains energy E_0 much greater than energy E_{B1} of the lowest photon mode in containing box B. For velocities not near the speed of light, B_0's two associated photons have energies at which the energy spacing of photon states in B is a very small fraction of the photon energy. What happens if B_0 has such high velocity that its associated down-shifted photon is in one of the low energy photon states in B?

Definition 13. *Call B_0 moving with velocity v 'fast' if it contains in its rest frame much more energy than the lowest photon energy level in containing box B, and v is such that B_0's associated down shifted photon is in a relatively low energy level in B.*

The following conditions ('$C1$', '$C2$') are used variously in Corollaries 8.6–8.16:

Condition 1. *In isolated box B, with linear size R_B, lowest photon energy E_{B1}; containing B_0 with energy E_0 in its rest frame, transporting energy E with velocity $v \approx c$ as seen by observer O_B substantially at rest relative to B; B_0 representing internally N bits of information*

Condition 2. $C1$; *possible photon energy levels E_{Bi}, i denoting the ith photon energy level; B_0's associated down-shifted photon having energy E_{Bj} and up-shifted photon having energy E_+*

Corollary 8.6. $C2$; *in domain of applicability of PODIC, B_0 has*

$$E_0(v) = 2E_{Bj}\frac{1+v/c}{\sqrt{1-v^2/c^2}}.$$

1. From PODIC and Cor 4.4 (a), in domain of applicability of PODIC, O_B sees B_0's down-shifted photon as having energy

$$E_- = \frac{E_0}{2}\frac{\sqrt{1-v^2/c^2}}{1+v/c}.$$

2. From EMT, for B_0's down-shifted photon to satisfy boundary conditions in B, $E_- = E_{Bj}$ for some j.

3. From 1 and 2, solving for $E_0(v)$,

$$E_0(v) = 2E_{Bj}\frac{1+v/c}{\sqrt{1-v^2/c^2}}.$$

●

In particular, *but with caution in view of Corollary* 8.5, *this being the case of* v_m,

Corollary 8.7. C2; *if formal method of Corollary* 8.6 *be applicable*, B_0 *has velocity upper limit, for which, if computational method of SR be approximately applicable*,

$$E_0(v) \sim 2E_{B1}\frac{1+v/c}{\sqrt{1-v^2/c^2}}.$$

1. From PODIC and Cor. 8.6, in domain of applicability of PODIC, B_0 has

$$E_0(v) = 2E_{Bj}\frac{1+v/c}{\sqrt{1-v^2/c^2}}.$$

2. From 1, B_0 has least $E_0(v)$ for same v iff E_{Bj} is as small as possible.
3. From EMT, for B_0's associated down-shifted photon to satisfy boundary conditions in B, $E_{Bj} \geqslant E_{B1}$.
4. From 1–3, if such formal method be applicable, B_0 has velocity upper limit, for which, if computational method of SR be approximately applicable,

$$E_0(v) \sim 2E_{B1}\frac{1+v/c}{\sqrt{1-v^2/c^2}}.$$

●

Definition 14. *Call a fast* B_0 *a 'fastest* B_0*' or '*fB_0*' if its associated down-shifted photon has lowest allowed photon energy in containing box B.*

Corollary 8.8. *C2; in domain of applicability of PODIC, for B_0,*

$$E(v)=2E_{Bj}\frac{1}{1-v/c}.$$

1. From Cor 4.4 (b),

$$E(v)=\frac{E_0}{\sqrt{1-v^2/c^2}}.$$

2. From Cor 8.6, and 1, factoring denominator,

$$E=2E_{Bj}\frac{1+v/c}{(1+v/c)(1-v/c)}.$$

3. From 2, cancelling,

$$E(v)=2E_{Bj}\frac{1}{1-v/c}.$$

●

Solving Corollary 8.8 for $v(E)$,

Corollary 8.9. *C2; in domain of applicability of PODIC, for B_0,*

$$v(E)=c\left(1-\frac{2E_{Bj}}{E}\right).$$

●

From Corollary 8.9, and composition of velocities in special relativity,

Corollary 8.10. *C2; in domain of applicability of PODIC, in any frame of reference seen by O_B as moving with velocity $v<c(1-\epsilon)$, $\epsilon\gg 2E_{Bj}/E$, fast B_0 would appear to have velocity $\approx c$, and its momentum would appear to transform substantially as does a photon's.*

●

Because so little of a fast B_0's energy is in its associated down-shifted photon, this result is not surprising. However, as I discuss hereinafter, a fast B_0 might interact quite differently than a photon.

Corollary 8.11. *C2; in domain of applicability of PODIC, for fast B_0, $E_0 \approx 2\sqrt{E\,E_{Bj}}$.*

1. From Cor 4.4 (d), $E_0^2 = 4E_+E_{Bj}$.
2. From Cor 4.4 (b), $E = E_+ + E_{Bj}$.
3. From Def 13, fast B_0 has $E_0 \gg E_{Bj}$.
4. From 1 and 3, $E_+ \gg E_{Bj}$.
5. From 2 and 4, $E_+ \approx E$.
6. From 1 and 5, solving for E_0, $E_0 \approx 2\sqrt{E\,E_{Bj}}$.

●

Fast B_0s seemed to have such large time dilation factors that it seemed interesting to examine the lifetime of a B_0 whose associated two photons are not stably coupled, and which exists for a time$\sim\tau_0$ in its own rest frame, for example, approximately equal to one period of its associated photons.

Corollary 8.12. *C2; in domain of applicability of PODIC, fast B_0, whose two associated photons are not stably coupled, having lifetime $\tau_0 \approx 2h/E_0$ in B_0's rest frame: O_B sees lifetime $\tau \approx h/2E_{Bj}$; for fB_0, if formal method of Corollary* 8.6 *and computational method of SR be approximately applicable, $\tau \approx h/2E_{B1} \approx \pi R_B/c$.*

1. From statement of corollary, $\tau_0 \approx 2h/E_0$.
2. From PODIC, $\tau = \tau_0 E/E_0$.
3. From 1 and 2, $\tau \approx 2hE/E_0^2$.
4. From Cor 8.11, and 3, $\tau \approx 2hE/4EE_{Bj}$.
5. From 4, $\tau \approx h/2E_{Bj}$.
6. From EMT and QM (from A-Def 17 and 22), $E_{B1} \approx hc/2\pi R_B$.
7. From Cor 8.7, and 5 and 6, if formal method of Corollary 8.6 and computational method of SR be approximately applicable, for fastest B_0, $\tau \approx \pi R_B/c$.

●

(cautionary note from Cor 8.7, for case of $v = v_m$ (see Cor 8.5))

Note that this result depends relatively weakly on E for $E \gg E_{Bj}$. With B as our universe, $\pi R_B/c$ is a long enough time to make it interesting to examine other properties of unbound fastest B_0s.

By how much would such fB_0 lag behind a photon in such 'lifetime'?

Corollary 8.13. $C1$; *if computational method of SR be approximately applicable,* fB_0, *in 'lifetime'* $\tau_0 \approx 2h/E_0$, *lags behind photon by approximately one wavelength of photon carrying same energy.*

1. From Def 14 and Cor 8.9, in domain of applicability of PODIC, $v(E) = c(1 - 2E_{B1}/E)$.
2. From Cor 8.12, if computational method of SR be approximately applicable, $\tau \approx h/2E_{B1}$.
3. Let $z = \tau(c - v(E))$. ($z \approx$lag as seen by observer substantially at rest in B)
4. From 1–3, in such case, $z \approx (h/2E_{B1})(c - c + c2E_{B1}/E)$.
5. From EMT and QM, and 4, if computational method of SR be approximately applicable, $z \approx hc/E = \lambda(E)$, wavelength of photon with energy E.

●

What do allowed energies of fast B_0s look like? There appear to be two substantially different cases. During formation, a fast B_0 might be given any energy which can be expressed as the sum of two photon energies in B. Were the fast B_0 required to have its velocity in some given direction, allowed photon energy levels would be substantially those in the 'linear' case, $E_{Bj} = jE_{B1}$, j (finite) integer>0.

Before I treated this subject axiomatically, I did not have formally a spectrum of photon states in B, and carried in formal analysis two spectra, chosen to define a range expected to include the true spectrum. For the finest energy level spacings, I took from EMT and QM the density of states presumed for 'blackbody' ('cavity') radiation, with the further assumption that every state, including states presumed in EMT and QM to differ only in polarization or direction, be of different energy. For the coarsest energy level spacings, I took restriction to a single direction, in which case energy levels would be integer multiples of some least energy greater than zero. Years later, I obtained from the axiomatic form that for the spectrum of photon states in our universe U, $E_j = jE_1$, j finite integer > 0, where E_1 is least photon energy > 0 possible in U. Because some results do not depend substantially on such choice of spectrum, and because this presentation is intended to be not only formal but also explicative, I have retained steps relating to spectra other than $E_j = jE_1$, but have placed square brackets around each such step or part of an expression to indicate that it is not part of the axiomatic form, and have added reference to A-Theorems 3 and 4 and A-Definition(s 21 and) 22 of the axiomatic form as a basis for $E_j = jE_1$. Therefore, B_0 could be formed with energies with separation $\Delta E \underset{[\leq]}{=} E_{B1}$. For a fast B_0, $E \gg E_{B1}$. Thus, its fractional energy spacing is $\ll 1$. Thus,

Corollary 8.14. *C1; fast B_0 may be formed with allowed fractional energy interval*

$$\frac{\Delta E}{E} \underset{[\leq]}{=} \frac{E_{B1}}{E} \ll 1.$$

•

The second case is for a fast B_0 already formed, and then subjected to conditions which might change E without changing E_0. Allowed energies then correspond to placing B_0's associated down-shifted photon in various low photon energy levels in B. These low levels have relatively larger fractional energy spacings than do higher energy levels. This leads to remarkable 'graininess' in momentum allowed for an already formed fast B_0.

Corollary 8.15. *C2; in domain of applicability of PODIC, fast B_0 formed with energy E; subjected to conditions to change E without changing E_0: B_0's next lower allowed energy E' has fractional energy spacing $\delta = \dfrac{E - E'}{E}$,*

$$\left[\frac{1}{\dfrac{4}{3\pi}\left(\dfrac{E_{Bj}}{E_{B1}}\right)^3 + 1} \lesssim \right] \delta \underset{[\lesssim]}{\approx} \frac{1}{\dfrac{E_{Bj}}{E_{B1}} + 1} = \frac{1}{j+1}.$$

1. From Cor 8.11, solving for E_{Bj}, $E_{Bj} \approx E_0^2/4E$.
2. From 1, with E_0 constant, decreasing E requires increasing E_{Bj}.
3. By definition, in B, next greater allowed photon energy is E_{Bj+1}.

4. From Lemma 4.1, Theorems 4 and 4′, and 2 and 3, B_0's next lower energy E' corresponds to its associated down-shifted photon having energy E_{Bj+1}.
5. From 1 and 4, solving for E', $E' \approx E_0^2/4E_{Bj+1}$.
6. Let $\delta = (E - E')/E$.
7. From 5 and 6, simplifying and substituting, $\delta = 1 - \dfrac{E'}{E} \approx 1 - \dfrac{\dfrac{E_0^2}{4E_{Bj+1}}}{\dfrac{E_0^2}{4E_{Bj}}}$.
8. From 7, simplifying, $\delta \approx 1 - \dfrac{E_{Bj}}{E_{Bj+1}}$.
9. From [EMT and QM, restricting B_0's motion to a given direction] A-Theorems 3 and 4, and A-Def 22, B_0's associated frequency-shifted photons have allowed energy levels $E_{Bj} \underset{[\approx]}{=} jE_{B1}$. ($B$ as U)
10. From 6, 8 and 9, for case in step 9, $\delta \approx 1 - j/(j+1) = 1/(j+1)$.

[11. From EMT and QM, allowed energy levels in case as in step 9 are at least as widely separated as photon energy levels in B.]

12. From 9 and 10 [and 11],

$$\delta \underset{[\lesssim]}{\approx} 1 - \frac{E_{Bj}/E_{B1}}{E_{Bj}/E_{B1} + 1} = \frac{1}{j+1}.$$

[13. In NRQM, by enumeration in phase space, density of photon states in volume V is

$$\rho(E_{Bj}) \approx \frac{V}{\pi^2} \frac{E_{Bj}^2}{(\hbar c)^3}.$$

(This is used in calculating black body radiation in NRQM.)]

[14. From 13, writing V as $(4/3)\pi R_B^3$, with $E_{B1} \approx \hbar c / R_B$,

$$\rho(E_{Bj}) \approx \frac{4}{3\pi} \frac{E_{Bj}^2}{E_{B1}^3}.$$]

[15. δ is smaller if energy levels in B are more closely spaced.]

[16. Closest energy level spacing in B corresponds to every photon state having different energy (including different polarization states).]

[17. From 14–16,

$$E_{Bj+1} - E_{Bj} \gtrsim \frac{3\pi}{4} \frac{E_{B1}^3}{E_{Bj}^2}.\Big]$$

[18. From 8 and 17,

$$\delta \gtrsim 1 - \frac{E_{Bj}}{E_{Bj} + \frac{3\pi}{4} \frac{E_{B1}^3}{E_{Bj}^2}}.\Big]$$

19. From 12 and 18, simplifying,

$$\left[\frac{1}{\frac{4}{3\pi}\left(\frac{E_{Bj}}{E_{B1}}\right)^3 + 1} \lesssim\right] \delta \underset{[\lesssim]}{\approx} \frac{1}{\frac{E_{Bj}}{E_{B1}} + 1} = \frac{1}{j+1}.$$

●

Corollary 8.15 has several remarkable consequences. A fast B_0's allowed energy levels are relatively widely spaced. This result is qualitatively different from particle behavior predicted by any other theory of which I am at the time of this writing aware. Questions in scattering theory relating to momentum changes, counting states in phase space, center of mass behavior, etc., might require treatment in this context different from what has previously been usual. In the extreme case, fB_0s, Cor 8.15 indicates that E' is substantially less than E. From Section 3, as noted hereinbefore, $E_2 = 2E_1$, so, for fB_0, *if computational methods of SR be approximately applicable*, $E' \approx \frac{1}{2}E$ [from the finest energy spacings in Cor 8.15, for fB_0, $E' < \sim 0.7725\, E$].

With Corollaries 8.5 through 8.15, we have a way to express a limitation from finiteness of B on separation of internal and external information. From Cor 8.15, an already formed B_0's allowed momenta become relatively widely spaced when its associated down-shifted photon is shifted into a relatively low energy level in B. In that case, an observer in a system moving with B_0 might see behavior different from (our) usual laws of physics, and thus be able to deduce external velocity. To avoid this, one would have to keep every B_0 in the observer's system sufficiently below its maximum velocity. Graininess of B_0's allowed energies would be unobservable if their spacings were much closer than the lowest photon energy level in the moving observer's system. [For all parts of a rapidly moving system to travel together, their velocities must be very nearly parallel.] Using [the case for] B_0's energy graininess [when restricted to momentum parallel to a given direction,] from Corollary 8.15, we can formulate conditions for PODIC to apply or fail *with respect to measurements made in a system (already) moving with uniform velocity*. In Section 3, the question of separability of internal and external information *in reaching such velocity* is considered formally (A-Theorem 6), and a relation between size and velocity, consistent with the following, is obtained.

Corollary 8.16. *C2; B_3 with linear size R_3 in its rest frame, moving with velocity v in B, containing system S of photons and B_0's, and containing observer O_3 substantially at rest relative to B_3: PODIC applies in B_3 (already) moving with uniform v with respect to current measurements by O_3 iff*

$$v < \approx c \frac{1-(R_3/R_B)^2}{1+(R_3/R_B)^2},$$

and fails by internal and external information being not separable if

$$v \geqslant \approx c \frac{1-(R_3/R_B)^2}{1+(R_3/R_B)^2}.$$

[1. From EMT and QM, for components of system S to travel nearly together with velocity $v \approx c$, all components must have nearly parallel velocities.]

2. From Cor 8.15, [and 1,] energy spacing ΔE for B_0 [traveling as seen by O_B substantially parallel to a given direction] with rest frame energy E_0, energy E, and down-shifted photon energy E_{Bj}, is given approximately by

$$\frac{\Delta E}{E} \approx \frac{1}{\dfrac{E_{Bj}}{E_{B1}}+1}.$$

3. From 2, for $E_{Bj} \gg E_-$, to first order in E_-/E_{Bj},

$$\frac{\Delta E}{E} \approx \frac{E_{B1}}{E_{Bj}}.$$

4. From Cor 4.4 (b).

$$\sqrt{1-v^2/c^2} = \frac{E_0}{E}.$$

5. From PODIC, and 4, lowest photon energy E_{31} in B_3 as seen by O_3 would appear to observer O_B as $E_{31}E/E_0$.
6. From PODIC, Theorem 6, Def 1, and 5, in domain of applicability of PODIC, O_3 would see representation of amount I bits of information in B_3 as requiring amount IE_{31} of energy.
7. From PODIC, Def 1, and 6, in domain of PODIC, O_3 would be able to observe in B_3 B_0's energy spacings as being different from as in step 6 iff such spacings were large enough that O_3 would see least amount $\Delta_3 I$ of information exchangeable with such B_0 representation as > 1 bit.
8. From PODIC, Theorem 6, Def 1, and 7, O_3 would be able to see $\Delta_3 I > 1$ iff $\Delta_3 I \geqslant 2$. (1 bit transcribed, and at least 1 bit to represent that such transcription involved an amount of energy greater than usual to represent 1 bit in B_3: there is not any virtual information)
9. From Lemma 8.1. through Cor 8.4, and 7 and 8, if O_3 cannot distinguish any effect on B_0's energy spacings due to finiteness of B, then O_3 cannot by observation of B_0 distinguish B_0's transformation properties from the case of B 'infinite'.
10. From Lemma 8.1 through Cor 8.4, definition of PODIC, and 2, 5, 6, 7, and 9, PODIC applies with respect to current measurements in B_3 moving iff $\Delta E < 2E_{31}E/E_0$ as seen by O_B.
11. From 3 and 10, PODIC so applies iff

$$\frac{EE_{B1}}{E_{Bj}} <\approx \frac{2E_{31}E}{E_0}.$$

12. From EMT; QM, and 5, noting definitions of E_{B1} and E_{31},

$$\frac{E_{31}}{E_{B1}} = \frac{R_B}{R_3}.$$

13. From 11 and 12, PODIC so applies iff

$$\frac{E_0}{E_{Bj}} <\approx \frac{2R_B}{R_3}.$$

14. From Cor 4.4 (a), solving for E_0/E_{Bj},

$$\frac{E_0}{E_{Bj}} = 2\frac{1+v/c}{\sqrt{1-v^2/c^2}}.$$

15. From 13 and 14, PODIC so applies iff

$$2\frac{1+v/c}{\sqrt{1-v^2/c^2}} <\approx \frac{2R_B}{R_3}.$$

16. From 15, solving for v/c, PODIC so applies iff

$$\frac{R_3}{R_B}(1+v/c) <\approx \sqrt{1-v^2/c^2}\ ;$$

squaring,

$$\left(\frac{R_3}{R_B}\right)^2 (1+v/c)^2 <\approx 1-v^2/c^2;$$

factoring,

$$\left(\frac{R_3}{R_B}\right)^2 \left(1+\frac{v}{c}\right)^2 <\approx \left(1+\frac{v}{c}\right)\left(1-\frac{v}{c}\right);$$

cancelling,

$$\left(\frac{R_3}{R_B}\right)^2 \left(1+\frac{v}{c}\right) <\approx 1-\frac{v}{c};$$

multiplying,

$$\left(\frac{R_3}{R_B}\right)^2 + \left(\frac{R_3}{R_B}\right)^2 \frac{v}{c} <\approx 1-\frac{v}{c};$$

transposing and factoring,

$$\left(\left(\frac{R_3}{R_B}\right)^2+1\right)\frac{v}{c}<\approx 1-\left(\frac{R_3}{R_B}\right)^2;$$

and dividing,

$$\frac{v}{c}<\approx\frac{1-\left(\frac{R_3}{R_B}\right)^2}{1+\left(\frac{R_3}{R_B}\right)^2}.$$

17. From 16, PODIC so applies iff

$$v<\approx c\frac{1-(R_3/R_B)^2}{1+(R_3/R_B)^2}.$$

18. From Theorem 6, and 6–8, if B_0's energy spacings due to finiteness of B are greater than or equal to $2E_{31}$ in B_3 moving, O_3 would be able to observe effects of such energy spacings on information exchange in B_3.
19. From 18, if B_0's energy spacings were thus $\geqslant 2E_{31}$, O_3 could obtain information about B_3's motion in B.
20. From 19, in such case, B_3's internal and external information would be not separate.
21. From 1 through 20, replacing $<$ in step 10 by $\geqslant$ in step 19, PODIC does not apply, by B_3's internal and external information being not separable, for

$$v\geqslant\approx c\frac{1-(R_3/R_B)^2}{1+(R_3/R_B)^2}.$$

22. From Cor 4.5 and Cor 8.5 step 2,

$$n_0=\frac{R_B}{R_1}.$$

23. From 22, for every B_0 contained in B_3,

$$n_0\geqslant\frac{R_B}{R_3}.$$

24. From Cor 4.5 (b), for B_0 in B,

$$\frac{v}{c}=\frac{n_+-n_-}{n_++n_-}.$$

25. From Cor 4.5, Cor 8.5 steps 14–16, and 24, if PODIC applicable, maximum possible velocity v_m in B of B_0 with n_0 is for $n_-=1$ and $n_+=n_0^2$.
26. From 24 and 25, for B_0 with n_0 in B, if PODIC applicable,

$$\frac{v_m}{c}=\frac{n_0^2-1}{n_0^2+1}=\frac{1-1/n_0^2}{1+1/n_0^2}.$$

27. From 23 and 26, solving 23 for $1/n_0^2$ and substituting in 26, if PODIC applicable, for every B_0 contained in B_3,

$$v_m \geqslant c\frac{1-(R_3/R_B)^2}{1+(R_3/R_B)^2}.$$

28. From Cor 8.5, PODIC applies to every B_0 in B_3 for $v<v_m$ respectively for each such B_0.
29. From 17, 27, and 28, for all velocities for which PODIC applies to S in B_3 with respect to current measurements by O_3 in B_3, PODIC applies to every B_0 in B_3.
30. From 21, 23, 27, 28, and 29, PODIC fails with respect to current measurements by O_3 in B_3 at a velocity $\leqslant$ least velocity at which PODIC fails for any B_0 contained in B_3.
31. From Cor 4.4, 4.5, and 8.5, and 17, 29, and 30, description in Corollaries 4.4, 4.5, and 8.5 of B_0 is useable in obtaining domain in step 17 in which PODIC so applies in B_3.
32. From EMT, QM, PODIC, Def 1 and 6, and 6, 7, and 8, measurement by O_3 as in step 8 involves a B_0 able to be in not lowest B_0 state in B_3 (not 'filling' B_3).
33. From Cor 4.4, 4.5, and 8.5, and 6, 7, 8, 21, 22, 23, 27, 28, and 32, PODIC can apply to a B_0 in B_3 useable by O_3 for measurement as in step 8 in a domain including velocity higher than that at which PODIC applies to S in B_3 in B.
34. From Cor 4.4, 4.5, and 8.5, and 1–21, 31, and 33, Corollaries 4.4, 4.5, and 8.5 are usable for description of B_0 in steps 1–21.

For $v \ll c$, this condition for PODIC to apply is approximately $(R_3/R_B)^2 < 1$, or approximately $R_3 < R_B$; for $R_3 \ll R_B$, neglecting terms of order $(R_3/R_B)^4$, this condition for PODIC to apply is approximately $v < c(1-2(R_3/R_B)^2)$.

PODIC becomes inapplicable by loss of separability of internal and external information. It was from this separability that I derived special relativity, so special relativity also does not apply in such cases.

This velocity limitation on special relativity appears interpretable in the following way: Special relativity (from PODIC) would lead one to calculate that, to an observer moving with B_3, B would appear to have linear size parallel to v of $R_B\sqrt{1-v^2/c^2}$. Writing x for R_3/R_B,

$$R_B\sqrt{1-v^2/c^2} \approx R_B\left(1-\left(\frac{1-x^2}{1+x^2}\right)^2\right)^{1/2}$$

$$= R_B\frac{\left((1+x^2)^2-(1-x^2)^2\right)^{1/2}}{1+x^2}$$

$$= R_B\frac{(1+2x^2+x^4-1+2x^2-x^4)^{1/2}}{1+x^2}$$

$$= R_B\frac{(4x^2)^{1/2}}{1+x^2} = R_B\frac{2x}{1+x^2}.$$

For $R_3 \ll R_B$, neglecting x^2 in comparison to 1,

$$R_B\sqrt{1-v^2/c^2} \approx R_B 2x = R_B 2\frac{R_3}{R_B} = 2R_3.$$

So, to an observer on B_3 watching the rest of B, special relativity would break down by graininess becoming observable at about the velocity at which such observer would see B as about twice as long parallel to v as B_3.

This limitation on domain of applicability of PODIC (and SR) appears reasonable, in that the concept of being inside the universe appears preserved.

Formal treatment of the process of acceleration and limitation associated therewith appears in Section 3 in relation to A-Theorem 6. The velocity and size relation so obtained is consistent with that obtained here. Such consistency is discussed hereinafter in relation to A-Theorem 6, in Subsection 3.3.6, and in Comment 4.1.

One might think of the small system PODIC velocity limit as corresponding to any entity which can fit into B_3 also being able to fit into the difference colinear with v between B_3 and B: if there were not any possibility of such entity being 'outside' B_3 along v, then, noting $\log_2 1 = 0$, what would it mean to say that such entity was 'inside' B_3? One might think of the low velocity PODIC size limitation in terms of filling a universe: in such case, with respect to what might one observe relative motion?

It is interesting that an information argument gives both Einstein's special theory of relativity and this limitation on its domain of applicability.

2.5 MOMENTUM

Using the preceding theorems, I next examine systems containing detectors and energy configurations in relative motion, to see if the momentum properties of inertia may be understandable in terms of information.

In Theorem 2, it appeared that all information contained in a box and encoded with electromagnetic energy is extracted by measuring photon state occupation numbers, without phase, with a perfect (QM) detector in infinite time. With infinite time, the energy uncertainty goes to zero, and each photon can be unambiguously assigned to its correct state count.

Suppose we wished to detect in finite time the contents of a box B_2, containing, for example, electromagnetic energy. In that case, it would be not sufficient to place a detector contiguous with the box, because, for any finite time, the residual electromagnetic energy in the box would have an expectation value greater than zero. Thus, photons might be missed, and not all the energy and information detected. To assure complete detection in finite time, the box must pass completely into the detector in that time. This process requires that the box and detector have nonzero relative velocity.

To avoid discussing external information, consider a detection process using a detector with just enough entry aperture to allow box B_2 to enter, which is turned on only while the box is entering it. Then, there is only one state of the box B_2 in the universe which will allow this process, so not any alternatives can be accepted, and thus not any information can be encoded external to B_2. This can be thought of as a limiting case of reasoning supporting extension as in Corollaries 5.3 and 6.1.

Suppose that the box moves with velocity v in the frame of reference of the detector during detection. Finite detection time makes the uncertainty in energy of detected photons greater than zero. What happens to information represented by photons in the box? In the frame of reference of the detector, the box of photons is moving, and the photon states are increased and spread apart in energy:

Theorem 9. *The increase and spread in energy of photon states in box B_2 moving relative to perfect (quantum mechanical) detector D is exactly sufficient to allow each detected photon to be unambiguously assigned to its correct state count by that detector in the finite time of detection ('unambiguous detection and assignment', or 'UDA').*

1. Put B_2, containing information coded as photon occupation numbers, and D in larger isolated box B. (which could (as in Cor 5.3) be 'the' universe)
2. From Theorem 1, and 1, within B, information is conserved.
3. From 2, during detection of B_2 by D, information is conserved (in B).
4. If the increase and spread of energy of photon states in B_2 as seen by D were insufficient to allow UDA, photon state occupation number would be ambiguous.
5. From 4, in case as in step 4, information would be lost.
6. From 3–5, photon states in B_2 are spread and raised in energy at least enough to allow UDA.
7. If photon states in B_2 were raised and spread more than necessary to allow UDA, then the energy uncertainty during detection could be larger without rendering assignment to true state ambiguous.
8. If larger energy uncertainty could be tolerated, the detection time could be partitioned into two or more (possibly overlapping) shorter times, each long enough to have small enough associated energy uncertainty to allow unambiguous assignment.
9. From EMT and QM, and 8, detection of photons as in step 8 would detect occupation numbers unambiguously, and would also detect 'phase' of detected photons.
10. From Theorem 2, all information present initially can be extracted by unambiguous detection of occupation numbers.
11. From Def 1, and 7–10, for case as in step 7 detected as in steps 8 and 9, information would increase.
12. From 3, 7, and 11, increase and spread of photon energy states is no greater than to allow UDA.
13. From 6 and 12, the photon energy increase and spread is exactly sufficient for UDA.

Note that total energy of a moving 'box' of photons is just the sum of the energies of its photons. In domain of applicability of PODIC, for relative velocity v, each such photon state i is raised to $E_i' = E_i/\sqrt{1-v^2/c^2}$. Thus, total energy $E_T' = E_T/\sqrt{1-v^2/c^2}$, and transforms the same way as does that of a moving mass.

We can transform to the rest frame of B_2, and apply Theorem 5, replacing the contents of B_2 with any energy configuration with the same total energy in B_2, and thereby represent the same amount of information. For example, we could replace photons with total energy E by one or more mass particles.

Intuitively, from the detector's viewpoint, one could think of the particle's external information in a position representation, with relatively sharply defined time and location of arrival at the detector, defining one out of all the possible times it could have arrived (while the detector was sensitive) and places (the detector entry aperture). Or, one could think of the particle's external information in a momentum representation, carrying a relatively closely defined momentum as energy and angle of entry, filling the detector aperture for its sensitive time (time not used to distinguish alternatives). The result is the same. The box becomes merely a convenient fiction.

Transforming back to the rest frame of the detector, we have the result sought:

Corollary 9.1. *All energy carried by a moving object can be fully attributed to its transport of information.*

●

This result is fully special relativistically covariant. The three spatial components have the form $m_{\mathrm{rel}}c^2\vec{v}$, and the fourth component $(d/dt)(Ect) = Ec$; representing, respectively, spatial and temporal flow of energy. These correspond respectively to the spatial transport of information and to its storage, or timelike transport. Note that the four vector formed is c^2 times the four momentum.

What we naively call inertia, with respect to momentum, is that you 'do work on' an object to give it speed relative to you. Thus, momentum of inertia appears in terms of information.

In particular, the rest mass of an object represents propagation of information from the past to the future, for example, by its presence at a 'place' in a position representation. Recalling the relativity of position, this might be thought of as defining location.

2.6 DETECTION AND TRANSCRIPTION

When information is detected, it must be transcribed into some storage system. Two kinds of transcription seem historically interesting. One of these involves recoding information in highly redundant form. For example, a small detector might have a large transcription box: a photon in a small box entering the detector could then be transcribed as a large number of photons in the corresponding state of the larger box. This large box could then be examined many times with more nearly negligible disturbance. This appears to correspond to the classical concept of detection:

Definition 15. *Call transcription of information from a box into highly redundant representation in a very much larger box 'classical transcription', or 'CT'.*

The other type of transcription involves creating a box of the same size as the detected box, emitted from the transcriber as the detected box enters the detector, into which entering information is transcribed with the same energy and redundancy of coding. I call this 'quantum mechanical transcription'. As a special case, any surface in 4-space may be thought of as such a system, with traversal of the surface regarded as a transcription process. This is in keeping with universality of coding constraints related to 'quantum mechanical' detector 'noise'. For example, a 2-surface in 3-space, existing for a time, could be such a detection and transcription system for things passing through it in 3-space. As another example, a volume in 3-space at one time could be such a surface in 4-space, acting as a detector and transcriber for the timelike propagation of objects which at that time were within that 3-volume, the largest such 3-volume being (a) universe.

Definition 16. *Call transcription of information from a box into a representation with the same redundancy, in a box with the same size, 'quantum mechanical transcription', or 'QMT'.*

When an object enters a detector, that part of its external information which is not specifically detected and transcribed appears as external information of the detector and transcription (box), added to their previous external information. This can be seen from conservation of energy and Corollary 6.1, or, in the axiomatic form, Postulate 1 and A-Definitions 21 and/or 22.

2.7 POSITION

Using the preceding theorems and intuitions, I next discuss position, distortion of a metric, relative position and nearness, and gravity. *Analysis in Theorem* 10-*Corollary* 13.5 *applies to systems containing only electromagnetic energy and/or mass particles.*

With relativity of position, as discussed hereinbefore, we can interpret presence of a timelike flow of information in a position representation as defining position. This information flow could appear, for example, as mass (energy) present in a 3-space region.

For the following Theorem 10 through Corollary 11.5 and discussion relating thereto, the following condition ('$C\,1$') is stated here:

Condition 1. *In isolated box B with linear size R_B; where v's and X's denote velocity(s) and linear size(s) respectively of each box contained in B at any stage of any process performed on such box(es),*

$$v < c\frac{1-(X/R_B)^2}{1+(X/R_B)^2},$$

v measured by observer O_B substantially at rest in B

Choice of

$$v < c\frac{1-(X/R_B)^2}{1+(X/R_B)^2}$$

is in conformity with Corollary 8.16, and with use of this portion of Section 2 as part of the axiomatic form in Section 3 hereinafter (from A-Theorem 6) which formally logically precedes this step in the axiomatic form. This constraint corresponds to restriction to domain of applicability of PODIC in finite enclosure (and adseparability). Note that such restriction to domain of applicability of PODIC in finite enclosure and adseparability replaces formal neglect of effects of finite size of B, as embodied hereinbefore in explicit neglect in Lemma 8.1 through Corollary 8.4 of terms of order $(X/R_B)^2$ in comparison to 1.

Using PODIC and PPR, we can obtain a relation between degree of localization of such 3-space region containing such information flow and amount of position information:

Theorem 10. $C\,1$; *if box B_1 with linear size L_1 containing amount I_1 of information is in proportionate box B_2 (containing only B_1) with linear size*

L_2, total amount of information represented by B_1 in B_2 is I_1L_2/L_1, and amount I_1' of external information of B_1 in B_2 is $I_1'=I_1(L_2/L_1-1)$.

1. Start with B_1 filling B_2, containing only I_1.
2. By gently scaling B_1, shrink it to size L_1.
3. From PODIC and Cor 8.16, step 2 does not change the amount or representation of I_1 in B_1. (see also Lemmas 8.8–8.11)
4. With PPR (Theorem 4 or 4′), we can equivalently represent any configuration of photons and/or mass particles by photons.
5. From EMT and QM, photon state energy in B_1 is proportional to $1/L_1$.
6. From 2–5, energy contained in B_2 is its initial energy times L_2/L_1.
7. From Theorem 6, and 1 and 6, $I_1+I_1'=I_1L_2/L_1$.
8. From 3 and 7, amount I_1' of external information of B_1 containing I_1 in B_2 is $I_1'=I_1(L_2/L_1-1)$.

●

With Corollary 8.5, choosing the case of B_1 containing two photons in the lowest photon state in B_1, (or from A-Theorem 6) this result is exact, and can be used in defining degree of localization without reference to physical sizes:

Definition 17. *Call the ratio of total information to internal information of a substantially timelike flow of information its 'localization'.*

Note that the smallest amount of contained information for which the information flow can be nearly entirely timelike is the least for which a nearly rest state in 3-space is defined, which is B_0 as defined before. In NRQM, one can think of localization in terms of reduction of position uncertainty by photon scattering experiments, which agrees with this definition, using B_0, in the nonrelativistic limit.

With this definition of localization and Theorem 10, allowing B_2 to be the universe, we obtain:

Corollary 10.1. C1; *detecting external information of box B without detecting internal information of B reduces localization of B.*

●

2.7.1 'Cosmological Red Shift'

In particular, equivalence of representation of position information by B_0 or freely propagating electromagnetic wave in finite enclosure B seems to offer some insight into electromagnetic propagation over macroscopic distances:

Corollary 10.2. *C1; in box B with constant linear size R_B, light with initial wavelength λ_0 appears after freely propagating distance $d \ll R_B$ to have wavelength*

$$\lambda(d) \approx \lambda_0\left(1 + \frac{d}{R_B}\right)$$

1. From EMT and QM, photon with wavelength λ_0 has energy $E = ch/\lambda_0$.
2. From EMT and QM, as geometric property of focus, photon with wavelength λ defines position to radius $r \approx \lambda/2\pi$.
3. From Theorem 6 (from A-Def 21 and/or 22), photon and B_0 representations of position information have same total energy iff they represent same amount of information in B.
4. From Cor 4.4(g), B_0 with total (mass) energy E has wavelength $\lambda_b = hc^2/Ev$, where v is velocity as seen by observer O_B.
5. From EMT, QM, Theorem 4′, Cor 4.4, and 4, least velocity v_1 of B_0 in B is given approximately by

$$v_1 \approx \frac{hc^2}{E 2\pi R_B}.$$

6. From 1, 3, and 5, v_1 of B_0 representing same position information as photon with wavelength λ_0 is given approximately by

$$v_1 \approx \frac{c\lambda_0}{2\pi R_B}.$$

7. From 2, 3, and 5, for time $t \ll R_B/c$, position initially defined by photon to radius r_0 is after time t defined to radius

$$r(t) \approx r_0 + v_1 t.$$

8. From 2, 6, and 7, after time $t \ll R_B/c$, freely propagating photon originally having wavelength λ_0 would appear to have wavelength $\lambda(t)$,

$$\frac{\lambda(t)}{2\pi} \approx \frac{\lambda_0}{2\pi} + \frac{c\lambda_0}{2\pi R_B} t.$$

9. For freely propagating photon, for $t \ll R_B/c$, $t \approx d/c$.
10. From 8 and 9, cancelling 2π's and factoring, for $d \ll R_B/c$,

$$\lambda(d) \approx \lambda_0 \left(1 + \frac{d}{R_B}\right).$$

●

From this, it appears that Maxwell's equations, whether or not applicable to description of normal modes in B, may not adequately describe 'locally propagative' transport of information by freely propagating electromagnetic waves over distances not $\ll R_B$.

For an enclosure of constant size, this 'red shift' may be approximately expressed in terms of a fictitious velocity of recession of sources:

Corollary 10.3. *C*1; *in box B with constant linear size R_B, light source at distance $d \ll R_B$ appears to have wavelengths of emitted light as if it were receding with (fictitious) velocity $v_r(d) \approx Hd$, $H \approx c/R_B$.*

1. From Cor 10.2, for $d \ll R_B$, $\lambda(d) \approx \lambda_0(1 + d/R_B)$.
2. From 1, for $d \ll R_B$, $(\lambda(d) - \lambda_0)/\lambda_0 \ll 1$.
3. From EMT and PODIC, for $(\lambda(v) - \lambda_0)/\lambda_0 \ll 1$, wavelength increase due to recessional velocity v_r is approximately $\lambda(v_r) \approx \lambda_0(1 + v_r/c)$.
4. From 1–3, by comparison of steps 1 and 3, light source at distance $d \ll R_B$ appears to have wavelengths of emitted light increased as if it were receding with (fictitious) velocity v_r,

$$\frac{v_r(d)}{c} \approx \frac{d}{R_B}.$$

5. From 4, for $d \ll R_B$, v_r may be approximately expressed as $v_r(d) \approx Hd$, $H \approx c/R_B$.

●

Note that this result depends on the nature of 'boundary conditions' for B: treating the longest wavelength for B_0 as $2\pi R_B$ (Cor 10.2, steps 4 and 5) may be thought of as a relatively 'soft' 'boundary condition'; 'harder' 'boundary conditions' might lead to estimating a smaller B_0 longest wavelength and correspondingly higher B_0 least velocity and larger 'red shift', with larger H for a given R_B. There appear to be conceptual and formal reasons for treating the 'boundaries' of B as in this sense 'soft'. This is treated more fully, as presented hereinafter, in 'Testing Postulates', in part of the axiomatic form logically preceding Corollary 10.2.

The formulation of Corollary 10.3 was chosen to simplify comparison with reports of observations as expressed by astronomers in terms of local value H for "Hubble's constant". Note that a universe with constant 'size' would appear to be 'expanding', and formal calculations of 'red shift' in a universe with 'variable size' might also include such a contribution to apparent 'red shift'. I have put quotes around 'variable size', because it is not clear to me what, if anything, it might mean to say that the entire universe had 'variable size', or that such 'changes' in such 'size' would be observable.

Note that a photon detector with unity efficiency cannot have phase sensitivity, in keeping with Theorem 2 and with reasoning in Theorem 9. Detecting direction of incidence of a photon involves phase comparison, for example, a telescope's focusing lens or mirror is an interferometer, and, therefore, photon source direction is not measured by a photon detector with unity efficiency. Thus, it is distance from the source that determines the angular spread of directions for a photon certain to be detected, and thereby determines the amount of information carried by the photon's momentum, and retained after detection as momentum of the detector, including, in the case of an extended detector, both linear and (if appearing in such momentum representation of information) angular momentum.

In the case of representing position information by a relatively localized mass particle or B_0, with increasing time the position so specified becomes less precise, and some position information appears as momentum information, with correlation between position and momentum of B_0 or mass particle. From equivalence of photon and B_0 representations of position information, it would seem that in the case of representation of position information by a spherically expanding freely propagating electromagnetic wave there would also be conversion of representation from position information to momentum information, conserving total information. The 'red shift' approximately calculated for $d \ll R_B$ in Corollary 10.2 seems to be an example of this effect, with that part of the photon's information that after red shift was no longer available as position information ($\lambda/2\pi$) having become available as momentum information, retained after detection of the red shifted photon as momentum information of a unity efficiency detector, represented in a way partly describable in terms of direction of 'recoil' of the detector, due to detecting the photon, defined more precisely the greater the distance from the source.

Although Corollaries 10.2 and 10.3 appear to describe division of a photon's information into position and momentum representations after detection, with the dependence of amount of information in each appearing to an observer as dependence of detected photon wavelength λ on distance d from the photon source, this result does not appear to require that λ be treated as a function of d if detection is not involved in defining such coding, as in calculating photon states (normal modes) satisfying EMT 'boundary conditions' in B.

In our universe, there appears to be a locally approximately linear 'red shift' with distance, previously approximately described by Hubble and attributed to a 'recessional velocity'. In view of this,

Correspondence Corollary 10.4. *Corollaries* 10.2 *and* 10.3 *appear, to the extent tested as presented hereinbefore, substantially allowed by knowledge prior to the time of this writing, with H in Corollary* 10.3 *corresponding to local "Hubble's constant".*

●

2.8 GRAVITY

How does GS of box B_1 in B look to an observer inside? It doesn't notice:

Theorem 11. C 1; *during GS of box B_1 in B, observer O_1 in B_1 does not see any change in the form and constants of the laws of physics.*

1. From PODIC and Cor 8.16, amount and representation of information in B_1 are unchanged by GS of B_1 in B.
2. Consider a system in B_1 including an experiment to measure every law and constant of physics.
3. From 1 and 2, all results of all measurements in system as in step 2 are unaffected by such GS.
4. From 3, O_1 in B_1 does not see any change in the form and constants of the laws of physics during GS of B_1 in B.

●

Using Definition 17, of localization, we obtain

Corollary 11.1. *C*1; *all laws and constants of physics as seen by an observer in box B_1 in B are independent of localization of B_1.*

Theorem 11 and Corollary 11.1 state more explicitly some consequences of PODIC mentioned hereinbefore as examples.

From Theorem 11 and Corollary 11.1, we expect that units of time, length, and mass look different to observers subjected to different amounts of GS. For example, suppose two observers are in a universe, and then one is placed into a box and shrunk by factor $\approx q$, by GS. For the shrunk observer not to notice reduction in size, his length standard must shrink in proportion, by factor $\approx q$. For the speed of light to appear constant, his time standard must shrink by factor $\approx q$; that is, his clocks run faster as seen from outside. Using PPR, his unit of mass is increased by factor $\approx 1/q$, corresponding to the higher energy of photon states in the smaller box. Collecting these results, we obtain:

Corollary 11.2. *C*1; *after GS of box B_2 by factor q, units of mass M, length L, and time T in B_2 appear to an unchanged observer (outside B_2) as $\approx M/q$, $\approx Lq$, and $\approx Tq$.*

(In the axiomatic form, '$\approx$' in the preceding paragraph and Cor 11.2 would be '$\hat{=}$'.)

For all of the following discussion, I have used quantum mechanical transcription (QMT) as restricted to transcription using the *same representation* as received.

Next, consider QMT of a small test box B_1 from outside B_2 to inside B_2 after B_2 has been shrunk by factor q by GS. During transcription, both energy and information are conserved. However, the process depends on the nominal sizes of the incident and transcription boxes, and these sizes are measured with different observers' units of length. Thus, QMT, by requiring the same nominal size for both boxes, corresponds to what the outside observer would otherwise describe as transcription into a box smaller by factor q.

Here we encounter an interesting difference between measurements of time and of length. Rates of two clocks can be compared by receiving timing pulses from one and comparing them to the other. In steady state, the number of pulses in transit is constant, so those arriving give true rate. For measurements of size, a relatively distant observer depends on a local observer, which codes size in terms of its local unit of length, and tells the more distant observer its result. In dealing with this, we proceed, in keeping with Principle 3, in a way which may be expressed as

Principle 4. *If an observer can not have any internal history that would tell it its unit of length is changed, its report of local length measurement is accepted as valid.*

This appears conspicuously in the limiting case of low velocity QMT. To represent the information from externally incident box B_i in internal transcription box B_t, with the same *nominal* size, requires $1/q$ times as much energy as in B_i. So, work must be done on B_i as it proceeds into the transcriber. That is, we have to push it in. More generally, we can treat the case of low velocity QMT of a box, between observers in regions subjected to different GS factors q:

Corollary 11.3. C1; *amount of work W done in slowly displacing by QMT test box B_1 from a region of space in which an observer O_1 has experienced GS by factor q_1 to a region in which an observer O_2 has experienced GS by factor q_2, q_1 and q_2 referenced to a common frame of reference, where E_1 is the total energy (including mass) of B_1, measured by O_1, appears to O_1 as $W_1 = E_1(q_1/q_2 - 1)$.*

1. From Def 11, the ratio of size of test box used by O_2 to that used by O_1 is q_2/q_1.
2. From EMT, QM, Theorem 6, Lemmas 8.8–8.11, PODIC, Cor 8.16, and 1, O_2 would require q_1/q_2 times as much energy as O_1 to represent the same internal information in B_1, for any representation using electromagnetic energy and/or mass particles.
3. From given, amount of energy available in B_1 as transmitted by O_1 is E_1.
4. From PODIC, Cor 6.3 and 8.16, and 2 and 3, amount of additional energy, as seen by O_1, required for transmission of B_1 from O_1's region to O_2's region by QMT is $W_1 = E_1(q_1/q_2 - 1)$.

Note that absolute values of q_1 and q_2 do not appear, only their ratio is used.

Or, B_i could enter transcription with a total of rest plus kinetic energy sufficient for transcription of its contained information I_i into B_t.

In that case, B_i would appear shortened to O_1, because of relative motion:

Corollary 11.4. C1; *two observers O_1 and O_2, respectively located in and substantially at rest relative to regions S_1 and S_2 subject to GS by factors of q_1 and q_2 with respect to a common frame of reference, $q_1 > q_2$, a box B_1 moving in S_1 relative to O_1 with exactly sufficient total energy for QMT from region S_1 into region S_2 appears to O_1 as shorter than nominal by factor q_2/q_1 parallel to its direction of motion.*

1. From Cor 11.3, and conservation of energy (in axiomatic form, from Postulate 1 and A-Def 21 and/or 22), B_1 is seen by O_1 as having total energy q_1/q_2 times its rest energy.
2. From PODIC and Cor 8.16 (SR), any object moving relative to an observer, with velocity such that its total energy is q_1/q_2 times its rest energy, appears to that observer as shortened by factor q_2/q_1 parallel to its direction of motion.

●

Note that absolute values of q_1 and q_2 do not appear, only their ratio is used.

We could compare the length of B_1 parallel to its velocity with a set of coordinates which had not been subject to GS:

Corollary 11.5. C1; *for two observers O_1 and O_2, respectively located in and substantially at rest relative to regions S_1 and S_2 subject to GS by factors q_1 and q_2, a box B_1, moving freely with at least sufficient total energy for QMT from S_1 into S_2, appears to have the same length parallel to its velocity, independent of QMT, compared to a superposed hypothetical coordinate system of observer O_0 which has not been subject to GS, relative to which GS factors q_1 and q_2 are defined.*

1. Suppose $q_1 \geqslant q_2$, and consider box B_2 moving in region S_1 relative to O_1 with least velocity sufficient for QMT into S_2.
2. From Cor 11.4, B_2 appears to O_1 as shortened parallel to its velocity by factor q_2/q_1.

3. From Cor 11.2, length units in S_1 appear to O_0 as shortened by factor q_1.
4. From 2 and 3, B_2 appears to O_0 as shortened by factor $q_1(q_2/q_1)=q_2$.
5. From Def 16, and 1, after QMT into S_2, in case as in step 1, B_2 appears to O_2 as having correct nominal length.
6. From Cor 11.2, and 4 and 5, in case as in step 1, B_2 in S_2 appears to O_0 as shortened by factor q_2.
7. From 4 and 6, if B_2 has exactly sufficient energy for QMT from S_1 into S_2, its length appears constant to O_0.
8. From PODIC and Cor 8.16, step 7 applies for B_2 containing any system.
9. Consider B_1 in B_2, moving with velocity relative to B_2 parallel to the motion of B_2 as seen by O_1, and appearing to O_1 as moving at least as fast as B_2.
10. From PODIC and Cor 8.16, and 9, B_1 appears to an observer O' in B_2 as shortened parallel to its apparent velocity by factor q'.
11. From Def 16, QMT, using the same representation for transcription as received, does not change the relative motion of B_1 in B_2 as seen by O' in B_2.
12. From 10 and 11, q' is not affected by QMT.
13. From 7 and 12, B_1 appears to O_0 to be shortened by the same factor, independent of QMT.
14. From 13, for $q_1 \geqslant q_2$, length of B_1 parallel to its motion appears constant to O_0, independent of QMT.
15. From Theorem 6, PODIC, Cor 8.16, and Cor 11.2, if $q_1 < q_2$, then B_2 in S_1 has sufficient total energy for transcription into S_2, for any velocity >0. (note that, for B_2, $v \neq 0$ in finite enclosure; see part of reasoning in A-Cor 2.2 steps 11–25, A-Cor 4.8, and A-Theorems 5 and 6, or QM)
16. Choose initial motion of B_2 in S_1 as 'quasistatic' displacement.
17. From PODIC, Cor 8.16, 11.3, and 11.4, and 16, in case as in step 16, to O_2, B_2 would appear after QMT as having total energy $\approx q_1/q_2$ times its rest energy.
18. From 2–14, and 17, by multiplying all velocities by -1, we can apply steps 2–14 as before.
19. From 14 and 18, length of B_1 parallel to its motion appears constant to O_0, independent of QMT.

●

Next, we could imagine many observers $O_{j,k,l}$ in different boxes $B_{j,k,l}$, arrayed contiguously, filling a 3-volume. Localization of each $B_{j,k,l}$ could be determined by amount of GS it had experienced, and thus, from Corollary 10.1, by extent to which external information had been encoded or detected for that box without changing its internal information.

For example, consider such an array of boxes, for which localization increases linearly with displacement positively along one coordinate, x, and is independent of displacement along perpendicular axes, y, z, and t. This would be a static, linear, 'one-dimensional', configuration of boxes $B_{j,k,l}$.

Suppose we explore this region of 3-space with a test box B_1, small compared to the $B_{j,k,l}$, and that the difference of GS factor between successive boxes along x is much smaller than the GS factor of each. If we slowly displace B_1 in the direction of increasing x (increasing localization) by QMT, we must keep doing work on it to allow the information contained within B_1 to be represented in what successive observers along x would regard as nominally the same size box.

Instead of displacing B_1 slowly, we could give it sufficient kinetic energy at an observer O_1, with q_1, for its total energy to just allow QMT to O_2, with $q_2 < q_1$. When released by O_1, B_1 has velocity $v_1 > 0$. As seen by O_1, v_1 must be just sufficient to give B_1 total energy increased by factor q_1/q_2 times its rest (mass) energy. Therefore, B_1 is seen by O_1 to be contracted by factor q_2/q_1 at its initial transmission. Successive observers from O_1 to O_2 along x would see B_1 as progressively longer and slower moving, until at O_2 its nominal length agreed with its observed length. This process could be reversed, with B_1 released by O_2 and accelerating toward O_1, seen as progressively shorter as it travelled in the $-x$ direction. For free motion by QMT, in either direction, transverse width of B_1 is seen by all intermediate observers as being its nominal width.

A clock at observer O_2 transmitting pulses to O_1 can be compared to a clock at O_1. In 'steady' state, presence of intermediate observers along x does not affect this comparison. Thus, we find, (Corollary 11.2), that the unit of time of O_2 is smaller than that of O_1, by factor q_2/q_1, that is, pulses arriving from O_2 at O_1 appear more closely spaced than nominal, compared to O_1's clock.

What we find is that an array of boxes as described displays many properties like a uniform, 'one-dimensional', static gravitational field. From preceding discussion of relation between information and momentum aspects of inertia, we know that mass and energy have the same properties for representing information. Thus, from dependence of effects of this array (of 'boxes') only on information in test box B_1, we obtain directly equivalence of inertial mass and gravitational mass.

Note that the 'potential' energy of B_1 in this system is directly attributable to B_1 itself (and, in domain of applicability of PODIC (adseparability), path independent): the concept of 'field energy' does not appear. A test box B_1 thrown freely upward in such a system could rise no higher than its energy allowed QMT to transcribe its contained information I_1; that is, it could not be transcribed to the 'box' of an observer who was so localized that such observer would require more energy to represent I_1 than was initially available.

Suppose that O_2 were formed by GS of a box B_2, starting at O_1. In that case, B_1 could have been placed in B_2, and subject to the same process of GS as O_2. Additional work would be required for GS of B_2 if it contained B_1, just equal to the work required to transport B_1 from O_1 to O_2 in the system described.

These effects on a test box, as seen locally, depend on ratio(s) of localization(s), also as seen locally. This can be used to formulate a representation for the effects in a way which depends only on local observations. For example, the effect of a small displacement along each local coordinate axis can be examined, and a table of such local transformations formed. If the dependence of transformations on position satisfies appropriate requirements, such as 'continuity', 'differentiability', etc., such a table, together with boundary conditions, can be represented as a differential manifold. Such representation (as from Postulates 1, 2, and 3, in domain of applicability of adseparability (PODIC)) might facilitate application of mathematical techniques developed for differential geometry in linking this conceptual picture with observation.

2.8.1 Nearness

Consider a large box B, for example, our universe, containing a much smaller box B_2 containing information I_2 and observer O_2. Choose a position representation for external information I_2' of B_2 in B. Suppose we were to form a box B_3 containing B_2 in B: how much of I_2' would be subsumed by the statement that B_2 is in B_3? If we had no knowledge of position of B_3 in B, then none of I_2' would be subsumed. If we knew exactly where, relative to B, B_3 was in B, then all of I_2' external to B_3 would be subsumed. However, location of B_3 in B represents information, so the amount of position information available to locate B_3 depends on its external information I_3' in B. For B_3 very much smaller than B, most of B_3's total information would be external information. Thus, B_3's external information would be substantially proportional to its total energy, which, for substantially timelike propagation of information in position representation, appears as substantially its rest mass.

What we usually mean by nearness is that two objects are at less than or equal to a specified distance from each other. For a test object with very little information, this is approximately equivalent to the statement that the test object is in a box whose position is known to an accuracy determined by the mass of another object.

Thus, the statement that two objects are near each other subsumes part of their external information. Therefore, proximating two objects decreases the total information represented by them, and, therefore, decreases their total remaining mass (energy). That is, energy and information are released when two objects are brought nearer each other. From this, I next construct a picture for gravity.

2.8.2 Gravity

Consider a large box B, containing a massive object M, representing information in B in a position representation. We could explore the region about M, by filling that region of 3-space with small, contiguous, observers, and then moving a much smaller test box among them.

For the following Theorem 12 through Corollary 12.2 and discussion relating thereto, the following condition is used:

Condition 2. *C*1 *from Theorem* 10, *and X is also each linear size or distance appearing at any step of any process in the system analyzed*

Start with many observers, all having common reference, and located near each other. Move these slowly to positions about M. As each observer is brought near M, part of each observer's external information is subsumed by nearness to M. (Please note convention, stated hereinbefore, that multiplication precedes division, in reading '/' fractions.)

Theorem 12. *C*2; *box B containing total information I_B, substantially isolated timelike information flow $I_M \ll I_B$ in position representation, and box B_1 with linear size $R_1 \ll R_B$, internal information I_1, and external information I_1' in position representation in B, $I_1 + I_1' \ll I_M$; for distance $r \gg R_1$ from I_M, $r \ll R_B$, nearness of B_1 to I_M subsumes amount of information*

$$\Delta I \approx (I_1 + I_1') \frac{R_B I_M}{I_B r} .$$

1. Consider B as contained in larger box B', representing information I_B' by position representation in B'.
2. From Theorem 6, energy E_B contained in B is proportional to I_B.

3. From Theorem 6, and 1 and 2, information represented in position representation by B in B' is proportional to E_B.
4. From Theorem 6, and 2, I_M contributes fraction $\approx I_M/I_B$ of E_B.
5. From Theorem 6, and 1–4, I_M contributes fraction $\approx I_M/I_B$ to I_B'.
6. From PODIC and Cor 8.16, position is relative. (PODIC example 2)
7. From 6, position information I_1' is defined (coded) relative to position of B.
8. From 5 and 7, I_M contributes fraction $\approx I_M/I_B$ to defining I_1'.
9. Consider a box B_3, with linear size r, in B, and containing B_1.
10. From Theorem 10, and 9, B_1 has external information $I_{1,3}'$ in B_3; and B_3, by containing B_1, has 'additional' external information I_3'' in B, 'added' to what it would have had without containing B_1.
11. From PODIC and Cor 8.16, $I_1' = I_{1,3}' + I_3''$.
12. From PODIC and Cor 8.16, determining position of B_3 constitutes detecting its external information, including I_3''.
13. From 12, determining that B_1 is within r of a fully known location in B constitutes detecting that part of I_1' external to a box of size r.
14. From 8 and 13, determining that B_1 is within r of a location known to fractional accuracy $\approx I_M/I_B$ of a fully known (relative to B) location in B constitutes detecting $\approx I_M/I_B$ of that part of I_1' external to a box of size r.
15. From Theorem 10, confining information $I_1 + I_1'$ from box B of size R_B to a box of size $r < R_B$ changes its total information in B to $I_1'' \approx (I_1 + I_1')R_B/r$.
16. From 14 and 15, fraction $\approx I_M/I_B$ of I_1'' is detected by proximating B_1 to r from I_M.
17. From 15 and 16, amount of information detected by proximating B_1 to r from I_M is $\Delta I \approx (I_1 + I_1')R_B I_M/I_B r$.

●

The '$\approx$' introduced in step 4 represents 'neglecting' 1 bit difference in accessibility of I_B in B' and in B, as (with $B = U$) in A-Definition 6 and A-Corollary 6.1, Section 3. The '$\approx$' of step 17 also represents neglecting $I_1 + I_1'$ in comparison to I_M.

(Sometimes on reading this I have had some reservations, as to whether treatment of B as being in a 'larger box' B' (steps 1–8) is meaningful and/or formally proper for B a universe, and share such reservations with you; after doing Section 3 through A-Theorem 8, which formally logically precedes this, I have had somewhat less reservation as to the usefulness of this derivation: note in relation to this that Postulate 3 (finite accessibility of information) might not appear to require that all observers' universes contain the *same* information and/or be coincident.)

PODIC led to scale invariance under size changes by GS, thus, it is reasonable that 'nearness' scales with size of box containing observer. Similarly, contained information is proportional to contained energy for given size of box, so it is reasonable, for independence of size scale, that effect of I_M in B is scaled by total information I_B in B.

Note that a substantially isolated timelike information flow in position representation appears as an isolated concentration of mass (energy). With this, Theorem 12 can be rewritten in terms of mass (energy):

Corollary 12.1. *C*2; *in box B with linear size R_B, containing total mass (energy) M_B, including substantially isolated mass $M \ll M_B$, and box B_1 with linear size $R_1 \ll R_B$ and total information I_T in B, $I_T \ll I_M$; for distance $r \gg R_1$ from M, $r \ll R_B$, nearness of B_1 to M subsumes amount of information*

$$\Delta I \approx I_T \frac{R_B M}{M_B r}.$$

1. From Theorem 6, in B, mass (energy) is proportional to information.
2. From Theorem 12, and 1, substituting M/M_B for I_M/I_B,

$$\Delta I \approx I_T R_B M / M_B r.$$

●

The detection process in Theorem 12 detects external information of B_1, but not internal information: B_1 and I_M can be proximated by steps of GA. For simplicity, assume $I_1' \gg I_1$, and neglect I_1 in comparison to I_1':

Corollary 12.2. *C*1; *in box B with linear size R_B, containing total mass M_B, including substantially isolated mass $M \ll M_B$, and box B_1 with linear size $R_1 \ll R_B$ containing mass $M_1 \ll M$; proximating B_1 from distance $\sim R_B$ to distance $r \gg R_1$ from M, $r \ll R_B$, reduces the localization of B_1 as by GS factor $q(r) \approx 1 + R_B M / M_B r$.*

1. Assume B_1 has internal information I_1 and external information I_1' in B.

2. From Theorem 10, $R_1 \ll R_B$ implies $I_1 \ll I_1'$.
3. Let $I_T = I_1 + I_1'$.
4. From Cor 12.1, proximating B_1 from $\sim R_B$ to $r \ll R_B$ from M subsumes information $\Delta I \approx I_T R_B M / M_B r$.
5. From Def 10, and 4, proximation as in step 4 can be performed by steps of GA of B_1.
6. From PODIC and Cor 8.16, GA of B_1 does not affect amount or representation of I_1.
7. From 1, 4, 5, and 6, I_1 is not changed by step 4.
8. From Cor 10.1, and 4 and 7, localization of B_1 is changed by proximating B_1 to distance r from M.
9. r is measured by an observer at r, subject to localization factor $q(r)$.
10. From Cor 11.2, and 9, observer at r sees M_B as $\approx q(r)M_B$, M as $\approx q(r)M$, and R_B as $\approx R_B/q(r)$.
11. From 4, 8, 9, and 10, total information $I_T(r)$ represented by B_1 at distance r from M would appear as
$$I_T(r) \approx (I_1 + I_1')(1 - R_B M / M_B r q(r)).$$
12. From 11, neglecting I_1 in comparison to
$$I_1', \quad I_1'(r) \approx I_1'(R_B)(1 - R_B M / M_B r q(r)).$$
13. From Def 17 (localization as I_T/I_1), and 4 and 12, and neglecting terms of order I_1/I_1' in comparison to 1, localization of B_1 is reduced as by GS factor $q(r) \approx 1/(1 - R_B M / M_B r q(r))$.
14. From 13, solving for $q(r)$,
$$q(r) \approx 1 + \frac{R_B M}{M_B r}.$$

Corollary 12.2 gives approximately the factor $q(r)$ by which observers located at distance r from M have been changed, as by GS, in relation to a set of observers were M not present. These observers are less localized near M than far from M. To find the nature of 'space' about M, an observer would have to rely on length measurements made by these changed observers. Thus, the reported changes would also depend on the nature of QMT.

A coordinate system could be formed by piecing together transformations relating adjacent observers. Or, a coordinate system could be derived from reports of such observers, by a single observer. Or, coordinates not subjected to GS might be assumed (O_0 hereinafter). To avoid discussing choice of coordinates, I next use the 'weak case' of $R_B M/M_B r \ll 1$ to display some features of an assembly of observers about M. In this limit, choice of coordinate system has relatively little effect.

Theorem 13 *and its lemmas are under the following conditions: with PPR, in B, with linear size R_B, containing total mass M_B, including substantially isolated mass M, and box B_1, with linear size $R_1 \ll R_B$ and contained mass $M_1 \ll M$; with a set of local observers about M; with hypothetical observer O_0 and superposed coordinates in a system not changed by GS, serving as reference for q factors of local observers; where X is each linear size or distance at any step of any process on system in B considered, each velocity v as seen by O_0 is $< c(1-(X/R_B)^2)/(1+(X/R_B)^2)$; every linear size of any part of said system $\ll$ separations between such parts for all steps of all processes thereon.*

Lemma 13.1. *B_1 moving by QMT freely radially toward M is seen by successive local observers as having its nominal width perpendicular to radius to M.*

1. From Def 16, QMT transcribes a received box into a transcription box of the same nominal size.
2. From PODIC and Cor 8.16, width perpendicular to velocity v is independent of v.
3. From 1 and 2, B_1 moving by QMT freely radially toward M is seen by successive observers as having its nominal width perpendicular to radius to M.

●

This result is independent of choice of coordinate r, and so is not limited to weak case.

Lemma 13.2. *For $R_B M/M_B r \ll 1$, B_1 moving by QMT freely radially toward M is seen by local observers as shorter than its nominal length by factor $\approx 1/q(r) \approx 1 - R_B M/M_B r$.*

1. From Cor 11.5, O_0 sees B_1 as having the same length, independent of r.
2. From Cor 11.2, each local observer has units of length which are $q(r)$ times O_0's unit of length.

3. From 1 and 2, local observers find $\approx 1/q(r)$ as many of their length units fit into the length of B_1.
4. From Cor 12.2, and 3, local observers see B_1 shortened by factor $1/q(r) \approx 1/(1+R_BM/M_Br)$.
5. From 4, for $R_BM/M_Br \ll 1$, $1/q(r)$ is given approximately by

$$1/q(r) \approx 1 - R_BM/M_Br.$$

6. From 3 and 5, for $R_BM/M_Br \ll 1$, B_1 moving by QMT freely radially toward M is seen by local observers as shorter than its nominal length by factor $\approx 1/q(r) \approx 1 - R_BM/M_Br$.

●

Lemma 13.3. *For $R_BM/M_Br \ll 1$, clocks near M appear to an observer far from M to run slower by factor $\approx q(r) \approx 1 + R_BM/M_Br$.*

1. From Cor 11.2, local observer's time unit is $\approx q(r)$ times as large as for O_0.
2. From 1, local clocks run slower by factor $\approx q(r)$ as seen by O_0.
3. From Cor 12.2, far from M, $q(r) \to 1$.
4. O_0 is defined as an unchanged observer.
5. From 3 and 4, a remote observer and O_0 would see substantially same clock rates.
6. From Cor 12.2, and 2 and 5, a remote observer would see clocks near M run slower by factor $\approx q(r) \approx 1 + R_BM/M_Br$.

●

Lemma 13.4. *For $R_BM/M_Br \ll 1$, slow displacement by QMT of B_1 with initial rest mass m_1, remote from M, releases energy as seen by O_0, in moving from distance r_1 to r_2, from M, approximately given by*

$$\Delta E \approx (m_1c^2R_BM/M_B)(1/r_2 - 1/r_1).$$

1. From Def 16, during slow displacement by QMT, information in B_1 and nominal size of B_1 are preserved.
2. From Theorem 6, and 1, each local observer sees B_1 as containing the same energy.
3. From Cor 5.1, with PPR (from A-Def 21 and/or 22), and 2, this energy is observed locally as m_1c^2.

4. From Cor 11.2, and 3, mass $m'(r)$ as seen by O_0 is $1/q(r)$ times the locally observed mass.
5. From Cor 12.2, and 4, to first order:

$$\begin{aligned} m'(r) &\approx m_1/q(r) \\ &\approx m_1(1 - R_B M/M_B r). \end{aligned}$$

6. From 3–5, energy released as seen by O_0 is approximately given by $\Delta E \approx c^2(m'(r_1) - m'(r_2))$.
7. From 5 and 6, such $\Delta E \approx (m_1 c^2 R_B M/M_B)(1/r_2 - 1/r_1)$.

●

Theorem 13. *For $R_B M/M_B r \ll 1$, behavior of a test box near M appears approximately as if a 'gravitational field' were present about M, with 'gravitational constant' $G \approx c^2 R_B/M_B$.*

1. From Lemma 13.4, comparison with Newton's law of gravitation, $\Delta E = G m_1 M(1/r_2 - 1/r_1)$, yields $G \approx c^2 R_B/M_B$.
2. From Lemma 13.3, and 1, with G as in step 1, effect of M on clock rates agrees with general relativity in the case examined.[1]
3. From Lemma 13.1, and 1, with G as in step 1, effect of M on lengths measured perpendicular to radius to M agrees with general relativity.[1]
4. From Theorems 6 and 10, extracting energy from B_1 without extracting internal information changes B_1's 'size'.
5. From Def 16, and 4, only a freely moving (QMT) box corresponds to transport of calibration standards for length.
6. From Lemma 13.2, and 1 and 5, with G as in step 1, effect of M on length measured parallel to radius to M agrees with effect in general relativity on a length calibration standard, in the case examined.[1]
7. From 1, 2, 3, and 6, for $R_B M/M_B r \ll 1$, behavior of a test box near M appears approximately as if a 'gravitational field' were present about M, with 'gravitational constant' $G \approx c^2 R_B/M_B$.

●

Note that choice of coordinates in step 9 of Corollary 12.2 led to a form for $q(r)$ which is singular only at $r = 0$ (note also Condition 2). This form is displayed explicitly in step 14 thereof: were step 9 omitted, the form would be $q(r) \approx 1/(1 - R_B M / M_B r)$, which diverges near $R_B M = M_B r$.

This conceptual picture for gravitational interaction might be enhanced by examining attraction of two equal, highly localized, masses. Unlike the treatment appearing in Theorem 13, effects of changes in units are not neglected in approximately describing magnitudes of such changes.

Corollary 13.1. *In box B, with linear size R_B, containing total mass M_B, including two substantially isolated equal masses M and $M' \ll M_B$, with hypothetical observers O and O′ respectively located at, but not affected by, M and M′; for M and M′ separated by r as observed by O,*

$$q(r) \approx \frac{1 \pm (1 + 4R_B M / M_B r)^{1/2}}{2}.$$

1. By symmetry, $q(r)$ is seen as the same by O and O'.
2. From PODIC, O sees M as unchanged.
3. From 1 and 2, O sees M' as unchanged.
4. From Cor 11.2, O sees R_B as $R_B/q(r)$, and M_B as $M_B q(r)$.
5. By choice of coordinates, O sees separation between M and M' as r.
6. From Cor 12.2 steps 1–13, and 1–5, noting difference between step 3 and Corollary 12.2 step 10, writing M for amount of mass of M,

$$q(r) \approx \frac{1}{1 - \dfrac{R_B M}{M_B r (q(r))^2}}.$$

7. From 6, solving for $q(r)$:

$$q(r) \approx \frac{1 \pm (1 + 4R_B M / M_B r)^{1/2}}{2}.$$

●

For the next corollary, I chose the positive square root.

Corollary 13.2. *In box B, with linear size R_B, containing total mass M_B, including two substantially isolated equal masses M and $M' \ll M_B$, with hypothetical observer O located at, but not affected by, M; starting with*

$r \sim R_B$, amount of mass (energy) $M_r(r)$ as observed by a remote observer, released in proximating M and M' to separation r, r as observed by O, is given approximately by

$$M_r(r) \approx 2M\left(1 - \frac{2}{1+\left(1+\dfrac{4R_BM}{M_Br}\right)^{1/2}}\right).$$

1. From symmetry, $q(r)$ is the same for M and M'.
2. From Cor 11.2, and 1, writing M for amount of mass of M, total mass remaining, as seen by a remote observer, is $\approx 2M/q(r)$.
3. From 1 and 2, mass released, as seen by a remote observer, is $M_r(r) \approx 2M(1-1/q(r))$.
4. From Cor 13.1, and 3,

$$M_r(r) \approx 2M\left(1 - \frac{2}{1+\left(1+\dfrac{4R_BM}{M_Br}\right)^{1/2}}\right).$$

●

Thus, proximating two equal, highly localized masses releases substantially all of their mass energy. For r small, remaining mass would vary approximately as $r^{1/2}$.

This process might be thought of as detection of position of one mass by coincidence with position of another. One mass might be thought of as a signal, coded as position, and the other mass as the decoding definition of a suitable match. Release of energy would signify a match. This energy could then be used to transmit other signals, or define other code and decoding matches (complementary information representations).

Were the energy released as photons, the expanding electromagnetic wave would be a photon representation of the position at which the interaction occurred. Or, one mass's trajectory could have been determined, and energy released by interaction used to represent time of intercept, thus decoding the other mass's position.

This conceptual picture of this interaction as decoding a code is like step 7 of Theorem 12.

For this mutual detection process to be total, the internal information carried by the two masses would have to be represented with complementary coding.

Note that this formulation does not lead to finite regions from which electromagnetic energy cannot escape. In part, this can be traced to the choice of coordinates in step 9 of Corollary 12.2. Also, a limitation on the domain of applicability of GS by $q(r)$ can be traced to Lemma 8.8, and suggests that the entire process by which metric effects about substantially localized mass energy M were derived is inapplicable in cases where $q(r)$ effects would lead to delocalization of a test body sufficient to include M or, even more strongly and with reference to Corollary 8.16, to more than fill B. Thus, this derivation suggests that the concept of 'distortion' of a metric about M does not properly include such cases, and, thus, that gravitational effects cannot lead to arbitrarily large 'red shift'. This also follows formally from A-Theorem 6. From these various considerations, or, in the axiomatic form, directly from Postulates 1 and 2,

Corollary 13.3. *Nearness to mass does not isolate any region in B with respect to information exchange.*

●

Because of use of nearness $q(r)$ as GS in deriving apparent 'curvature' of spacetime near mass, which might not appear in describing B itself as a whole, this approach might not yet contain any explicit concept corresponding to some notion of 'global' curvature of all space in B itself as a whole.

Although one might previously have construed SR as implying that information not be transportable at velocity $v > c$, and that this would apply also to communication by Theorem 13 interaction, such argument does not appear sufficient, and, although we have SR from PODIC, for the next corollary I use part of the axiomatic form formally logically preceding Theorem 13 to avoid such informality:

Corollary 13.4. *Theorem* 13 $q(r)$ *effect due to* M *at time* t *if observable as locally propagated to distance* $\approx r$ *from* M *is there so observable at time* $t' \approx t + r/c$.

1. From A-Theorem 1, A-Cor 1.1 and 1.3, and A-Def 4, 5, 7, and 8, velocity of communication is locally c. ('linear coefficient of local communication'; see also note following A-Theorem 1)

2. From 1, velocity v of local communication by Theorem 13 $q(r)$ effect is $\leqslant c$.
3. From A-Theorem 6 and restriction to system velocity $< c(1-(X/R_B)^2)/(1+(X/R_B)^2)$, adseparability applies.
4. From A-Theorem 1, A-Cor 4.8, A-Theorem 5, and 3, any entity in such system transporting information locally at velocity $v' < c$ is adequately describable in terms of B_0 ψs.
5. From Def 1 (or A-Def 1 and 2 and Postulates 1–3), and 4, in such system, local transport of information at $v' < c$ is adequately describable in terms of B_0 ψs.
6. From A-Cor 4.8, A-Def 2 and 12, Theorem 13, and 3, local transport in such system of information by Theorem 13 $q(r)$ effect is not adequately describable in terms of B_0 ψs.
7. From 5 and 6, local $v \nless c$.
8. From 2 and 7, local $v = c$.
9. From A-Cor 1.3, A-Def 7 and 8, and 8, Theorem 13 $q(r)$ effect due to M at t if observable as locally propagated to distance $\approx r$ from M is there so observable at $t' \approx t + r/c$. ('local propagation' as 'piecing together' of successive local transport of information)

(Note that Cor 13.4 might not adequately describe communication, if any, by Theorem 13 interaction not adequately describable in terms of successive local transport ('local propagation'; 'adequately' as in A-Def 2, Section 3).)

(In Section 6, question 26 is related to 'gravitons'.)

Corollary 13.4 is stated with restriction to 'locally propagated to' and 'if observable' because such effect on observer at such distance might be adequately (as in A-Def 2) describable as zero.

Note that the gravitational 'constant' in Theorem 13 is expressed in terms of R_B, M_B, and c. This offers a way to compare this conceptual picture with observation.

The theory as presented depends on B being isolated with respect to information flow. Although prior to the time of this writing there does not appear to have been any direct observation of the 'size' of our universe, a red shift with distance has been reported (Hubble's red shift). Using the relation in Corollary 10.3 between c, R_B, and local value H for Hubble's constant, mean density ρ of mass (energy) in our universe can be approximately expressed in terms of gravitational constant G, c, and H:

$$R_B \approx c/H \qquad \text{(from Cor 10.3)},$$

$$M_B \approx c^2 R_B / G \qquad \text{(from Theorem 13, solving for } M_B\text{)},$$

$$\rho \approx 3M_B / 4\pi R_B^3 \qquad \text{(using isotropy from A-Cor 1.3)};$$

substituting,

$$\rho \approx \frac{3H^2}{4\pi G}.$$

A. R. Sandage[2] reported H as $\sim 15 < H < \sim 40$ km/sec megalightyear, or $H \sim (1.587 \text{ to } 4.231) \times 10^{-18}$/sec. With $G \approx 6.672 \times 10^{-11}$ m^3/kg sec^2, this range for H yields a calculated density range of $\sim 9.007 \times 10^{-27}$ to $\sim 6.405 \times 10^{-26}$ kg/m^3. The smaller density corresponds to $\sim 5.053 \times 10^9$ eV/m^3. This range appears roughly one to two orders of magnitude higher than reported values for the density of luminous matter in our universe, and appears allowed by reports, of which I am at the time of this writing aware, of measurement prior to the time of this writing.[3] The objective here was to test whether a conceptual picture is compatible with such prior knowledge, and this appears sufficient for such purpose. Thus,

Correspondence Corollary 13.5. *The conceptual picture for gravity as an interaction of substantially timelike flows of information in position representation, and formal structure based thereon, presented in formal sequence hereinbefore, appears, to the extent tested as presented in formal sequence hereinbefore, at the time of this writing, allowed by knowledge prior to the time of this writing.*

●

This conceptual picture appears to offer insight into why any form of matter or energy should act as a *source* for a gravitational 'field'.

It appears that what corresponds in general relativity to a hypothetical undistorted metric, to which distortions could be compared, is replaced in this approach by an assembly of observers *who do not have any internal record of having been distorted*, and whose length measurements are the only way a remote observer has of learning the length of an object near them.

Note that gravitational 'field energy' does not appear explicitly in this picture. Gravitational radiation has not been specifically formally included in detail herein. (See Section 6, question 26.)

I suspect that ability to accelerate small systems with negligible effect has in it some aspects which might lead to a generally covariant formulation. Time dependent systems also do not appear precluded.

An observer in a test system freely falling in a central gravitational field would be unaware of the gravitational field iff his test system were small enough. In this regard, it is interesting to note that the down-shifted component of a B_0 might not be spatially small enough for the 'principle of equivalence' to apply. I discuss this effect quantitatively in 'Fast B_0s' (Subsection 2.10).

2.9 B_0 AND 'ELEMENTARY PARTICLES'

All calculations in this Subsection 2.9 *are in, or construed as if in, domain of applicability of PODIC* (*adseparability*). ('construed' re Cor 14.3 'historical solid' filling *B*, and re Cor 14.5 as in Cor 14.3)

The 'softest' B_0 which can be formed in box *B* consists of an association of two photons in states of *B*. This has an interesting consequence:

Theorem 14. *In box B, with linear size R_B; for every form of interaction of photons in B, the lowest bound to amount of mass* (*energy*) *which can form by such interaction a bound B_0 in B is at least that with which such form of interaction associates amount of energy of lowest energy photon in B.*

1. From Cor 8.5 step 23, least energy which can represent occurrence of coupling between two photons in *B* is energy of lowest energy photon in *B*.
2. From 1, for every form of interaction of photons in *B*, there is a lowest bound to amount of mass (energy) for two photons so coupled in forming bound B_0 in *B*, such that said form of interaction associates with coupling of said two photons the amount of energy of lowest energy photon in *B*.

●

The form of interaction might provide for forming B_0s with other than such least mass (energy), or might accommodate several discrete values of mass (energy), for a B_0 in B, at which it just provides the lowest photon energy in B.

Corollary 14.1. *Lowest momentum state of bound B_0 in B has $v \neq 0$, and is the same as for corresponding mass particle P.*

1. From Cor 8.5 step 30, B_0 in lowest velocity state of bound B_0 in B has wavelength equal to the wavelength of the lowest frequency photon possible in B.
2. From NRQM, and 1, the lowest momentum state of P in B is the same as the lowest momentum state of bound B_0 in B.

●

This (or A-Cor 2.2 steps 13–23) justifies restriction of Theorems 4 and 4′ to $v \neq 0$. Theorem 14 has an interesting consequence: it offers a reason to expect that, for each form of interaction between photons, B might contain 'mass particles' with not less than a certain mass, and that all particles with approximately that least mass will have exactly that least mass. This offers a context within which one might search for 'elementary' particles.

Information associated with coupling of two photons to form a bound B_0, that is, a B_0 which can be stable when substantially at rest and 'free' in B, although separately calculated, appears to be merely an alternative way of describing part of the information represented by the photons if not coupled. In such B_0's rest frame, such B_0's associated two photons have (formally) equal frequency. For these reasons, it appears improper to assume phase independence for photon amplitudes in a representation of such information, so energy in a photon state may be expected to be substantially proportional to the square of amount of information so represented. For convenience, 'lowest' in describing photon states may be used herein to denote 'having lowest energy' or, for a set of states, a set satisfying any conditions given and having lowest energy for doing so.

The following approximate treatment of this appears sufficient to proceed:

Corollary 14.2. *For every form of photon interaction which allows a few (N) bits of information to be represented internally in a two photon B_0 substantially stable when substantially at rest and 'free' in isolated B, such interaction releases approximately enough energy to place one photon in each of the N different lowest photon states in B.*

1. From Theorem 1, information is conserved in B.
2. From 1, total information represented in B by photons forming B_0, before forming B_0, is reduced by the amount of information in B_0, after forming B_0. (total (amount) of information represented in B_0 and by such photons after forming B_0 is the same as before; an alternative way of describing representation of such total (amount) of information (see preceding discussion))
3. From Theorem 6, energy and information are proportional.
4. From Theorems 4′ and 14, least energy which can represent information in B is that amount required to represent it in lowest photon states in B.
5. From Theorem 2, this information can be fully represented by occupation numbers for photon states.
6. From Def 6, in B_0's rest frame, both of B_0's two associated photons have equal frequency.
7. From EMT, and 6, amplitudes of photons to represent information of the coupling of photons to form B_0 as in statement of this Corollary 14.2 are not independent. (see also preceding discussion)
8. From EMT, Theorem 6, and 7, amount of energy required to represent I bits of information in one photon state is proportional to I^2.
9. From EMT, and 8, and energy relations of low photon states in 'reasonable' boxes, less energy is required to represent a few bits if each photon state contains at most approximately one photon (e.g., in spherical or cubic boxes).
10. From 2–5, and 9, least energy which can be released in forming B_0 containing N bits in B is approximately that which allows each of the N lowest different photon states in B to contain one photon.

●

With Corollary 14.2, the following function appears useful:

Definition 18. *Let* $Y(N)=\Sigma_{i=1}^{N} E_{Bi}/E_{B1}$, *where the* E_{Bi} *are energies of lowest* N *photon states in* B, *of which* E_{B1} *is least.*

(Note that more than one photon state might have the same energy.)

The only fundamental form of interaction which has appeared so far in this theory appears to correspond to gravity. Using this interaction, I next present derivation for a least mass for B_0 in B.

The following derivation uses a calculation of $q(r)$ self-interaction for two photons together. The longest photon path in a cavity of radius r is in the 'whispering' mode, with path length 2π times the cavity radius. Similarly, the common travel for the photons with their accompanying $q(r)$ is 2π times the cavity radius. Although for geometric simplicity, the following calculation treats B as nested shells with substantially constant angular distribution for all values of shell 'radius', this factor of 2π is retained. (Isotropy and homogeneity appear formally in A-Cor 1.3, which, in the axiomatic form, formally precedes this step.) This approximate calculation relies on the choice of time and position as seen by an observer at the center of such converging nested shells, which observer would, at the time of arrival of the incoming 'signal', see with it all of its prior history. That such a result is calculated with respect to a time selected for convenience appears to not affect the validity of the result: the coupling in such coupled 'mode' of 'electromagnetic energy' in B is what is so sought, *not* 'time development' of such 'mode'. Freedom in choosing description(s) for convenience (e.g., in calculation) is discussed more fully hereinafter in A-Theorem 5, A-Corollary 5.1, and text near after A-Corollary 5.1, in the axiomatic form formally preceding Corollary 14.3.

Corollary 14.3. *In box B, with linear size R_B, containing total mass M_B, including B_0 containing N bits of information; an upper bound for least mass m_0 for which B_0 in B is defined by the interaction stated in Theorem* 13 *is given approximately by*

$$m_0\left(32\pi^2(\ln(R_B 2m_0c/\hbar))^2\right)^{1/3} \approx \left(\frac{\hbar^2 M_B Y(N)}{c^2 R_B^2}\right)^{1/3}$$

1. Consider a shell of electromagnetic energy converging as at speed c of light.
2. From PODIC and Theorem 13, and 1, such electromagnetic energy has associated with it Theorem 13 $q(r)$ effects.
3. From Cor 13.4, and 1 and 2, converging shell moves in as with its own $q(r)$ from further out.
4. From Theorem 13, and 3, inside shell, $q(r)$ is greater than outside shell.
5. From Cor 11.2, and 4, for radial direction, c inside shell appears slower to observer outside shell.
6. From 1, 2, and 5, electromagnetic energy is moving into region with lower velocity of propagation.
7. From 6, shell is kept 'sharp'. (as 'shock' wave)

8. Photons contain total mass energy m.
9. Path length in shell is 2π times shell radius.
10. From 8 and 9, shell has effective mass $2\pi m$. (see also preceding discussion)
11. From 3, 7, and 10, $q(r)$ behaves as for a set of nested shells, each with thickness $\approx y$ and mass $2\pi m$. (y depends on degree of 'sharpening' of incoming electromagnetic energy.)
12. From 10 and 11, m behaves as a body of 'density' $\approx 2\pi m/4\pi r^2 y = m/2r^2 y$, from the 'wall' of B at R_B, in to a radius r_1.
13. From 7 and 12, for Theorem 13 $(1/r)$ interaction, m behaves approximately as:

$$\int_{r_1}^{R_B} 4\pi r^2 dr(m/2r^2 y)/r \approx \frac{2\pi m}{y} \ln(R_B/r_1).$$

(from point of view of observer seeing arriving wave and its past together; 'historical solid')
14. From 7, interaction between two such masses is approximately describable as confined to a shell of thickness y, sweeping inward as at velocity c.
15. From 13 and 14, an overall km_1m_2/r interaction appears as

$$\approx ky\left(m\frac{2\pi}{y}\ln(R_B/r_1)\right)^2 = km^2 4\pi^2 (\ln(R_B/r_1))^2/y.$$

16. From EMT and QM, and 1–15, for strongest interaction as in step 15 (least mass required), least value for both r_1 and y may be estimated as $\approx \hbar/mc$ (shortest radial distance definable by photon with ('tangential') wavelength h/mc; see also Comment 4.8, p. 309).
17. From EMT, QM, PODIC, and Theorem 13, in B_0's 'rest frame', for strongest Theorem 13 interaction (least mass required), the state of B_0's two photons is that for which the photons are superposed in phase.
18. From PODIC and Def 6, sum of mass energies of B_0's two photons is m_0.
19. From EMT, QM, PODIC, and 17 and 18, $m = 2m_0$.
20. From Theorem 13, and 15, $k \approx R_B/M_B$. (This yields a mass change Δm as result of interaction.)
21. Substituting 16, 19, and 20 in 15, mass change Δm is given approximately by

$$\Delta m \approx \frac{R_B}{M_B} 8m_0^3 c 4\pi^2 (\ln(R_B 2m_0 c/\hbar))^2/\hbar.$$

22. From EMT, QM, and PODIC (in axiomatic form, from A-Cor 1.5 and A-Def 11 and 18), lowest photon mode in B has wavelength $\lambda_1 = 2\pi R_B$, and mass $\hbar / cR_B$.
23. From Theorem 14, Cor 14.2, and Def 18, and 21 and 22, for N bits of information in B_0,

$$\frac{Y(N)\hbar}{cR_B} \approx \frac{R_B}{M_B \hbar} 32 m_0^3 c\pi^2 (\ln(R_B 2 m_0 c/\hbar))^2.$$

24. From 23, isolating dependence on m_0,

$$m_0 \left(32\pi^2 (\ln(R_B 2 m_0 c/\hbar))^2\right)^{1/3} \approx \left(\frac{\hbar^2 M_B Y(N)}{c^2 R_B^2}\right)^{1/3}.$$

25. From Theorem 13, all interaction by interaction stated in Theorem 13 is of the same sign.
26. From 1–24, for additional interaction with the same sign, m_0 would be smaller.
27. From 24–26, m_0 appearing in step 24 is an upper bound for the least mass m_0 of B_0 in B bound by interaction stated in Theorem 13.

●

One might think of Corollary 14.3 in the following general way: Although there is not hereinbefore any definition for an observer with velocity c, and so does not appear to be any way to discuss Theorem 13 interaction in such an 'observer''s 'frame of reference' in which the 'photons' would 'appear' to have 'velocity' $<c$ (as required in statement of Theorem 13), this does not appear to prevent use of a 'historical' approach: from the point of view of a centrally located observer, the concept of 'velocity' of the 'photons' does not appear directly, and is subsumed in the simultaneous availability of the 'photons' and all their prior history from R_B 'on in', ('arriving') together as a 'historical solid' as in steps 12 and 13. The factor 'ky' in the left side of the expression in step 15 might be thought of as in some sense 'taking the place' of an 'observer' 'moving' 'with' the 'photons', by selecting for inclusion in 'km_1m_2/r' interaction only effect 'due to' a 'shell' of 'thickness' y, taken 'over' the entire 'historical solid'.

Note that this result depends on spectrum of photon state energies in B, for which is taken (from axiomatic form formally), as $\lambda_1 = 2\pi R_B$, in some sense relatively 'soft' 'boundary conditions' for B. This is discussed further hereinafter in Subsections 3.3.4 and 3.3.5, where formal derivation of such spectrum is presented (A-Theorems 3 and 4). As noted hereinbefore, such result is formally available here in the axiomatic form of derivation herein.

Note that Corollary 14.3 appears to use an approximate calculation of a 'mode' of electromagnetic energy in B, and such B_0s might be regarded as 'excitations' of such 'mode'. A mode of an enclosure can, in EMT, be excited in less time than a signal would take to traverse such enclosure; similarly, it would seem that transcription of information to be represented by such B_0s would not require times as large as R_B/c.

The factor of 2π in step 9 (and 10) might be thought of in terms of tangential direction of propagation observed radially, so that the distance swept by the photons would be 2π times the radius r to the observer, for which sweep of $2\pi r$ the photons and $q(r)$ effects would be as at distance r from the observer. This appears consistent with choices in the axiomatic derivation (Section 3) of globally 'flat' spacetime coordinates (A-Def 8) and R_U (here R_B) as $\lambda_1/2\pi$, where λ_1 is the longest photon wavelength possible with amplitude >0 in universe U (here B); and absence of boundary conditions on U, with photon state 'energy' spectrum $E_j = jE_1$, where E_1 is the lowest photon state energy >0 and j is a finite integer >0 (A-Theorems 3 and 4, and A-Def 21 and 22).

At the time of this writing, I am not sure that I have stated clearly here and in Corollary 14.5 the physical picture which I have had in mind for this form of interaction, which reservation as to clarity of statements has, at the time of this writing, left me with reservation as to the clarity of what seems to be my understanding of this. At the time of this writing, I have not (yet?) made a derivation not involving the 'coordinate' transformation which might seem to underlie this factor of 2π, to information arriving at such a centrally located observer: see introduction to Section 6, and question 25 (Section 6) concerning this. However, with respect to such transformation, there appear to be conceptual and formal reasons (Postulates 1 and 2, A-Theorem 5, A-Cor 5.1, A-Def 5 and 6 (Section 3)) for regarding treatment of such information arriving at such centrally located observer as adequate (in the sense of A-Def 2 and A-Cor 4.1 (Section 3)) for description of amount of information represented in such way in U, and, thus (A-Def 21, 22 (Section 3)), of mass energy in such information representation.

Physically, what might it mean for electromagnetic energy to be in the 'extended interaction' state approximately described in Corollary 14.3? This state was defined by effects of two superposed photons. However, if it contains less energy than that corresponding to such superposed amplitude, energy would not be conserved if it interacted as such two in an energy exchange. Thus, it might seem that electromagnetic energy in the extended state might seem 'hidden' from electromagnetic interaction, and to interact only through $q(r)$. (see also question 26, Section 6, about 'gravitons')

Corollary 14.4. *There are two types of B_0 as in Corollary* 14.3, *distinguished by* 1/4 *cycle relative phase lead or lag between photons in extended interaction and less coupled states.*

1. From Theorem 4′, the two photons in B_0 moving have the same frequency measured in B_0.
2. From Theorem 13 and Cor 14.3, the interaction in Corollary 14.3 depends on energy density in the electromagnetic field.
3. From EMT, QM, and 2, the interaction in Corollary 14.3 depends on the square of the electromagnetic field.
4. From EMT, QM, Theorems 6 and 13, and 3, time dependence of $q(r)$ has twice the frequency of the electromagnetic field.
5. From Cor 8.5 step 25 ('normal modes' 'split' by coupling), Theorem 13, Cor 14.3, and 4, decomposing B_0's electromagnetic energy into such modes, the phase relation for least effect due to $q(r)$ is $\pm 1/4$ cycle relative to the electromagnetic field producing $q(r)$.
6. From 1 and 5, there are two types of B_0 coupled by interaction as in Corollary 14.3, corresponding to + or − 1/4 cycle between photon in extended interaction state ('source' of $q(r)$) and photon in less coupled state.
7. From 1 and 6, these relative phases are constant in time, independent of motion of B_0 in B.
8. From 6 and 7, two types of B_0 are defined by relative phase $\pm 1/4$ cycle between photons in extended interaction and less coupled states.

●

Note significance of equality of photon frequency as observed in B_0: with this, the two types of B_0 are distinct in domain of applicability of PODIC, independent of velocity.

How does presence of 1 bit represented by ± relative phase $\pi/2$ affect interaction in extended interaction state of Corollary 14.3?

Corollary 14.5. *In isolated box B, with linear size R_B, containing total mass M_B, including B_0 representing internally N bits of information of which* 1 *bit is represented by $\pm$ relative phase $\pi/2$; B_0's associated two photons coupled* *by extended interaction as in Corollary* 14.3; *an upper bound for least mass* m_0 *for which* B_0 *in B is bound by interaction stated in Theorem* 13 *is given approximately by*

$$m_0\left(32\pi^2(\ln(R_B 2m_0c/\hbar))^2\right)^{1/3} \approx \left(\frac{\hbar^2 M_B Y(N)}{c^2R_B^2\left(1+\frac{1}{N^2}\right)}\right)^{1/3}.$$

1. From Theorems 4′ and 6, and PODIC (from A-Def 21 and/or 22), information represented by mass m is proportional to m.
2. From 1, representation of 1 bit of a total of N bits of information contained in B_0 corresponds to appearance in such representation of an amount of mass $1/N$ of B_0's mass.
3. From Cor 14.4, $\pm$ relative phase $\pi/2$ represents 1 bit of information in B_0 bound by extended interaction as in Corollary 14.3.
4. From 2 and 3, m in Corollary 14.3 appears multiplied by factor $(1\pm(1/N)^2)^{1/2}$, with $+$ or $-$ sign depending on the relative phase of parts of representation for each state of lead or lag.
5. From Cor 14.3 steps 13 and 16, and 4, limit r_1 of integration in Corollary 14.3 step 13 appears unchanged by such effect. (expression for wavelength not algebraically changed)
6. From Cor 14.3 step 19, and 4 and 5, m in Corollary 14.3 step 19 becomes effectively

$$\approx\left(1\pm\frac{1}{N^2}\right)^{1/2} 2m_0.$$

7. From Cor 14.3 steps 13–24, and 4–6, noting distinction between m appearing in Theorem 13 interaction and m appearing in limit of integration, Corollary 14.3 step 24 becomes

$$m_0\left(32\pi^2(\ln(R_B 2m_0c/\hbar))^2\right)^{1/3} \approx \left(\frac{\hbar^2 M_B Y(N)}{c^2R_B^2\left(1\pm\frac{1}{N^2}\right)}\right)^{1/3}.$$

8. From 7, m_0 appears smaller for $+$ sign than for $-$ sign in denominator of expression in step 7.

9. From 7 and 8, as an upper bound for least such m_0,

$$m_0\left(32\pi^2(\ln(R_B 2m_0c/\hbar))^2\right)^{1/3} \approx \left(\frac{\hbar^2 M_B Y(N)}{c^2 R_B^2\left(1+\frac{1}{N^2}\right)}\right)^{1/3}.$$

10. From Cor 14.3 steps 25–27, and 9, m_0 appearing in step 9 is an upper bound for least mass m_0 of B_0 bound by Theorem 13 interaction in B and representing internally N bits of information of which 1 bit is represented by $\pm$ relative phase $\pi/2$ in extended interaction state as in Corollary 14.3.

●

(See also question 46, Section 6, in relation to step 4.)

Note that choice of '+' sign in step 9 corresponds to seeking least mass for this form of interaction, as part of looking for a stable bound system.

Corollary 14.6. *Corollary* 14.3 *interaction allows substantially exactly two different relative polarizations of B_0's associated two photons.*

1. From EMT and QM, a photon has exactly two possible states of polarization.
2. From 1, by enumeration, B_0's associated two photons have exactly two relative polarizations.
3. From Theorem 13 and Cor 14.3, $q(r)$ as in Corollary 14.3 interaction does not substantially interchange such relative polarization states.
4. From 2 and 3, Corollary 14.3 interaction allows substantially exactly two different relative polarizations of B_0's associated two photons.

●

Corollary 14.7. *For B_0 in B, using the interaction in Corollary* 14.3, *B_0 contains at least* 3 *bits of information.*

1. From Theorem 14, formation of B_0 subsumes at least 1 bit of information.

2. From Cor 14.4, relative phase of the two photons associated with B_0 represents 1 bit of information.
3. From Cor 14.6, relative polarization of the two photons associated with B_0 represents 1 bit of information.
4. From 1–3, using the interaction in Corollary 14.3, B_0 contains at least 3 bits of information.

●

Such third bit not coded as relative phase or relative polarization is discussed further in Chain 17 and its links, in Section 3.

Next, compare Corollary 14.5 interaction with interaction based on relatively localized electromagnetic energy, phase coherent with a representation as in Corollary 14.5. As in Corollary 14.5, Theorem 13 interaction is used.

Corollary 14.8. *In isolated box B, with linear size R_B, containing total mass M_B, including B_0 with mass $m_0 \ll M_B$; compared to relatively locally interacting mass energy m_0, mass energy m_0 in Corollary* 14.5 *extended interaction appears more effective for releasing information by factor* $F \approx (32\pi^2(\ln(R_B 2m_0c/\hbar))^2)^{1/3}$.

1. From EMT and PODIC, in smallest size to which an amount m of electromagnetic mass (energy) can be localized, its wavelength fits around once in a circle of that size.
2. From EMT, electromagnetic energy density is proportional to square of electromagnetic field.
3. From 1 and 2, wave as in step 1 has two diametrically opposed relatively localized mass (energy) portions.
4. From EMT and QM, least wavelength for electromagnetic mass (energy) m is h/mc.
5. From 1, 3, and 4, localized electromagnetic mass (energy) m appears as two masses $m/2$, separated by distance $\approx 2\hbar/mc$.
6. From Theorem 13, and 5, mass (energy) Δm released by these interacting mass portions is given approximately by

$$\Delta m \approx \frac{R_B\,(m/2)^2}{M_B 2\dfrac{\hbar}{mc}}.$$

7. From Cor 14.3 and 14.5, as in Corollary 14.5 steps 1–4, for such relatively localized interaction to represent with same relative photon

phases as in Corollary 14.5 N bits of information internal to B_0, of which 1 bit is represented by $\pm$ relative phase $\pi/2$, m appears larger in Theorem 13 interaction by factor $(1+1/N^2)^{1/2}$.

8. From Def 6, two photons are associated with B_0.
9. From Def 6 and Theorem 4′, both photons associated with B_0 have equal energy in B_0's rest frame.
10. From Theorem 13, and 6, Δm is largest for m largest.
11. From EMT and QM, and 8, B_0's two photons have greatest mass (energy) when superposed in phase, substantially equal to twice their mass energy if summed.
12. From 7, 8, 9, and 11, m appears largest in Theorem 13 interaction as $\approx 2(1+1/N^2)^{1/2}m_0$.
13. From 10 and 12, noting as in Corollary 14.5 step 5 that factor $(1+1/N^2)^{1/2}$ does not appear in wavelength, largest Δm is given approximately by

$$\Delta m \approx \frac{R_B m_0^3 c}{M_B \hbar}(1+1/N^2).$$

14. From Cor 14.2, and 13, as in Corollary 14.3 and 14.5, for B_0 to contain N bits of information, of which 1 bit is represented by $\pm$ relative phase $\pi/2$, least m_0 is given approximately by

$$\frac{Y(N)\hbar}{cR_B} \approx \frac{R_B m_0^3 c}{M_B \hbar}\left(1+\frac{1}{N^2}\right).$$

15. From 14, solving for m_0,

$$m_0 \approx \left(\frac{\hbar^2 M_B Y(N)}{c^2 R_B^2\left(1+\frac{1}{N^2}\right)}\right)^{1/3}.$$

16. From Cor 14.5, and 15, comparing, mass m_0 in Corollary 14.5 extended interaction appears more effective in releasing information than m_0 in the relatively localized interaction of this Corollary 14.8 by factor

$$F \approx \left(32\pi^2(\ln(R_B 2 m_0 c/\hbar))^2\right)^{1/3}.$$

Note that there appears to be not enough information present in relatively localized system as in steps 1–15 for description dividing 3-space more finely than for description of interacting 'parts' as 'lumps'. (See also Comment 4.8, p. 309.)

Note that use of '+' sign in step 7 was determined by its use in Corollary 14.5, with this relatively more localized mode appearing as copossible with such extended interaction mode.

Although this derivation uses (in step 14) the same spectrum of photon state energies ('boundary conditions') in B as was used in Corollary 14.3 steps 22 and 23 and in Corollary 14.5, such dependence is very nearly removed in step 16, and F appears to depend on 'boundary conditions' in B only logarithmically through variation of R_B by roughly $1/2\pi$ of one wavelength of photons coupled to form B_0. 'Boundary conditions' in B are discussed more fully hereinafter, in 'Testing Postulates'.

What happens when two B_0's are proximated?

Corollary 14.9. *For two B_0's, with equal mass m, in B, as in Corollary* 14.4, *interacting at separation $r \gg \hbar/mc$; like types (lead or lag) repel each other, and unlike types attract each other.*

1. From Cor 14.3 and 14.4, phase of photon in extended interaction state determines phase of $q(r)$.
2. From Cor 14.4, phase of photon in less coupled state is referenced to $q(r)$.
3. From Cor 14.3, 14.5, and 14.8, when two B_0's with equal mass are proximated, strongest coupling is between their respective extended interaction states (rather than less coupled states).
4. From EMT and QM, when two photons are superposed in phase, their total energy is greater than for any other relative phase.
5. From Cor 14.3 step 15, and 1 and 4, coupling of photons in respective extended interaction states of two B_0's is greatest for photons in phase.
6. From 3 and 5, when two B_0's with equal mass are proximated, photons in their extended interaction states are in phase.
7. From 2 and 6, when two B_0's with equal mass are proximated, relative phase between photons in the B_0's respective less coupled states is the difference between their respective phases relative to their respective extended interaction state photons.
8. From Cor 14.4, phase of less coupled state photon is $\pm 1/4$ cycle relative to extended interaction state photon.
9. From 7 and 8, like types of B_0 have their respective less coupled state photons in phase, and unlike types have their respective less coupled state photons $1/2$ cycle out of phase.

10. From EMT and QM, two at least partially superposed equal photons in phase together represent more energy than the sum of their separate energies.
11. From EMT and QM, two at least partially superposed equal photons with opposite phase together represent less energy than the sum of their separate energies.
12. From EMT and QM, and Def 6, proximating two B_0's more closely superposes more nearly completely their respective electromagnetic fields. (their 'photons')
13. From 9, 10, and 12, proximating like (equal mass) B_0's increases their total energy.
14. From 13, like (equal mass) B_0's repel each other.
15. From 9, 11, and 12, proximating unlike (equal mass) B_0's reduces their total energy.
16. From 15, unlike (equal mass) B_0's attract each other.

●

Corollary 14.9 appears to offer some insight into the meaning of one of the bits of internal information, represented by photon relative phase, in a B_0 based on photon interaction as in Theorem 13 and Corollaries 14.3 and 14.5.

What does the internal bit represented by photon relative polarization mean?

Corollary 14.10. *An assembly of B_0s as in Corollaries* 14.3 *and* 14.5, *of the same type, each with mass m, with maximum kinetic energy per $B_0 \ll mc^2$, is nearly described by Fermi–Dirac statistics for an assembly of like particles with exactly two internal states ('spin $\frac{1}{2}$').*

1. Consider box B, containing one B_0, B_{01}, in a momentum state in B.
2. From Theorem 4′, B_{01}'s two associated photons are in photon states in B.
3. From Cor 14.4, phase of B_0's less coupled state photon is fixed as + or $-1/4$ cycle relative to its extended interaction state photon.
4. From Cor 14.6, polarization of B_0's less coupled state photon is fixed in one of two allowed relations to that of its extended interaction state photon.
5. Consider a second B_0, B_{02}, of the same type as B_{01}, in B, in a different momentum state than B_{01}.

6. From Theorem 4′, B_{02}'s two associated photons are in photon states in B.
7. From EMT, Theorems 2 and 4′, and 5 and 6, B_{02}'s photons have no phase relation to B_{01}'s photons.
8. From EMT, different photon states in B are independent.
9. From 7 and 8, B_{01}'s and B_{02}'s photons do not interfere in B.
10. Consider a third B_0, B_{03}, of the same type as B_{01}, in B, in the same momentum state as B_{01}.
11. From EMT, QM, Theorem 6, Cor 14.3 and 14.5, and 10, as in Corollary 14.9, electromagnetic field in respective extended interaction states of B_{01} and B_{03} interacts by $q(r)$ most strongly for identical phase and polarization.
12. From 10 and 11, photons in respective less coupled states of B_{01} and B_{03} have the same phase.
13. From 4 and 11, photons in respective less coupled states of B_{01} and B_{03} have fixed relative polarization.
14. From EMT and QM, for two photons in phase, being more than $\pm\pi/2$ radians apart in polarization is equivalent to phase reversal.
15. From Cor 14.4, for B_{01} and B_{03} to be of the same type, phase reversal does not occur.
16. From 14 and 15, relative polarization between photons in respective less coupled states of B_{01} and B_{03} is $\pm \leqslant \pi/2$ radians.
17. From EMT and QM, and 1 and 10, photons of B_{01} and B_{03} are in photon states in B.
18. From EMT and QM, in B, with relative polarization $\pm \leqslant \pi/2$ radians, there are exactly two different states of photon relative polarization, 0 and $\pi/2$ radians relative polarization.
19. From 17 and 18, relative polarization of photons in respective less coupled states of B_{01} and B_{03} is either 0 or $\pi/2$ radians.
20. From EMT and QM, if two photons in respective less coupled states of B_{01} and B_{03} have $\pi/2$ radian (perpendicular) relative polarization, they are in different photon states in B.
21. From EMT and QM, and 19 and 20, in case as in step 20, photons of B_{01} and B_{03} respectively do not interfere electromagnetically in B.
22. From 12, 17, and 18, if two photons in respective less coupled states of B_{01} and B_{03} have the same polarization, they are in the same state of B with the same phase.
23. From Theorem 4′, for B_0 kinetic energy $\ll mc^2$, its two associated photons have nearly equal energy.

24. From EMT and QM, and 22 and 23, in case in step 22, two photons in respective less coupled states of B_{01} and B_{03} superpose to total mass (energy)$\sim 2m$.
25. From 22 and 24, additional mass (energy)$\sim m$ is required for case in step 22.
26. Consider many B_0's of the same type, in B, with kinetic energy per $B_0 \ll mc^2$.
27. From 9, B_0's of the same type in different momentum states of B_0 in B substantially do not interfere.
28. From (EMT, and) 21 and 25, for energy per $B_0 \ll mc^2$, almost exactly two B_0s of the same type are allowed in each momentum state of B_0 in B.
29. From QM, and 27 and 28, neglecting multiple scattering which might result in a few B_0s having kinetic energy $\sim mc^2$, many B_0s as in Corollaries 14.3 and 14.5 of the same type, in B, with kinetic energy per $B_0 \ll mc^2$, are nearly described by Fermi–Dirac statistics for an assembly of like particles with exactly two internal states ('spin $\frac{1}{2}$').

●

Note that, with enough energy, B_0s can be in exactly the same state. Thus, an assembly of B_0s whose Fermi level were highly relativistic might contain more than one B_0s in each relatively low momentum state. From superposition of photon amplitudes, total mass (energy) per multiply occupied state might be expected to vary approximately as the square of the number of B_0s in the same state.

Corollary 14.10 appears to offer some insight into the meaning of the internal information bit represented by relative polarization of the two photons of B_0 of Corollaries 14.3 and 14.5.

What happens if two such B_0s, of unlike types, are together in B?

Corollary 14.11. *For two B_0s as in Corollary 14.5, with equal mass m, in isolated box B, with different relative phase of their respective associated two photons; if the B_0s are in different momentum states in B, their respective associated photons do not interfere electromagnetically; if the B_0s are in the same momentum state in B, with their respective less coupled state photon polarizations perpendicular to each other, their respective associated photons do not interfere electromagnetically; if the B_0s are in the same momentum state in B, with their respective less coupled state photon polarizations parallel* *to each other, both B_0s mutually annihilate, releasing their mass (energy) as propagating electromagnetic energy.*

1. From and as in Cor 14.10 steps 1–9, for unlike B_0s with equal mass m in different momentum states in B, their respective associated photons do not interfere electromagnetically.
2. Consider two B_0s as in Corollary 14.5, of opposite type and equal mass m, B_{01} and B_{02}, in the same momentum state in B.
3. From EMT, QM, Theorem 6, Cor 14.3 and 14.5, and 2, as in Corollary 14.9, electromagnetic fields in respective extended interaction states of B_{01} and B_{02} interact by $q(r)$ most strongly for identical phase and polarization.
4. From Cor 14.4, and 2, photons in respective less coupled states in B_{01} and B_{02} have phase $\pm 1/4$ cycle relative to their respective extended interaction state photons.
5. From 3 and 4, unlike B_0s in the same momentum state in B have relative phase $1/2$ cycle between photons in their respective less coupled states.
6. From and as in Cor 14.10 steps 13–19, relative polarization of photons in respective less coupled states of B_{01} and B_{02} is either 0 or $\pi/2$ radians.
7. From and as in Cor 14.10 steps 20 and 21, and 6, if relative polarization as in step 6 is $\pi/2$ radians, associated photons of B_{01} and B_{02} respectively do not interfere electromagnetically.
8. From Theorem 4′, for both B_0s with equal mass in the same momentum state in B, photons in their respective less coupled states have equal energy.
9. From EMT, QM, and 5, 6, and 8, with relative polarization 0, photons in respective less coupled states of B_{01} and B_{02} exactly cancel.
10. From Lemma 4.1 and Theorems 4 and 4′, B_0 can have velocity $< c$ only if it has both its associated photons.
11. From 6, 9, and 10, with relative polarization 0 as in step 6, two unlike B_0s in the same momentum state in B mutually annihilate each other, with remaining electromagnetic energy propagating at the speed of light.
12. From PODIC, and 11, energy and information are conserved in B during the process in step 11.
13. From 2, 11, and 12, propagating electromagnetic energy remaining after such mutual annihilation contains all mass energy of B_{01} and B_{02}.

●

From the definition of 'complementary' (Def 2), and Corollaries 14.10 and 14.11, from conditions for mutual annihilation,

Corollary 14.12. *For two B_0s as is Corollary* 14.5, *with equal mass, in isolated box B; two such B_0s with the same relative phase of their respective associated two photons are not able to represent information in complementary codes, and two such B_0s with different relative phase of their respective associated two photons can represent information in complementary codes.*

●

Next, I evaluate the approximate magnitude and form of the attractive and repulsive interaction whose existence was set forth in Corollary 14.9.

With Corollary 6.4, we obtain from Corollary 14.8:

Corollary 14.13. *Electromagnetic field in B_0 relatively localized state has $1/F^{1/2}$ times amplitude in extended interaction state, and has phase $\pm 1/4$ cycle relative to electromagnetic field in extended interaction state, with F as in Corollary* 14.8.

1. From Cor 14.8, compared to relatively locally interacting mass energy, mass energy m_0 in extended interaction appears more effective for releasing information by factor $F > 1$.
2. From PODIC, existence of representation in relatively localized mode or in extended mode requires release of same N bits of information. (copossibility)
3. From Def 6 and PODIC, all mass (energy) of B_0 is electromagnetic.
4. From Theorem 6, Cor 6.4, and 1–3, electromagnetic amplitude in relatively localized state is $1/F^{1/2}$ that in extended interaction state.
5. From Cor 14.4, and 1, relatively localized state is (a) less coupled state, as in Corollary 14.4.
6. From Cor 14.4 and 14.9, and 5, as in Corollary 14.9, electromagnetic field in relatively local state has phase $\pm 1/4$ cycle relative to that of extended interaction state.

●

Corollary 14.14. *In isolated box B with linear size R_B, containing total mass M_B, including two B_0s as in Corollary* 14.5, *with equal mass $m_0 \ll M_B$,* *representing external information in position representation in B; with separation* $\approx r$; *like types of* B_0 *(as in Cor* 14.4*) repel* ('+') *and unlike types attract* ('−'), *for* $r \gg \hbar/m_0c$, *with energy change* $\Delta E(r) \approx \pm \hbar c/Fr$, *with F as in Corollary* 14.8.

1. Consider two B_0s as in Corollary 14.5, B_{01} and B_{02}, with equal mass m_0, representing external information in position representation in B, with separation $\approx r$. (r is not defined more closely than provided by position information of B_{01} and B_{02} in B.)
2. From Cor 14.3, electromagnetic field in respective extended interaction states of B_{01} and B_{02} couples most strongly by $q(r)$ for the same phase in both.
3. From Cor 14.13, and 2, electromagnetic fields in respective local states of B_{01} and B_{02} have relative phase 0 for like B_0s and π for unlike B_0s.
4. From Cor 14.13, electromagnetic amplitude in local state of B_0 is $\approx 1/F^{1/2}$ times electromagnetic amplitude in extended interaction state.
5. From given, electromagnetic mass energy in extended interaction form is m_0 as in Corollary 14.5.
6. From EMT, QM, PODIC, and 4 and 5, in suitable units, electromagnetic amplitude in local state is $\approx (m_0/F)^{1/2}$.
7. From EMT, QM, PODIC, and 5, electromagnetic amplitude in local state may be approximately described as having wavelength $\hbar/m_0c$.
8. From EMT, Cor 14.8, and 7, electromagnetic energy in local state is localized to radius $\approx \hbar/m_0c$.
9. From EMT, Cor 14.3 and 14.8, and 8, electromagnetic amplitude in B_{01}'s local state has spatial distribution $\approx \hbar/m_0cr$ about B_{01}'s position. (Note that B_{01} is located no more accurately than its external information provides.)
10. From EMT, and 9, change of angle of orientation of electromagnetic field by more than $\pi/2$ radians is equivalent to reversing phase.
11. From Cor 14.4, and 10, for B_{01} to remain of the same type, orientation of its local state electromagnetic field is in a hemisphere. (all 'states' representing the same information appear in calculation; see also Comment 4.10, p. 312)
12. From 9 and 11, integrating over this hemisphere, local state electromagnetic field amplitude at distance r from B_{01} has form $\approx \hbar/2m_0cr$.
13. From 6 and 12, local state electromagnetic field amplitude about B_{01} is $\approx (m_0/F)^{1/2}\hbar/2m_0cr$.
14. From and as in steps 10 and 11, orientation of B_{02}'s local state electromagnetic field is in a hemisphere.

15. From 14, integrating over this hemisphere, B_{02}'s local state receives a contribution of $\approx\frac{1}{2}$ B_{01}'s presented local state electromagnetic field.
16. From 3, 10, and 15, B_{02}'s local state electromagnetic amplitude is increased for like B_0s, and decreased for unlike B_0s, by $\approx\frac{1}{2}$ B_{01}'s presented local state electromagnetic field.
17. From 13, 15, and 16, B_{02}'s local state electromagnetic amplitude is approximately

$$\left(\frac{m_0}{F}\right)^{1/2} \pm \left(\frac{m_0}{F}\right)^{1/2} \frac{\hbar}{4m_0cr},$$

with '+' for like B_0s and '−' for unlike B_0s.
18. From EMT, and 6 and 17, in units as in step 6, total electromagnetic energy in B_{02}'s local state is approximately

$$\frac{m_0}{F}\left(1 \pm \frac{\hbar}{2m_0cr} + \frac{\hbar^2}{16m_0^2c^2r^2}\right),$$

with '+' for like B_0s and '−' for unlike B_0s. (mass units)
19. From 18, change Δm_2 in mass of B_{02}'s local state is approximately

$$\Delta m_2 \approx \pm \frac{\hbar}{2Fcr} + \frac{\hbar^2}{16m_0c^2r^2F},$$

with '+' for like B_0s and '−' for unlike B_0s.
20. Steps 1–19 also approximately describe B_{02}'s effect on B_{01}.
21. From 19 and 20, total mass change for two interacting B_0s is $\Delta m \approx 2\Delta m_2$.
22. From PODIC, total energy change $\Delta E = c^2\Delta m$.
23. From 19, 21, and 22,

$$\Delta E\,(r) \approx \pm \frac{\hbar c}{Fr} + \frac{\hbar^2}{8m_0r^2F},$$

with '+' for like B_0s and '−' for unlike B_0s.
24. For $r \gg \hbar/m_0c$, $\Delta E\,(r)$ is approximately describable as $\Delta E\,(r) \approx \pm\hbar c/Fr$, with '+' for like B_0s and '−' for unlike B_0s.

Note that, for r large enough, ΔE depends on m_0 only through F. Thus, were another form of interaction used to form B_0s, which gave the same phase relations and value of F, but different m_0, repulsion and attraction approximately calculated in Corollary 14.14 might be unchanged. (See also Section 6 question 80 in relation to step 23 $\hbar^2/8m_0r^2F$ term.)

Note that polarization of less coupled state electromagnetic field A_L is fixed relative to extended interaction state electromagnetic field A_E, in one of two states. As in Corollary 14.10 steps 18 and 19, only two relative polarizations appear possible between respective A_Ls in two like B_0s. Were polarization of A_L relative to A_E to be labeled as 0 or $\pm\pi/2$ radians, then two like B_0s might have three labels of relative polarization of their A_Ls, 0, $\pi/2$, and π radians. I suspect that polarization of A_L relative to A_E might better be labeled as $\pm\pi/4$ radians. This also might seem to offer relatively simple symmetry properties (under parity reversal, charge conjugation, and time reversal).

Noting that the only interaction formally presented hereinbefore before Corollary 14.9 is Theorem 13 interaction, using PPR, and noting that $1/(1+1/3^2)=0.9$, we can collect results of Theorem 14 through Corollary 14.14, as

Corollary 14.15. *In isolated box B, with linear size R_B, containing total mass M_B; least (rest) mass particle P_0, mass $m_0 \ll M_B$, contains 3 bits of internal information; P_0 is of two types, of which like types repel and unlike types attract, and unlike types mutually annihilate with all contained mass (energy) released as propagating electromagnetic energy; an assembly of P_0's of a single type, with kinetic energy per $P_0 \ll m_0c^2$, is nearly described by Fermi–Dirac statistics for an assembly of like particles with 'spin $\frac{1}{2}$'; like types are unable, and unlike types are able, to represent information with complementary coding; an upper bound on least m_0 is given approximately by*

$$m_0F \approx \left(\frac{0.9\hbar^2 M_B Y(3)}{c^2R_B^2}\right)^{1/3}$$

with $F=(32\pi^2(\ln(R_B 2m_0c/\hbar))^2)^{1/3}$; and, for two P_0's separated by distance $\approx r \gg \hbar/m_0c$, repulsion or attraction is of magnitude $\Delta E \approx \pm\hbar c/Fr$, with '+' for like P_0's and '$-$' for unlike P_0's.

(See discussion following Cor 14.3 and 14.8 in relation to dependence of these results on spectrum of photon state energies ('boundary conditions') in B.)

We can compare this with our universe, by calculating the least allowed mass for a particle which can exist substantially at rest. This is to be compared uniquely with the mass and properties of the least massive (rest) mass particle in our universe.

Of data in the formulae for F and m_0, M_B is least well known. From Theorem 13, $G \approx c^2 R_B / M_B$, so M_B can be replaced using R_B and G. From Corollaries 10.3 and 10.4, $R_B \approx c/H$. Substituting, we obtain

$$F \approx \left(32\pi^2\left(\ln\left(c^2 2m_0/\hbar H\right)\right)^2\right)^{1/3} \qquad \text{(a)}$$

and

$$m_0 \approx \left(\frac{0.9\hbar^2 H Y(3)}{Gc}\right)^{1/3} \Big/ F \qquad \text{(b).}$$

To calculate m_0 and F, we need $Y(3)$. In the axiomatic derivation of photon states in our universe U, I show explicitly that 'energy' E_2 of the second photon 'energy level' is twice energy E_1 of the lowest 'energy' photon state in U. Although such derivation appears subsequently herein, in Section 3, it formally precedes this part of the derivation, and the spectrum of states so obtained may properly be used here.

Using such spectrum, for $Y(3)$, only two photons (with different polarization) can be placed in the lowest 'energy level', with energy E_1, and 1 is placed in the second 'energy level', with $E_2 = 2E_1$, so $Y(3) = (2\times 1) + (1\times 2) = 4$.

From the 'adjusted' values recommended for physical constants by E. R. Cohen and B. N. Taylor[4] ('CT', 1973), we take for this calculation $G = 6.672(41)\times 10^{-11}$ m^3/kg sec^2, $\hbar = 1.0545887(57)\times 10^{-34}$ J sec, and $c = 299792458(1.2)$ m/sec, where digits in parentheses were stated by CT as 'error' respectively of the last two digits of the numbers stated, as discussed further hereinafter.

Using these values for G, $\hbar$, and c, and A. R. Sandage's[2] values for range of H, $\sim(1.587$ to $4.231)\times 10^{-18}$/sec, solving numerically for m_0 and F by successive approximation, I obtained

H (sec^{-1})	m_0 (kilogram)	F (number)
1.587×10^{-18}	10.7528×10^{-31}	136.7101
4.231×10^{-18}	14.9821×10^{-31}	136.0518

m_0 appears remarkably close to the mass of the electron and positron, which at the time of this writing are the least massive particles of which I am aware reported as observed substantially at rest in our universe; and F appears remarkably close to α^{-1}, describing electrostatic interaction between pairs of electrons and/or positrons.

R_U, radius of ('a') universe as 'box' $B(R_B)$, was expressed in (a) and (b) as c/H, and prior reports of which I am at the time of this writing aware appear to provide value(s) for H only much more roughly than for the other physical constants appearing in those expressions.

With the closeness of correspondence of mass and charge calculated with such rough values for H, it seemed reasonable to test this apparent correspondence more sharply, by:

(1) setting $m_0 = m_e$ (mass of electron) and $F = \alpha^{-1}$, combining (a) and (b) to eliminate H, and thereby obtaining a(n approximate) relation among G, $\hbar$, c, m_e, and α; and
(2) if such relation appeared to be consistent with observation, calculating a value for H from G, $\hbar$, c, and m_e, and then, from such calculated value for H, calculating both m_0 and F, to see if they agreed with m_e and α^{-1}, as a way of checking to see if such consistency in (1) might have been due to cancellation in the combined expression of discrepancies in the separate approximate expressions for m_0 and F.

CT stated the error of their recommended value for G as more than 100 times their stated error for c, $\hbar$, m_e, or α. Combining (a) and (b) to isolate G:
from (a), solving for H,

$$H \approx \frac{c^2 2m_0}{\hbar \exp(F^3/32\pi^2)^{1/2}} \tag{c}$$

from (b), noting that $Y(3) = 4$, cubing both sides, and isolating G,

$$G \approx \frac{\hbar^2 H \times 4 \times 0.9}{m_0^3 F^3 c} \tag{d}$$

substituting (c) in (d),

$$G \approx \frac{4 \times 0.9 \times 2\hbar^2 c^2 m_0}{m_0^3 F^3 c\hbar \exp(F^3/32\pi^2)^{1/2}} \tag{e}$$

collecting terms, and setting $m_0 = m_e$ and $F = \alpha^{-1}$,

$$G \approx \frac{7.2\hbar c}{m_e^2 \alpha^{-3} \exp(\alpha^{-3}/32\pi^2)^{1/2}} \quad \text{(f)}$$

or, isolating dependence on F,

$$F^3 \exp(F^3/32\pi^2)^{1/2} \approx \frac{7.2\hbar c}{m_e^2 G} \quad \text{(g)}.$$

This appears to be a ('locally') new relation among G, $\hbar$, c, m_e, and α: is it compatible with prior knowledge, as expressed in prior reports of observation?

It seems, at the time of this writing, that the most recent set of values for these constants is in the set of 'adjusted' values for physical constants recommended by CT[4] (1973) based on what purports to be a meticulous study of then available experimental data, partly by means of relationships among such data provided by prior theory.

CT listed values for these constants, and listed errors in parts per million ('ppm') respectively as c, 0.004 ppm; α^{-1}, 0.82 ppm; $\hbar$, 5.4 ppm; m_e, 5.1 ppm; and G, 615 ppm. However, they note that such 'errors' are standard deviations computed on the basis of internal consistency of data which they chose to use, after eliminating data which they regarded as placing excessive 'strain' on consistency of their adjusted system of values for physical constants and weighting remaining data.

Such definition of error does not include possible systematic error, and does not indicate spread in reported measurements. Moreover, because the process used involved both selection to reduce 'strain' and weighting of values used, one might imagine circumstances in which new data, near data which had been discarded, might form a consistent set 'strained' by data on which various recommended adjusted values had been based, so that the processes of selection to reduce 'strain' and of weighting might then involve discarding data on which some 'old' adjusted value had been based and/or reducing 'weight' accorded it; in which case, such 'new' adjusted value might differ from the 'old' adjusted value by many times the standard deviation, of either the 'old' or the 'new' adjusted value, as computed in each case on the basis of internal consistency of the data not discarded.

The notion of 'strain' depends not only on spread of measurements of a single constant, but also on 'strain' due to calculated results based on measurements of other constants. Which constants are interrelated, and in what ways, depends also on the theory or theories used. A new theory might relate constants previously regarded as independent, or relate constants in new ways, and newly computed values obtained through such new relation(s) might, by entering into such weighting and averaging process, lead to different 'adjusted' values; and such process of selection to reduce 'strain' might then discard or keep different data: such adjusted value might be so changed to an extent not indicated by an 'error' computed on the basis of consistency of either the formerly used or the then remaining data.

For various reasons mentioned, 'adjusted' values based on weighting and averaging data selected to reduce 'strain' in sets of values for constants as related by prior theory, and 'errors' computed on the basis of internal consistency of data then remaining, appear questionable as a basis for testing compatibility of a new theory with observation.

It seems to me that, for the purpose of comparing a new theory with observation, 'error' in recommending measured values should, if feasible, be defined in such way that further refinement of any measured quantity not remove recommended values from their stated error ranges.

In an earlier compilation, B. N. Taylor, W. H. Parker, and D. N. Langenberg[5] made several reasonable suggestions relating to reporting of experiments (p. 485 in ref. 5). I would like to make two additional suggestions here:

(1) A report of experimental data on physical constants might reasonably include a mathematical expression or computational procedure (e.g., a computer program) formally describing the apparatus and data analysis together in detail, with, for each parameter appearing therein, explicit statement of how the value used and the 'error' stated therefor were obtained; and

(2) A report recommending a set of values for physical constants might reasonably include, for each value proposed, besides an 'error' defined in terms of internal consistency, two other statements of error: a probable range of absolute error, stated in terms of some specified confidence level (e.g., 95% or higher) that the true value lies within that range; and a range for use in logical study of a new theory, defined by the criterion that if a calculated value for such quantity were to lie outside that range it would explicitly contradict all known (and plausible?) observation, with a level of certainty comparable to that sought by a cautious competent mountain climber in inspecting his or her rope.

TPL noted, in relation to their Figures 1 and 8, graphically displaying changes in various recommended values and 'standard deviations' from 1952 to 1969, and to their Table XXXVI, that various recommended values had changed by more than would be expected in view of probabilities calculated from such 'standard deviations'. Such situation, of recommended values changing by several times their stated 'error's, seems to me to indicate that statement of a single number for 'error', calculated solely from internal and mutual consistency of data selected for inclusion at least partly on a criterion of increasing such mutual consistency, has not been sufficient as a description of the state of knowledge of such value recommended for such constant. With incomplete theory(s), precision (repeatability, narrowness of variation) is not a formally sufficient test for accuracy (agreement with 'true' value). Use of these two suggested additional characterizations of 'error' might help to avoid such variations greater than stated 'error's of recommended values, not apparently related to changes of actual values, and might provide a relatively simple and convenient way for compilers of such reviews and/or sets of recommended values to share significantly more of the fruits of their extensive efforts.

To test compatibility with prior observational data, I solved (g) with F as α^{-1}, $\alpha^{-3}\exp(\alpha^{-3}/32\pi^2)^{1/2}\approx 7.2\hbar c/m_e^2G$, for α^{-1}, by successive approximation (using a Hewlett-Packard Company Model 65 stored program 'pocket' calculator (for specificity as to read only memory programs, serial number 1333A04145), with internal registers storing 10 decimal digit 'floating point' numbers with 2 decimal digit exponent of 10).

I tested this relation by approximately calculating F, using values for c, G, m_e, and $\hbar$ recommended by E. R. Cohen and J. W. M. DuMond[6] ('CD', 1963 adjustment), Taylor, Parker, and Langenberg[5] ('TPL', 1969), and Cohen and Taylor[4] ('CT', 1973), and comparing these values for F with CD, TPL, and CT's recommended values for α^{-1}. Recommended values, approximate calculated values for F, and comparisons with recommended values for α^{-1} are listed in Table 1. Entries in the 3×3 comparison matrix are $\approx(F-\alpha^{-1})/F$. SI units are used throughout (meter, kilogram, second). (Please see Table 1 on next page.)

Note that G was not 'adjusted' together with c, m_e, $\hbar$, or α^{-1}, there apparently having not been used therefor any theoretical relation between G and such quantities.

I obtained numerically approximate values for partial derivatives of F, using increment of 0.0001 times each independent variable, for CT 1973 input values for c, G, m_e, and $\hbar$, of

$$\partial F/\partial c\approx +7.2\times 10^{-3}$$

$$\partial F/\partial G\approx -7.2\times 10^{-3}$$

$$\partial F/\partial m_e\approx -14.4\times 10^{-3}$$

$$\partial F/\partial \hbar\approx +7.2\times 10^{-3}$$

Table 1. Comparisons of F and α^{-1} for Various Sets of Recommended Values

	CD[6] 1963	TPL[5] 1969	CT[4] 1973
c (m/sec)	299792500(100)	299792500(100)	299792458(1.2)
G(m^3/kg sec^2)	$6.670(5)\times10^{-11}$	$6.6732(31)\times10^{-11}$	$6.6720(41)\times10^{-11}$
m_e (kg)	$9.10908(13)\times10^{-31}$	$9.109558(54)\times10^{-31}$	$9.109534(47)\times10^{-31}$
$\hbar$ (J sec)	$1.054494(25)\times10^{-34}$	$1.0545919(80)\times10^{-34}$	$1.0545887(57)\times10^{-34}$
$F\approx$(number)	137.0390240	137.0385373	137.0387173
CD 1963 $\alpha^{-1}\approx$ 137.0388(6)	$\sim+1.634\times10^{-6}$	$\sim-1.917\times10^{-6}$	$\sim-0.6035\times10^{-6}$
TPL 1969 $\alpha^{-1}\approx$ 137.03602(21)	$\sim+21.92\times10^{-6}$	$\sim+18.369\times10^{-6}$	$\sim+19.683\times10^{-6}$
CT 1973 $\alpha^{-1}\approx$ 137.03604(11)	$\sim+21.77\times10^{-6}$	$\sim+18.223\times10^{-6}$	$\sim+19.537\times10^{-6}$

(entries in 3×3 comparison matrix are $\approx(F-\alpha^{-1})/F$)

From these, and Table 1, it appears that differences among the three approximate values for F are due almost entirely to different input values for G. All three F values are consistent with CD 1963 recommended α^{-1} value, and substantially inconsistent with TPL 1969 and CT 1973 recommended values for α^{-1}.

For F versus CT's recommended value for α^{-1}, G is only recommended to 615 ppm, which, with $\partial F/\partial G \sim -7.2\times10^{-3}$, would yield F only to ~4.428 ppm per 'standard deviation' ('σ'). All other inputs for F have error stated as $<1\%$ of, and independent of, that of G, and 'error' propagated therefrom may be neglected in this estimate. For comparison with α with stated error of 0.82 ppm, noting that G and α^{-1} were not co-adjusted by CT, 'error'~4.5 ppm/σ, and the $\sim+19.537$ ppm discrepancy appears as ~4.34σ, and to such extent might appear as likely statistically significant.

The relatively large change in values recommended for α^{-1} between CD 1963 and TPL 1969 appears as more than four times CD 1963's stated α^{-1} error. This change appears traceable to TPL's reliance on Josephson effect measurements for $2e/h$, which carried stated errors small enough, and were mutually 'consistent' enough, to dominate their selection, weighting, and averaging process.

If we look to the data on which TPL based their studies before selection and weighting on the basis of consistency and prior theory (see for examples TPL's Table XXVIII and Figures 6 and 7, ref. 5), there appears to be a considerable range of values reported for, or expressed in terms of, α^{-1}, bracketing the approximate values obtained for F.

Calculating an approximate value for H by solving (d) for H and substituting CT's recommended values for $\hbar$, c, and m_e, and the calculated approximate value for α^{-1} as F, I obtained $H \approx Gm_e^3(\alpha^{-1} \text{as} F)^3 c/3.6\hbar^2 \approx 9.719106719 \times 10^{-19}$/sec, retaining ('extra') digits for error checking (last two digits suspect, but not critical for this test). Since it was first proposed, the lower limit offered by astronomers[2] for H has decreased by a factor ~ 10, and, at the time of this writing, it would seem that this calculated approximate value for H could not reasonably be excluded by such prior knowledge.

Substituting this value for H in (a) and (b), and solving by successive approximation, I obtained, in the accuracy of execution of the calculation, $F \approx 137.038717$ and $m_0 \approx 9.1095340 \times 10^{-31}$ kg, from which it would seem likely that the result obtained for α^{-1} did not arise spuriously by cancellation in the combined expression for G of any substantial errors in the separate expressions for m_0 and F.

As is discussed more fully hereinafter, I am not, at the time of this writing, aware of any formal reason within this conceptual and formal structure to regard interaction between e^+ and e^- as having the same 'strength' as that between p^+ and e^-. In their 1973 compilation, Cohen and Taylor stated (discussion following expression 22.12, p. 695, ref. 4) that (preliminary) interpretation of positronium 'hyperfine structure' measurements yielded an approximate value for α^{-1} of $\approx 137.0387(17)$ (13 ppm), and that most of the 'uncertainty' was from lack of detail in the theory used. Such approximate value for α^{-1} appears to substantially allow the approximate value for F, of $\approx 137.0387173(\sim 4.43 \text{ ppm}/\sigma)$, obtained as presented herein from CT's 1973 values for G, c, $\hbar$, and m_e, and to avoid questions of relation between e^+e^- and p^+e^- interaction 'strengths'. (Cohen and Taylor stated (ref. 4, p. 695) that such value for α was not selected for use as an input to their weighting and averaging process.) Because F was obtained for interaction between $e^\pm$ and $e^\pm$, such comparison might seem more directly relevant than measurements of p^+e^-, or $p^+\mu^-$, interaction, as in some spectroscopic measurements.

CT's 1973 recommended values included effects due to choice as a standard of $2e/h$ from Josephson effect measurements, and thus substantially retained changes introduced in TPL 1969. Josephson effect measurements appear to have been interpreted using J. Bardeen, L. N. Cooper, and J. R. Schrieffer ('BCS')'s theory of superconductivity.[7] In BCS's theory, 'charge' carriers appear to have been treated as having 'charge' $2e$, where 'e' denotes the charge of one substantially isolated electron.

Restriction to forming 'bosonic' carriers appears to allow an even number of electrons to be coupled to form a 'charge' carrier. For this reason, one might expect that, in an expansion in powers of α for the value of α^{-1} obtained from Josephson effect measurements, coefficients of odd orders in α might have much smaller magnitude than those of even orders. Thus, a first 'correction' to such value for α^{-1} might be of order α^2. The $\sim$19.537 ppm discrepancy with CT's 1973 recommended value for α^{-1} appears to be roughly of order $\alpha^2/(2.7 \pm \sim 23\%)$, which might suggest nonzero contribution(s) from higher orders than electron pairs as treated in BCS's theory (e.g., four (or six, or more) electrons). (Please see also Section 6 questions 12 and 38.)

Also, at the time of this writing, I am not certain as to whether the $\sim$615 ppm/σ 'error' stated by CT for G should be relied on as a statement of accuracy, or construed only as a statement of precision (repeatability, narrowness of spread), and/or questioned as to magnitude; and have had corresponding reservations as to whether what appeared as a(n) '$\sim$ 4.34σ' discrepancy should be regarded as significant. (Please note also items 1 and 2 of Table 6.)

In view of the foregoing, the relationship derived, among G, $\hbar$, c, m_e, and α^{-1}, appears, to the extent tested as presented hereinbefore, substantially allowed by the ranges of values for such constants as reported, before 'adjustment' involving prior theory:

Correspondence Corollary 14.16. *Theorem* 14 *through Corollary* 14.15 *appear, to the extent tested as presented hereinbefore, substantially allowed by prior knowledge, with the two types of P_0 corresponding to the electron and positron, F corresponding to α^{-1}, and $\alpha^{-3}\exp(\alpha^{-3}/32\pi^2)^{1/2} \approx 7.2\hbar c/m_e^2 G$.*

●

Note that this analysis is directed toward testing whether a physical picture is allowed by prior knowledge: Corollary 14.16 conforms to this purpose.

This correspondence, Table 1, and discussion hereinbefore relating to BCS's theory, might raise the question of whether the values for $2e/h$ relied on by TPL and CT might be substantially erroneous, and all values based thereon and/or adjusted therewith substantially distorted thereby. This also might introduce some difficulty in finding a 'clean' set of input values for calculations presented herein. Note that values recommended for G were substantially independent of such values for $2e/h$.

In SI units, by definition, $\alpha = \{\mu_0 c^2/4\pi\} e^2/\hbar c$, where, by definition, $\mu_0 = 4\pi \times 10^{-7}$ and e is electron charge magnitude in coulombs. Substituting F^{-1} for α, and solving for e,

$$\frac{\mu_0 c^2}{4\pi}\frac{e^2}{\hbar c}\approx F^{-1},$$

$$10^{-7}\frac{e^2 c}{\hbar}\approx F^{-1},$$

$$e^2\approx\frac{10^7\hbar}{Fc},$$

$$e\approx\left(\frac{10^7\hbar}{Fc}\right)^{1/2}.$$

Substituting $\hbar$ and c from CT (1973) (ref. 4) (see Table 1), and F as obtained from CT 1973 input values (see Table 1), I obtained $e\approx 1.602173569\times10^{-19}$ coulomb. From CT's 1973 stated 'error' for G, as noted hereinbefore, F would contain propagated 'error'$\sim$4.5 ppm/σ, primarily due to input 'error' of G, and thus likely not strongly correlated with error in $\hbar$. However, as discussed hereinbefore, the numerical value recommended by CT (1973) for $\hbar$ might contain substantially larger error, which might be $\sim$20 ppm, from interpretation and compilation of measurements. Because of the square root in calculating this approximate value for e, it would seem likely that error propagated in such value for e would be <50 ppm, but such error might be not statistical. In order to avoid conflict of units in using tabulated data from CT and other sources (e.g., ref. 8), I have used for energy units conversions CT's 1973 recommended value of $1\text{eV}\approx1.6021892(46)\times10^{-19}$ joule (parenthetical digits stated by CT as 1σ 'error' in last two decimal places, as discussed hereinbefore). The absolute value of the fractional difference between this approximate calculated value for e and CT's 1973 recommended value for e of $1.6021892(46)\times10^{-19}$ coulomb is <10 ppm, and, for reasons given hereinbefore in relation to F, appears substantially allowed by comparisons presented hereinbefore between approximate value for F and knowledge prior to the time of this writing.

Using $\rho\approx3H^2/4\pi G$, derived hereinbefore, and H as approximately calculated by solving (d) with F as approximately calculated, I obtained average density of our universe U $\rho_U\approx3.3799\times10^{-27}$ kg/m^3, and, with from CT c and 1 eV, $\rho_U c^2\approx1.896\times10^9$ eV/m^3. From (d), H is roughly proportional to G. CT stated for G 'error' 615 ppm/σ, more than 100 times the 'error' which they stated for m_e, $\hbar$, or c. Thus, ρ_U approximately calculated in this way would contain 'error' propagated substantially as G, or $\sim$615 ppm/σ.

With Corollaries 10.3 and 10.4, $R_B \approx c/H \approx 3.084568 \times 10^{26}$ m, containing propagated 'error' substantially as G^{-1}, $\sim$615 ppm/σ. From EMT and QM (exact in the axiomatic form, from A-Def 21 and/or 22), lowest energy photon in B has energy $E_1 = \hbar c/R_B$. Substituting, from Corollary 10.3, $R_B \approx c/H$, $E_1 \approx \hbar H$. With H as obtained hereinbefore, and $\hbar$ (and eV/J, from CT 1973, $E_1 \approx 1.024966 \times 10^{-52}$ J $\approx 6.397284 \times 10^{-34}$eV.

From Theorem 13, $G \approx c^2 R_B / M_B$. Substituting, from Corollary 10.3, for R_B, and solving for M_B, I found, for total mass M_U in our universe (universe 'U' as 'box' 'B'), $M_U \approx c^3/HG \approx 4.155 \times 10^{53}$ kg, containing propagated error substantially as G^{-2}, $\sim$1230 ppm/σ.

From PODIC, total energy E_B in B is $E_B = M_B c^2$. From Theorem 6, total amount I_B of information in B is proportional to E_B, as $I_B \approx E_B/E_1$. Combining these (exact in the axiomatic form, directly from A-Def 22), $I_B = M_B c^2/E_1 \approx c^5/H^2 G\hbar \approx 3.64344 \times 10^{122}$ bits, containing propagated error substantially as G^{-3}, $\sim$1845 ppm/σ.

The amount of mass (energy) in a 1 m^2 column from where we are to the 'edge' of 'our' 'universe(s?)' would be $\approx R_B \rho_U$. Substituting,

$$R_B \rho_U \approx \frac{c}{H} \frac{3H^2}{4\pi G} = \frac{3c}{4\pi} \frac{H}{G}.$$

Solving (d) for H/G, and substituting, with $m_0 = m_e$, and F as α^{-1} as approximately calculated hereinbefore,

$$R_B \rho_U \approx \frac{3c}{4\pi} \frac{m_e^3 \alpha^{-3} c}{3.6\hbar^2} = \frac{c^2 m_e^3 \alpha^{-3}}{4.8\pi\hbar^2} \approx \frac{c^2 m_e^3 F^3}{4.8\pi\hbar^2}$$

$\approx$1.0425632 kg, which result, because it depends on G substantially only through F, might be expected to be substantially more accurate than approximate results obtained hereinbefore for ρ_U, R_B, M_B, E_1, H, and I_B, and covariance of error in terms appearing therein might be significant in estimating error of $R_B \rho_U$.

For ρ_U, R_B, M_B, E_1, H, and I_B, where propagated 'error' is dominated by 'error' in G, such 'error' is traceable to mechanical properties of macroscopic apparatus which might seem relatively well understood, and such 'error's might have usefulness as describing widths of distributions which, within a few standard deviations of the value stated, might reasonably be approximated by 'normal' ('Gaussian') distributions. With respect to $R_B \rho_U$, although propagated 'error' appearing therein might at the time of this writing reasonably be expected to be substantially less than for ρ_U, R_B, E_1, M_B, H, and I_B, likely less than 100 ppm/σ, I have reservations as to the sufficiency of stating one standard deviation of a supposedly 'normal' statistical distribution as a means for describing, at the time of this writing, the state of knowledge of error propagated in calculating such quantity, and do not state herein such an 'error' for such quantity, only a general comment as to an approximate upper bound for such 'error'.

For convenience, various of these results are collected in Table 2, with expressions for recalculating them using other input values:

Table 2. For Our Universe U as 'Box' B

Item	Expression	Approximate Value	Approximate 'Error' ppm/σ	Units
1. α^{-1}as F	$F^3\exp(F^3/32\pi^2)^{1/2}$ $\approx 7.2\hbar c/m_e^2G$	137.0387173	4.43	number
2. e	$(10^7\hbar/Fc)^{1/2}$	1.6021736×10^{-19}	<50[a]	coulomb
3. H	$H\approx Gm_e^3F^3c/3.6\hbar^2$	9.7191×10^{-19}	615	$\sec^{-1}$
4. R_U	$R_U\approx c/H$	3.084568×10^{26}	615	meter
5. E_1	$E_1\approx\hbar H$	1.024966×10^{-52}	615	joule
		6.39728×10^{-34}	615	eV
6. M_U	$M_U\approx c^3/HG$	4.15508×10^{53}	1230	kg
7. I_U	$I_U\approx c^5/H^2G\hbar$	3.64344×10^{122}	1845	bit
8. ρ_U	$\rho_U\approx 3H^2/4\pi G$	3.3799×10^{-27}	615	kg/m^3
		1.896×10^9	615	eV/m^3
9. mass in 1m^{2}column to 'wall'	$R_U\rho_U\approx\dfrac{c^2m_e^3F^3}{4.8\pi\hbar^2}$	1.0425632	<100[a]	kg

[a]: might be not 'statistical'

Input values (from Cohen and Taylor 1973, ref. 4):

$c\approx 299792458(1.2)$ m/sec $G\approx 6.6720(41)\times10^{-11}$ m^3/kg sec^2

$\hbar\approx 1.0545887(57)\times10^{-34}$ J sec $m_e\approx 9.109534(47)\times10^{-31}$ kg

With this closeness of correspondence between P_0 of Corollary 14.15 and the electron and positron, we apply Theorem 13 to electrons and positrons. Thus, we might expect that electrons and positrons would be affected by gravity in exactly the same way, and be equally effective as sources of gravity.

As in Definition 2, in relation to Corollary 13.2, and in Corollary 14.12, for two information representations to together be fully convertible to any other representation, they must be complementary, including that their representations of internal information must be complementary. From closeness of correspondence between Corollary 14.15 3 bit Theorem 13 B_0s and electrons and positrons, it would seem that an electron and positron in an 'annihilation' or 'formation' process have complementary representations of internal information. This seems to offer some insight into why the three internal bits, which appear to correspond to 'charge' sign, 'spin', and 'lepton number', appear conserved in annihilation and creation of electron and positron pairs.

The closeness of correspondence between properties of the least mass B_0 bound by Theorem 13 interaction and the electron and positron suggests that this may be a useful way to think about 'elementary' particles.

Definition 19. *Call a least mass B_0 allowed by an interaction between photons a 'photonic particle'. ('photicle'?)*

For example, the electron (positron) seems to be a 3 bit gravity photicle.

One can consider other photicles. Note that B_0s are capable of representing substantially timelike flows of information, in their lowest allowed momentum states in B. One can also examine properties of least timelike B_0s, for example, B_0 with one photon in the lowest photon state in B, and the rest of its energy in the other photon. Rapidly moving systems might be interesting in relation to neutrinos. Also, one might examine 'boxes' containing more than two photons; these might be interesting in relation to more massive and/or unstable particles. Various of these considerations are pursued further in Section 3, in Chain 17 and its links. In the next subsection (2.10), I present some approximate descriptions of B_0s moving with velocity $\approx c$.

2.10 FAST B_0s

I have collected into this subsection some derivations relating to various properties of fast B_0s with respect to Theorem 13 interaction, relatively localized interaction, 'graininess', coding, and formation. Some discussion of fast B_0s was included in Corollaries 8.5–8.16, because use in evaluating the domain of applicability of PODIC seemed to warrant dividing discussion herein of fast B_0s.

Following Theorem 13 and its corollaries, I noted that fast B_0s might behave in special ways in gravity. From Theorem 13 and its corollaries, one might expect that, in a small enough test system in free fall, an observer would not observe effects of a gravitational field. This is usually called the 'principle of equivalence'. From Corollary 8.11, the lower photon energy of B_0 moving with velocity $v \approx c$ is $E_{Bj} \approx E_0^2/4E$, and this lower photon's wavelength could be quite large. For a fastest B_0, $\lambda_{j=1}$ fills B.

For Theorem 15–Corollary 15.11, I have used various conditions:

Condition 1. *In isolated box B, with linear size R_B, lowest photon energy E_{B1}; containing B_0 with energy E_0 in its rest frame, transporting energy E with velocity $v \approx c$ as seen by observer O_B substantially at rest relative to B; B_0 representing internally N bits of information*

Condition 2. $C1$; *B containing total mass energy M_B; allowed photon energy levels E_{Bi}, i denoting ith photon energy level; B_0 with its associated down-shifted photon having energy E_{Bj} and its up-shifted photon having energy E_+; with 'm_b' representing least mass (energy) to form bound B_0 using interaction in Theorem* 13, *with charge magnitude a_b; 'm_p' representing rest mass of least massive particle which is stable when substantially at rest*

Condition 3. $C2$; *B containing substantially isolated relatively localized mass $M \ll M_B$; observer O with mass negligible but sufficient for substantially localizing such observer, at approximate distance $R \ll R_B$ from M, O substantially at rest relative to M and freely falling; B_0 with energy $E \ll Mc^2$; $R_B M / M_B R \ll 1$*

To simplify the following discussion, I considered the 'weak field' case, in which various coordinate systems about M are approximated by a 'flat' metric.

I discuss first applicability of principle of equivalence for two cases: where it appears to be a relatively good approximation, and where it appears to be substantially irrelevant:

Theorem 15. $C3$; *for O to see B_0 behaving with respect to Theorem* 13 *interaction with M as being in region S with linear size r about O, $r \ll R$, within which region the principle of equivalence is presumed to be a good approximation, $E_{Bj} \geqslant \hbar c/r$ and $E - 2E_{Bj} \geqslant \hbar c/r$; for fast B_0 and $E_0 \gg \hbar c/r$, $E \leqslant\approx E_0^2 r/4\hbar c$.*

1. From Theorems 12 and 13 and their corollaries, it is B_0's information that interacts with M.

2. From EMT, QM, Theorems 4 and 4′, and Cor 4.1, for B_0's information to be represented within S, B_0's associated two photons must be able to satisfy boundary conditions in S.
3. From EMT and QM, for B_0's down-shifted photon to be able to satisfy boundary conditions in S, $E_{Bj} \geqslant \hbar c/r$.
4. From EMT, for both B_0's photons to satisfy boundary conditions in S, the spatial period λ_b of the difference frequency between B_0's up- and down-shifted photons must fit into S.
5. From Cor 4.4 (b), B_0's up-shifted photon has energy $E - E_{Bj}$.
6. From Cor 4.4 (c), and 5, λ_b can fit into S iff $E - 2E_{Bj} \geqslant \hbar c/r$.
7. From 1–4, and 6, O could see B_0 as in region S with respect to Theorem 13 interaction with M iff $E_{Bj} \geqslant \hbar c/r$ and $E - 2E_{Bj} \geqslant \hbar c/r$.
8. From Cor 8.11, solving for E_{Bj}, for fast B_0, $E_{Bj} \approx E_0^2/4E$.
9. From Def 13, and 8, for fast B_0, $E_0 \gg \hbar c/r$ iff $E - 2E_{Bj} \gg \hbar c/r$.
10. From 7–9, for fast B_0 with $E_0 \gg \hbar c/r$, O could see B_0 as in S with respect to Theorem 13 interaction with M iff $E \leqslant \approx E_0^2 r/4\hbar c$.

●

Next, consider the case where one of B_0's two photons has much longer wavelength than the distance R to M.

Corollary 15.1. $C3$; *'principle of equivalence' is substantially irrelevant for O's description of B_0's behavior with respect to Theorem* 13 *interaction with M, if $E_{Bj} \ll \hbar c/R$; for fast B_0, if $E \gg E_0^2 R/4\hbar c$.*

1. From Theorems 12 and 13 and their lemmas and corollaries, it is B_0's information that interacts with M.
2. From EMT, QM, Theorems 4 and 4′, and Cor 4.1, B_0's information is not entirely representable in any region V with linear size x smaller than to allow B_0's associated two photons to satisfy boundary conditions in V.
3. From EMT, and 2, for B_0's down-shifted photon to be able to satisfy boundary conditions in V, $x \geqslant \hbar c/E_{Bj}$.
4. From 3, for $E_{Bj} \ll \hbar c/R$, $x \gg R$.
5. From 4, for $E_{Bj} \ll \hbar c/R$, B_0 interacts with M as in Theorem 13 as spread over a region much larger than the distance from O to M.
6. From Theorem 13, and 5, for $E_{Bj} \ll \hbar c/R$, B_0 interacts with M substantially differently than does O.

7. From Cor 8.11, for fast B_0, solving for E_{Bj}, $E_{Bj} \approx E_0^2/4E$.
8. From 7, for fast B_0 and $E \gg E_0^2 R/4\hbar c$, $E_{Bj} \ll \hbar c/R$.
9. From 1–6, and 8, for fast B_0 and $E \gg E_0^2 R/4\hbar c$, B_0 interacts with M substantially differently than does O.
10. From 6 and 9, if $E_{Bj} \ll \hbar c/R$, or for fast B_0 if $E \gg E_0^2 R/4\hbar c$, 'principle of equivalence' is substantially irrelevant for O's description of B_0's behavior with respect to Theorem 13 interaction with M.

●

The interaction examined in Theorems 12 and 13 and their corollaries appears as a relation between relatively localized information in position representations. As B_0's 3-momentum is increased for the same total energy, a larger portion of B_0's total information appears in momentum representation, and less in position representation. Thus, although a fast B_0 may have the same total information and mass energy as a B_0 whose information is relatively localized, it does not participate in Theorem 13 interaction in the same way. In particular, a fastest B_0's information fills B, and so cannot serve in a Theorem 13 interaction to localize further anything already in B. Thus,

Corollary 15.2. *C*1; *fastest B_0 does not have any position information in B.*

●

For x (Cor 15.1 step 2) very much larger than R, B_0's information is spread over a region which becomes more nearly symmetric about M as x is increased. Thus, it seems that there is not any reason to expect a fastest B_0 to 'fall' toward M. In some sense, B_0's information being spread over a large part of our universe seems like the notion of a photon from a distant star filling a telescope aperture, and much of the surface of a sphere at that distance from the star, before interacting with a single electron in a detector.

This 'breakdown' of the 'principle of equivalence' appears as a consequence of one component of B_0 being too large, despite what one might formerly have thought of as B_0's localizability based on its (de Broglie) wavelength. This breakdown might offer some insight into the remarkable behavior so suggested for fast B_0's in a gravitational field.

From Theorem 13's symmetry as to being affected and being a source, we see next that for conditions for which the principle of equivalence is substantially irrelevant, B_0 and M are substantially unattracted by each other.

Corollary 15.3. *C3; observer O sees B_0 and M as substantially unattracted by each other by Theorem 13 interaction if $E_{Bj} \ll \hbar c/R$; for fast B_0, if $E \gg E_0^2 R/4\hbar c$.*

1. From Cor 8.11, for fast B_0 with $E \gg E_0^2 R/4\hbar c$, $E_{Bj} \ll \hbar c/R$.
2. From and as in Cor 15.1 steps 1–4, for $E_{Bj} \ll \hbar c/R$, B_0's information cannot be localized within a region V with linear size $x < \hbar c/E_{Bj}$.
3. From 1 and 2, if $E_{Bj} \ll \hbar c/R$, or for fast B_0 $E \gg E_0^2 R/4\hbar c$, $x \gg R$.
4. From Theorems 12 and 13 and their corollaries, and $R_B M/M_B R \ll 1$, B_0 gives up an amount of information nearly inversely proportional to its distance from M.
5. From 3 and 4, B_0 cannot give up a larger fraction of its information than $R_B M/M_B x$.
6. From 3–5, compared to O, such B_0 gives up a substantially negligible fraction of its information in Theorem 13 interaction with M.
7. From Theorem 6, and 6, O sees such B_0 as substantially unattracted by M.
8. From Theorems 12 and 13 and their corollaries, Theorem 13 interaction between such B_0 and M is symmetric.
9. From 7 and 8, for B_0 with $E_{Bj} \ll \hbar c/R$ or fast B_0 with $E \gg E_0^2 R/4\hbar c$, O sees B_0 and M as substantially unattracted by each other by Theorem 13 interaction.

●

In particular, a fastest B_0's information fills B, so

Corollary 15.4. *C3; fB_0 and M do not attract each other by Theorem 13 interaction.*

1. From Theorem 12 and Cor 12.1 and 12.2, interaction between B_0 and M cannot make B_0 less localized than to fill B.
2. From Cor 15.1 and Def 14, as in Cor 15.1 steps 1–4, fB_0's information already fills B.

3. From 1 and 2, Theorem 13 interaction with M cannot reduce B_0's localization.
4. From Theorems 12 and 13 and their corollaries, and 3, Theorem 13 interaction with M cannot subsume any position information from fB_0.
5. From Theorems 12 and 13 and their corollaries, and 4, noting symmetry of Theorem 13 interaction, fB_0 and M do not attract each other by Theorem 13 interaction.

●

Note that Theorem 13's symmetry allows an equivalent approach: M's interaction with fB_0 does not detect any more of M's position information than fB_0's information is localized, so M still is merely known to be in B, as before interacting with fB_0. Therefore, there is not any Theorem 13 attraction between fB_0 and M.

Next, I discuss interaction with fast B_0s. With Corollary 6.4, B_0's rest frame energy E_0 would be partially in the relatively localized form approximately calculated in Corollary 14.8, displayed in step 15 thereof. This leads to a fraction $1/F_f$ (subscript 'f' for 'fast' (or 'free')) of B_0's energy in relatively localized form. From this, one might approximately describe interaction with B_0 in terms of an effective 'charge' a_f.

Corollary 15.5. *C2; for $m_pc^2 > E_0 < m_bc^2$, $1/F_f$ of B_0's rest frame energy E_0 appears in relatively localized form, with*

$$F_f \approx \frac{1}{E_0}\left(\frac{M_B Y(N)\hbar^2 c^4}{R_B^2}\right)^{1/3}.$$

1. From Cor 14.3–14.5, 14.8, and 14.9, representation of 1 bit of internal information by $\pm$ relative phase $\pi/2$ as in Corollaries 14.3–14.5, 14.8, and 14.9 corresponds to representation in terms of two states between which distinction arises from the interaction coupling photons to form such B_0.
2. From Theorem 14, for $m_p > m_0 < m_b$, B_0's associated two photons are not bound.
3. From 1 and 2, for $m_p > m_0 < m_b$, B_0 does not represent any internal information by relative phase of its associated two photons.
4. From Cor 14.5 and 14.8, the term $(1 + 1/N^2)$ appearing in the denominator of Corollary 14.8 step 15 corresponds to representation of

1 bit, of a total of N bits, of internal information by $\pm$ relative phase $\pi/2$ of B_0's associated two photons.

5. From 3 and 4, for $m_p > m_0 < m_b$, term $(1+1/N^2)$ does not appear in said denominator.
6. From PODIC, Cor 8.16, and Def 6, in domain of applicability of PODIC, for two photon B_0, in B_0's rest frame, B_0's two photons have equal frequency and coherent phase.
7. From Cor 14.2, Def 18, and 6, representation internally of a few (N) bits of information by such B_0 involves energy $\approx Y(N)E_{B1}$.
8. From PODIC and Cor 14.8 step 15, and 5 and 7, for $m_p > m_0 < m_b$, in isolated B with linear size R_B and containing total mass energy M_B, relatively localized two photon state representing internally N bits of information has energy

$$\approx c^2\left(\frac{M_B Y(N)\hbar^2}{c^2R_B^2}\right)^{1/3}.$$

9. From EMT, PODIC, Cor 6.4, and 8, as in Corollary 14.13, noting that electromagnetic energy is proportional to amplitude squared, for $m_pc^2 > E_0 < m_bc^2$, fraction $1/F_f$ of E_0 appears in relatively localized form, with

$$F_f \approx c^2\left(\frac{M_B Y(N)\hbar^2}{c^2R_B^2}\right)^{1/3}\Big/ E_0.$$

10. From 9, simplifying, for $m_pc^2 > E_0 < m_bc^2$,

$$F_f \approx \frac{1}{E_0}\left(\frac{M_B Y(N)\hbar^2c^4}{R_B^2}\right)^{1/3}.$$

●

Substituting Corollary 8.11 into Corollary 15.5, F_f can be approximately expressed in terms of E and E_{Bj}:

Corollary 15.6. *C2; in domain of applicability of PODIC, for* $m_pc^2 > 2\sqrt{E\,E_{Bj}} < m_bc^2$, $1/F_f$ *of fast* B_0*'s rest frame energy appears in* *localized form,*

$$F_f \approx \frac{1}{2\sqrt{EE_{Bj}}}\left(\frac{M_B Y(N)\hbar^2 c^4}{R_B^2}\right)^{1/3}.$$

●

Conventionally, one expresses electrostatic interaction between two electrons in terms of their charges. The result in Corollary 14.14, $\Delta E \approx \pm \hbar c/Fr$, might be so expressed (in suitable units) as $\Delta E \approx \pm e^2/r$, where e is electron charge. Thus, $|e| \approx (\hbar c/F)^{1/2}$. Thus, e is proportional to electromagnetic field amplitude in localized form. Interaction with localized form of a fast B_0 might similarly be expressed in terms of amplitude in localized form, as 'charge' a_f:

Definition 20. *Call $(\hbar c/F)^{1/2}$ 'charge magnitude' associated with presence in localized form of $1/F$ of electromagnetic energy associated with B_0 in its rest frame.*

Corollary 15.7. *C2; in domain of applicability of PODIC, for $m_p c^2 > 2\sqrt{EE_{Bj}} < m_b c^2$, fast B_0 has charge a_f related to charge a_b of least massive bound B_0 using interaction in Theorem 13, mass m_b:*

$$|a_f| = |a_b|\left(\frac{F_b}{F_f}\right)^{1/2} \approx |a_b|\sqrt{2}\, E^{1/4}\frac{E_{Bj}^{1/4}}{m_b^{1/2}c}\left(\frac{0.9\,Y(3)}{Y(N)}\right)^{1/6}.$$

1. From Def 20, $|a_b| = (\hbar c/F_b)^{1/2}$ and $|a_f| = (\hbar c/F_f)^{1/2}$.
2. From 1, $|a_f| = |a_b|(F_b/F_f)^{1/2}$.
3. From Cor 14.5, 14.7, 14.8, and 14.15,

$$F_b \approx \frac{1}{m_b}\left(\frac{0.9 M_B Y(3)\hbar^2}{c^2 R_B^2}\right)^{1/3}.$$

4. From Cor 15.6, for $m_b c^2 > 2\sqrt{EE_{Bj}} < m_p c^2$, in domain of applicability of PODIC, factoring out c^2,

$$F_f \approx \frac{c^2}{2\sqrt{EE_{Bj}}}\left(\frac{M_B Y(N)\hbar^2}{c^2 R_B^2}\right)^{1/3}.$$

5. From 3 and 4, for $m_b c^2 > 2\sqrt{EE_{Bj}} < m_p c^2$,

$$\frac{F_b}{F_f} \approx \frac{2\sqrt{EE_{Bj}}}{m_b c^2} \left(\frac{0.9\, Y(3)}{Y(N)} \right)^{1/3}.$$

6. From 2 and 5, for $m_b c^2 > 2\sqrt{EE_{Bj}} < m_p c^2$, in domain of applicability of PODIC,

$$|a_f| \approx |a_b| \sqrt{2}\, E^{1/4} \frac{E_{Bj}^{1/4}}{m_b^{1/2} c} \left(\frac{0.9\, Y(3)}{Y(N)} \right)^{1/6}.$$

●

I next discuss N for unbound fast B_0s, which seems to be an interesting case.

Corollary 15.8. *C2; in domain of applicability of PODIC, for $m_p > E_0/c^2 < m_b$, number N of bits of internal information $\geqslant 2$.*

1. Consider a set of photons being detected: from EMT and QM, the first several (many) photons' polarizations can be used to define orientation for subsequent detection.
2. From EMT and QM, and 1, each subsequent photon's polarization can be compared with this orientation.
3. From EMT and QM, and 1 and 2, each subsequent photon's two polarization states are distinguishable.
4. From Def 1, and 3, each subsequent photon's polarization represents 1 bit of information. (or, from A-Theorem 1, each photon's polarization represents 1 bit of information)
5. From PODIC and Theorem 14, for $m_p > E_0/c^2 < m_b$, B_0's two photons are not bound. (this is the use of PODIC in this corollary)
6. From 4 and 5, polarizations of B_0's two photons represent 2 bits of information.
7. From Def 1, for B_0's two photons not bound, there is not any information released, by associating B_0's two photons, to represent binding.
8. From 5 and 7, total information represented by energy of each of B_0's two photons is unchanged by their association.
9. From Theorem 1, and 5–8, in domain of applicability of PODIC, for $m_p > E_0/c^2 < m_b$, $N \geqslant 2$ bits of information.

●

An fB_0's associated down-shifted photon represents only 1 bit of information in B (such photon's 'existence' bit). Although a few bits of information represented in B by a fast B_0's associated down-shifted photon might appear to be represented internally by such fast B_0, for fB_0, writing E_0/c^2 as m_0 to avoid use of PODIC, from Corollary 15.8 we obtain

Corollary 15.9. *C2; for $m_p > m_0 < m_b$, for fB_0, number N of bits of internal information* $= 2$.

Next, under what circumstances might one expect to see peculiar behavior associated with 'graininess' of fast B_0's allowed energies? Briefly, any particle which interacts more weakly than a photon must transport information saying that its interaction is less, to which information corresponds structure involving at least two photons and, suitably decomposed, an associated (with PODIC, rest frame) energy E_0. With Corollary 6.4, E_0 can appear co-possibly in substantially localized form, so the 'particle' has 'charge'. For small enough charge compared to energy transported, with PPR, one has a fast particle which resembles a fast B_0. Thus, from energy and interaction 'strength', one can estimate energy graininess. Note that this theory does not so far explicitly formally include gravitational radiation. (Brackets are used as noted herein preceding Cor 8.14.)

Corollary 15.10. *C2; in domain of applicability of PODIC, with PPR; for any particle P, with energy E and velocity $v \approx c$, interacting differently than photon with energy E, with rest frame mass m_0, $m_p > m_0 < m_b$, and interaction strength corresponding to rest frame charge magnitude a_f; P's next lower allowed energy E' is approximately related to E, a_f, and allowed photon energy levels E_{Bj} in B by*

$$\frac{E}{1+\dfrac{4E_{B1}E}{m_b^2c^4}\left|\dfrac{a_b}{a_f}\right|^4\left(\dfrac{0.9\,Y(3)}{Y(N)}\right)^{2/3}} \underset{[\lesssim]}{\approx} E'$$

$$\left[\lesssim \frac{E}{1+\dfrac{48\pi E_{B1}^3E^3}{m_b^6c^{12}}\left|\dfrac{a_b}{a_f}\right|^{12}\left(\dfrac{0.9\,Y(3)}{Y(N)}\right)^{2}}\right], \qquad N \geqslant 2.$$

1. From Cor 4.1 and Def 7 (or A-Cor 5.1), with PPR, for every particle there is a corresponding photon configuration representing same internal and external information and having same allowed states.
2. From given, P interacts differently than photon with same energy.
3. From EMT and QM, and 1 and 2, P's PPR has more than one associated photon.
4. From Theorem 3, any configuration of photons having same allowed states as P will have same energy as P's PPR.
5. By decomposing electromagnetic energy of P's PPR into forward and backward components along P's line of motion, form from P's PPR two component system B_0'.
6. From Theorems 1 and 4′, and 3–5, P's allowed states correspond to those of its corresponding B_0'.
7. From PODIC and Theorem 4′, B_0' has $E_0 = m_0 c^2$.
8. From Theorem 14, and 7, noting $m_p > m_0 < m_b$, B_0''s associated components are not bound.
9. From EMT and Def 6, B_0''s associated components must satisfy boundary conditions in B.
10. From EMT and QM, and 9, B_0''s associated components must be integer numbers of photons in photon states in B.
11. From EMT, QM, Theorem 4′, and 9 and 10, for same total energy, B_0' has finest energy spacings if integer number of photons in each of its associated two components is 1.
12. From given, $v \approx c$.
13. From Def 13, and 8, 11, and 12, P's energy spacings are finest if P's PPR is fast B_0.
14. From Cor 15.8, and 13, for finest energy spacings, $N \geqslant 2$. (for fastest P, with PPR, from Cor 15.9, $N = 2$, but note $v = v_m$ for fB_0, and limitation $v < v_m$ on domain of applicability of PODIC)
15. From Cor 15.7, and 13, for P's energy spacing to be finest,

$$|a_f| \approx |a_b| \sqrt{2}\, \frac{E^{1/4} E_{Bj}^{1/4}}{m_b^{1/2} c} \left(\frac{0.9\, Y(3)}{Y(N)} \right)^{1/6}.$$

16. Solving 15 for E_{Bj},

$$E_{Bj} \approx \left| \frac{a_f}{a_b} \right|^4 \frac{m_b^2 c^4}{4E} \left(\frac{Y(N)}{0.9\, Y(3)} \right)^{2/3}.$$

[17. From Cor 8.15, for P's finest energy spacings,

$$E-E' \gtrsim E\left(1-\frac{1}{1+\frac{3\pi}{4}\left(\frac{E_{B1}}{E_{Bj}}\right)^3}\right).$$

(writing as 1 − fraction to simplify solution)]

[18. Substituting 16 in 17,

$$E-E' \gtrsim E\left(1-\frac{1}{1+\frac{3\pi}{4}E_{B1}^3\left|\frac{a_b}{a_f}\right|^{12}\frac{64E^3}{m_b^6c^{12}}\left(\frac{0.9\,Y(3)}{Y(N)}\right)^2}\right).$$]

[19. From 18, simplifying,

$$E' \lesssim \frac{E}{1+\frac{48\pi E_{B1}^3E^3}{m_b^6c^{12}}\left|\frac{a_b}{a_f}\right|^{12}\left(\frac{0.9\,Y(3)}{Y(N)}\right)^2}.$$]

20. From [Cor 8.15, for upper bound of coarseness of P's finest energy spacings] A-Theorems 3 and 4 and A-Def 22,

$$E-E' \underset{[\lesssim]}{\approx} E\left(1-\frac{1}{1+\frac{E_{Bi}}{E_{Bj}}}\right).$$

21. From 16 and 20, substituting 16 in 20,

$$E-E' \underset{[\lesssim]}{\approx} E\left(1-\frac{1}{1+E_{B1}\left|\frac{a_b}{a_f}\right|^4\frac{4E}{m_b^2c^4}\left(\frac{0.9\,Y(3)}{Y(N)}\right)^{2/3}}\right).$$

22. From 21, simplifying,

$$E' \underset{[\gtrsim]}{\approx} \frac{E}{1+\dfrac{4E_{B1}E}{m_b^2c^4}\left|\dfrac{a_b}{a_f}\right|^4\left(\dfrac{0.9\,Y(3)}{Y(N)}\right)^{2/3}}.$$

23. From 14, 19, and 22,

$$\frac{E}{1+\dfrac{4E_{B1}E}{m_b^2c^4}\left|\dfrac{a_b}{a_f}\right|^4\left(\dfrac{0.9\,Y(3)}{Y(N)}\right)^{2/3}} \underset{[\lesssim]}{\approx} E'$$

$$\left[\lesssim \frac{E}{1+\dfrac{48\pi E_{B1}^3E^3}{m_b^6c^{12}}\left|\dfrac{a_b}{a_f}\right|^{12}\left(\dfrac{0.9\,Y(3)}{Y(N)}\right)^2}\right], \qquad N \geqslant 2.$$

●

Are there processes in which 3 bit Corollary 14.15 B_0s and 2 bit fast B_0s are related? (For the following corollary, reference to a result obtained in the axiomatic form formally preceding this is used in step 10.)

Corollary 15.11. *C2; In domain of applicability of PODIC; in transcription of information to be represented by ('making') B_0 as in Corollary 14.15, 'B_{03}'; by a process supplying least possible internal information; supplying, or involving making an entity representing information with coding complementary to ('complementing'), coding for exactly 1 bit of B_{03}'s internal information: said 1 bit is coded as relative phase of B_{03}'s associated two photons; such process involves making a fast B_0, 'B_{02}', representing internally exactly 2 bits of information, in coding complementary to B_{03}'s other 2 bits of internal information; B_{02} represents internally 1 bit of information by relative polarization of its associated two photons; B_{02} does not represent any internal information by relative phase of its associated two photons; and such process involves transcription of 5 bits of information to be represented internally by B_{02} and B_{03}.*

1. From Def 1 (or Postulates 1 and 2), information is accessible.
2. From Theorem 1, in B (or from Postulate 1), information is conserved.

3. From Cor 14.15, B_{03} represents internally 3 bits of information.
4. From Def 2, and 3, in domain of applicability of PODIC, in making B_{03}, coding for each of B_{03}'s 3 bits of internal information must be supplied or complemented for information so represented to be accessible.
5. From 1–4, coding for each bit of B_{03}'s 3 bits of internal information must be supplied or complemented in making B_{03}.
6. Suppose that, in domain of applicability of PODIC, B_{03} were made by a process supplying least possible amount of internal information, which supplied or complemented coding for only 1 bit of B_{03}'s internal information.
7. From 3 and 6, in such process, coding for 2 bits of B_{03}'s internal information would not be supplied.
8. From 5 and 7, in such process, coding for 2 bits of B_{03}'s internal information must be complemented.
9. From Cor 1.2 (or A-Theorem 7), and 8, such process must make an entity representing exactly 2 bits of information internally with coding complementary to the 2 bits of B_{03}'s internal information for which coding was not specifically supplied or complemented by the process in step 6, 'B_{02}'.
10. From A-Theorem 1 step 15, two photons are necessary and sufficient to represent as internal information 2 bits of information.
11. From 9 and 10, B_{02} is describable in terms of two photons. ('adequately', as in A-Def 2 (Section 3))
12. From PODIC and Theorem 14, in domain of applicability of PODIC, for B_{02}'s associated two photons to be coupled, such coupling must be designated by 1 bit of information represented 'internally' by B_{02}.
13. From 9–12, B_{02}'s associated two photons would be not coupled.
14. From 13, B_{02} would be unable to represent information by coding, or complement coding of information represented, by means of states distinguished in a way derived from interaction coupling photons.
15. From Cor 14.3–14.5, in representation of B_{03}'s internal information, states derived from coupling of B_{03}'s associated two photons are distinguished by $\pm$ relative phase $\pi/2$ of photon in extended interaction state and photon in less coupled state.
16. From 14 and 15, B_{02} would not represent any internal information by photon relative phase, and would be unable to complement such coding.
17. From 5, 6, and 16, process as in step 6 would supply or complement coding by relative phase of B_{03}'s associated two photons.

18. From Cor 14.6, and 9 and 17, B_{02} would complement B_{03}'s coding of 1 bit of information by relative polarization of B_{03}'s associated two photons.
19. From EMT, QM, PODIC, Cor 14.11 and 14.12, and Def 2, in domain of applicability of PODIC, for B_{02} to complement the coding of 1 bit of B_{03}'s internal information by relative polarization of two photons, B_{02} would represent 1 bit of internal information by relative polarization of two photons.
20. From 18 and 19, B_{02} would represent 1 bit of internal information by relative polarization of B_{02}'s associated two photons.
21. From 9, 10, and 20, B_{02}'s associated two photons would have together only 1 bit of internal information to describe existence of both. ('adequately', as in A-Def 2 (Section 3))
22. From 21, B_{02}'s associated two photons would not have separate existence.
23. From Cor 8.12, Def 13, and 13 and 22, B_{02} would be describable as a fast B_0. ('adequately', as in A-Def 2 (Section 3))
24. From 3, 6, 9, 16, 17, 20, and 23, in domain of applicability of PODIC, process as in step 6 would supply or complement coding of 1 bit of B_{03}'s internal information by relative phase of associated photons; such process would involve making a fast B_0 representing internally exactly 2 bits of information in coding complementary to B_{03}'s other 2 bits of internal information; B_{02} would not represent any internal information by relative phase of its associated two photons; B_{02} would represent 1 bit of internal information by relative polarization of its associated two photons; and such process would involve transcription of 5 bits of information to be represented internally by B_{02} and B_{03}.

●

Note that internal information, and coding thereof, in some sense the 'nature', of the fast B_0 produced depend on the entity whose internal information coding it partly complements, and not otherwise on the process making both.

With reference to Corollaries 14.3, 14.4, 14.9, 14.13, 14.14, and 14.16, the bit of internal information coded by photon relative phase appears to correspond to 'charge' sign. With reference to Corollaries 14.10 and 14.16, the bit of internal information represented by photon relative polarization appears to correspond to 'spin $\frac{1}{2}$'. As noted following Corollary 14.16, B_{03}'s other bit of internal information appears to correspond to 'lepton number'.

Thus, with Corollary 15.11, it appears that, in domain of applicability of PODIC, any process which made an electron or positron, and which supplied or complemented coding for 1 bit of internal information thereof, would supply or complement coding for 'charge' sign, and would also make a fast B_0 with 2 bits of internal information, 'spin $\frac{1}{2}$', and 'lepton number' complementary to such electron's or positron's, and without 'charge' sign; and that such process would involve transcription of 5 bits of information to be represented internally by such electron or positron and such fast B_0.

In view of the foregoing, such fast B_0 would seem to correspond to an electron neutrino or antineutrino.

In some respects, all information represented by a fast B_0's associated down-shifted photon might be regarded as represented 'internally' by such B_0. With such interpretation, for a fast B_0 to represent internally only 2 bits of information, its associated down-shifted photon would have to represent only 1 bit of information in B, and, therefore, be in the lowest energy photon state in B, so such B_0 would be a fastest B_0.

One can make some numerical comparisons with our universe (see also Table 2, p. 155):

From Corollary 8.10, in our universe, a 1 MeV fB_0's momentum would appear to transform substantially like a photon's in frames of reference moving with velocity $< c(1-\epsilon)$, $\epsilon \gg 2E_{B1}/E \approx 1.279\times 10^{-39}$.

From Corollary 8.11, for fast B_0, $E_0 \approx 2\sqrt{EE_{Bj}}$. For fastest B_0, $j=1$. For a 1 MeV fB_0, $E \gg E_{B1}$, and subject to the question noted hereinbefore as to applicability of computational methods of SR in case of fB_0 (see discussion near Cor 8.7), $E_0 \approx 5.05857\times 10^{-14}$ eV.

From Corollary 14.16, least massive bound B_0 appears to correspond to electron and positron. Taking m_bc^2 as m_ec^2, from CT 1973 (ref. 4)$\approx$ 511003.4 eV, and writing a_b as electron charge $|e|$, with $N=2$, using $Y(3)=4$ from Section 3 as noted hereinbefore, from Corollary 15.7, and Definition 20, if computational methods of SR be applicable, for fB_0, $|a_f| \approx |e|\times 1.09736\times 10^{-11}E^{1/4}$, E in eV. For a 1 MeV fB_0, $|a_f| \approx 3.4701\times 10^{-10}|e|$. Note that a_f is in B_0's rest frame, which, in our universe, has $v \approx c$. Note also that interaction cross sections might depend on graininess of B_0's allowed momenta.

From Corollaries 15.2–15.4, an fB_0 near earth would not be attracted by the earth by Theorem 13 interaction.

With reference to Table 2 and Corollary 8.16 in relation to domain of applicability of PODIC and, with it, SR, SR would appear applicable for systems with linear size(s) X moving with velocity satisfying the relation for PODIC to be applicable, $< \approx c(1-(X/R_U)^2)/(1+(X/R_U)^2)$. In view of the magnitude of R_U, SR appears applicable in all domains in which I am at the time of this writing aware of its having been explicitly tested prior to the time of this writing. (but see question 7 and others, Section 6)

At the time of this writing, these various results for fast B_0s appear allowed by prior knowledge of which I am at the time of this writing aware, with correspondence between fast B_0s and neutrinos and antineutrinos:

Correspondence Corollary 15.12. *Corollary* 8.6 *through Corollary* 8.16 *and Theorem* 15 *through Corollary* 15.11 *appear, to the extent tested as presented hereinbefore, allowed by knowledge prior to the time of this writing, with correspondence between fast* B_0*'s and neutrinos and antineutrinos.*

●

This correspondence might offer some insight into particle processes involving neutrinos.

For example, if one were to assume that protons ('p') and neutrons ('n') represent internally approximately the same amount of information, then the relatively long half-life of free neutrons might be interpreted in terms of transcription of external information to be represented as internal information of the 'decay' products, as giving some notion of a lower limit of the 'degree of separability' of amounts of internal and external information in such process, in this case, fractional rate of transcription of external to internal information. From $E_{B1} \approx 6.39728 \times 10^{-34}$ eV (see Table 2), neutron mass energy $m_n c^2 \approx 939.573$ MeV (from N. Barash-Schmidt+, ref. 8), and (also from ref. 8) neutron mean life $\sim 918 \pm 14$ sec; for e and ν_e together representing 5 bits of information internally; with such assumption, it would seem that, in $n \rightarrow p^+ + e^- + \bar{\nu}_e$ or $\bar{n} \rightarrow p^- + e^+ + \nu_e$, part (5 bits) of the neutron's or antineutron's external information would be transcribed to be represented as internal information of the electron or positron and electron antineutrino or neutrino, at an average 'rate' (for each bit of the original amount of external information) of not more than about 3.708×10^{-45}/sec (cancelling units of (bits internal)/(bits external)).

2.11 INFORMATION AND INERTIA

I have discussed four aspects of 'inertia': momentum; position; gravity; and discreteness of mass, charge, and statistics. From Corollaries 9.1, 10.4, 13.5, 14.16, and 15.12,

Correspondence Theorem 16. *A physical picture of inertia in terms of information appears, to the extent tested as presented hereinbefore, allowed by knowledge prior to the time of this writing.*

●

2.12 COMPATIBILITY WITH PRIOR KNOWLEDGE

In view of all development of these ideas as presented in this Section 2, it appeared that, in each case in which a formal result was obtained therein which was compared with that part of prior knowledge of which I am at the time of this writing aware, such result seemed, to the extent so tested, substantially allowed by such prior knowledge, in some cases with comparison to qualitative and/or quantitative features of reports of prior observation, in other cases with absence of such reported prior observational data where such formal result differed substantially from various prior theory(s).

Thus, the physical picture presented herein, in which physical phenomena are viewed in terms of information, appeared to that extent substantially compatible with prior knowledge.

In view of this apparent compatibility, it seemed reasonable to seek an axiomatic form based on information, into which this treatment might be merged.

So, on to Section 3.

Section Three

Axiomatization

I wish to in this section display more starkly the nature of the physical picture whose compatibility with prior knowledge was explored hereinbefore, by finding a set of postulates from which such picture can be formally and conceptually derived. If this is done, then the reader may be guided by an esthetic sense regarding such conceptual and formal structure.

This Section 3 is under the interpretation stated on p. 2.

(Because they form a sequence independent of the derivation appearing in Section 2, various definitions, lemma, theorems, and corollaries appearing in this axiomatic form are designated by prefix 'A-'.)

3.1 LOOKING FOR POSTULATES

In this subsection, I summarize part of the thinking by which I came to propose the postulates presented hereinafter, partly because such summary might give some perspective for subsequent formal sufficiency tests, and partly because you, the reader, might wish to make some postulate sets yourself, in which pursuit you might find use for some of this approach. One interesting aspect of theoretical formalism for a physical picture is induction thereby of equivalence classes of postulate sets: comparison of such equivalent postulate sets and corresponding formalisms may help elucidate relations between physical content and formalism, as, for example, appears to have happened to some extent in quantum mechanics with wave and matrix formalisms.

In Theorem 1 through Theorem 8 hereinbefore, quantum mechanics (QM) entered primarily for information conservation; modeling was from electromagnetic theory (EMT), including mass particles by photon particle representation (PPR) shown in Theorems 4 and 4′ (with lemmas and corollaries) to have the same external information as mass particles. Corollary 13.3 indicates that formal structure in Section 2 with respect to Theorem 13 interaction is consistent with information conservation.

Inspection of the formal structure hereinbefore suggests that, by using PPR, one could obtain QM from EMT, $E = hf$, and information conservation: the mechanics of B_0s appears to correspond, within special relativity (SR)'s domain of applicability, to relativistic QM, and thus to include as special cases nonrelativistic QM, Einstein's special relativistic classical mechanics, and Newton's nonrelativistic classical mechanics. This appears especially in Theorem 4′, Corollaries 4.4 and 8.16, Theorem 9, Corollary 9.1, and Theorem 14 and its corollaries.

In view of these considerations, information conservation seemed interesting as a postulate candidate.

Although EMT and QM have previously been formulated without regard for boundary conditions from 'the' universe, and although I attempted in earlier parts of this work to explore certain hypothetical aspects of 'infinite enclosures' (e.g., Cor 5.3 and 6.1), it became apparent that many interesting aspects of the theory depended on finite accessibility of information and finiteness of enclosure linear size, as for 'cosmological red shift' (Cor 10.2–10.4), gravity (Theorem 13 etc.), ('etc.' may be read here as 'and its lemmas and/or corollaries'), electron and positron (Theorem 14 etc.), and limitations on domains of applicability of PODIC, SR, and one notion of 'principle of equivalence' (Cor 8.16, Theorem 15 etc.). Finite accessibility of information appears to imply finiteness of linear size of (accessible) enclosure, because being in any finite subregion of an infinite accessible enclosure would designate one out of infinitely many possible alternatives, and thus represent infinite information. In this respect, finite accessibility of information and finiteness of accessible information appear equivalent; but the concept of finite accessibility appears to not exclude the possibility of a conceptual picture in which different observers might have different (not entirely coincident) universes. (See also Comments 4.9 and 4.11, and Section 6 question 47.)

In view of these considerations, finite accessibility of information seemed interesting as a postulate candidate.

Because specifying any characteristic as occurring in a region requires more information for smaller regions (more alternative regions), one remarkable consequence of such postulate would be that every property of a physical system would be describable, for a small enough region, by mathematical functions which were continuous and, in a small enough region, linear.

Only part of EMT was used: Maxwell's equations without sources, to which, absent sources, must in such theory be added a definition of energy as the volume integral of the square of field amplitude, and, from QM, quantization rule $E = hf$ for photons, where h is Planck's constant and f is photon frequency. Dependence of energy on square of field amplitude was essential to the concept of 'gentle acceleration' (GA) and 'gentle scaling' (GS) not substantially disturbing states in a small box, underlying separability of internal and external information and special relativity (Theorem 8 etc., and PODIC). Transversality of wave equations was used in various of the corollaries of Theorems 14 and 15.

But, what is such transversality? What it seems to mean is that photons carry 1 bit of information, called 'polarization' or 'helicity', or, saying it more formally, a mathematical function space in which electromagnetic waves can be decomposed may be thought of as a product of 2-spaces, and orthogonal basis functions of each such 2-space may optionally be chosen to correspond to two perpendicular linearly polarized waves or 'right' and 'left' helicity states. By considering decomposition into helicity states, one can see by inspection that a photon carries one bit of information as 'polarization'. Because 'linear polarization' spans the same function 2-space (e.g., one can express 'linear polarization' as equal 'contributions' of opposite 'helicity' with $\pm\pi/2$ relative phase), linear polarization represents the same one bit of information per photon.

This can be 'turned around', in that one can use a requirement that quanta exchanged with a wave always represent internally 1 bit as a way of requiring that the wave equation be decomposable in a function space itself a product of 2-spaces whose basis 'vectors' (functions) differ only with respect to that 1 bit, that is, that the wave equation be 'transverse' (as two coupled waves). However, that such quanta exist requires 1 bit of information corresponding to existence, so quantization of such wave equation requires that existence bit, and thus requires 'transversality'.

Thus, quantization, a wave equation and squared amplitude, and postulates that information be conserved and finitely accessible should be enough to provide a basis for all of the theory hereinbefore.

Finite accessibility of information implies finite 'enclosure', as noted hereinbefore, and a discrete spectrum of solutions of the wave equation, only finitely many of which can have amplitude different from zero, and for each such solution only finitely many discrete amplitudes and only finitely many possible discrete combinations with relative phases and amplitudes; and, therefore, for each solution in that spectrum there is a least amplitude A_1 greater than zero. From this comes a form of quantization, so one can thus get 'transversality' of the wave equation.

In adding to an allowed solution initially having zero amplitude an amount of information sufficient to bring its amplitude to the least amount greater than zero which it can have, A_1, one is adding amplitude without relative phase. By repeating this process n times, one obtains an amplitude $n^{1/2}A_1$, which, by postulated information conservation, represents n times as much information as was represented by A_1 in that state. Thus, one can eliminate as redundant the separate statement that information (energy) be proportional to amplitude squared, and energy becomes merely a construct proportional to information.

Where can we get a wave equation? Well, what *is* a wave equation? To me, one of the most striking features of a wave is found at its back: as the wave passes, the 'medium' is returned to substantially its original condition and *relatively little or none of the wave is left behind*. We already have as a postulate candidate information conservation, so, if we add a postulate that information be *communicable*, we will have a process which requires that there be an information representation in which availability of information at various positions in 3-space is describable as a function of time, and that none of the representation be left behind; and, from the postulate candidate that information be finitely accessible, such propagation must in a sufficiently small region be adequately describable by linear equations and be characterizable by a velocity.

The mathematical function describing such residue-free propagation is not monotonic, and, because linearity in a small enough region implies possibility of superposition of another propagating signal in such region, the basis functions of a mathematical function space in which the function describing such residue-free propagation can be completely expressed must be both positive- and negative-going, and must be combinable with relative phase and amplitude. From finite accessibility of information, for a sufficiently small region of space and time, such a function space is spanned by a finite set of solutions of a (as noted hereinbefore, 'transverse') 'wave equation'.

Thus, postulates of information conservation, information communicability, information finite accessibility, and 3-space and time, would be enough. Where can we get 3-space and time?

Communication requires existence of two events, or, stated differently, of a transmitter and a receiver, or a (coded (for local propagation)) message and a receiver, or an event and an observer. Existence of each requires 1 bit, and existence of both requires a total of 2 bits. Therefore, basis functions of a mathematical function space in which every mathematical description of observed events, or of communication, can be completely expressed are functions of four independent variables; which, from postulate candidate that information be finitely accessible can, for a small enough region of those four variables, be adequately described as linear. Communication in such region is describable by linear relations among four variables, completely expressable in a function space spanned by a finite set of solutions of two (coupled) transverse wave equations in those four variables. Choose an orthonormal basis, x_0, x_1, x_2, x_3, for those four variables, and express the wave equation solutions in terms of such basis; for specificity, let x_0 denote the variable describing availability of information at regions labeled by x_1, x_2, and x_3.

As discussed in example 5 following PODIC (p. 70), from consideration of information which an observer has as a function of x_0, it would seem that, for any x_0, an observer O can regard x_0 as unidirectional, and can order events by x_0 increasing only. x_0 thus seems to correspond to our previous notion of time, and O's criterion is the one, if you think you really know the future, of asking yourself if you have ever in the past correctly predicted an event in what was then yet the future *and had your ability to so predict already confirmed by the occurrence without your causing it of that event*: for the absolute foreknowledge of reversed observation, the answer is always *no*, because you do not now have information of the occurrence in your past of reversed observation. The problem might not be knowing the future, but rather knowing that you know the future.

By comparison of adjoining regions, local choice of variable to describe communication appears to extend to (a) global choice, in that the special status of x_0 in describing events only monotonically distinguishes it from x_1, x_2, and x_3, in terms of which events may be described subject to any coordinate rotation, but x_0 may not be fully interchanged with any linear combination of variables x_1, x_2, and x_3.

Thus, the variable used to describe communication (x_0) appears to correspond formally and conceptually to the notion of time usually used at the time of this writing, and the 3-space spanned by x_1, x_2, and x_3 appears to correspond similarly to such notion of space.

From the preceding analysis, the four independent variables formally required by the two existence bits for functions describing observed event(s) or received communication(s) appear to correspond to space and time coordinates as previously used in physical theory. Note that communicability of information appears to be an essential part of any theory which describes observation, and, therefore, of physics (for observer $< U$).

Thus, postulating that information be conserved, communicable, and finitely accessible appears to be enough:

3.2 POSTULATES

A-*Definition 1.* *Amount I in 'bits' of information equals $\log_2$ of number of possibilities from which one is designated.*

('information' may occasionally be used herein to mean 'amount of information')

Postulates: *1. Information is conserved.*
2. Information is communicable.
3. Information is finitely accessible.

3.3 TESTING POSTULATES

A-*Definition 2.* *With respect to description by mathematical function, let 'adequate' and/or 'adequately' mean 'with any and all error less than* 1 *bit of information for the description and/or any observable dependence thereon'; let 'is adequately describable as' be denoted by '*A*' above the corresponding relational symbol; let 'ad(a)' denote '(is)(are) adequately describable (as)'; and let 'irad' denote '(an) information representation adequately describable (as)(by)(in such way)'.*

(Usefulness of 'adequate' as a criterion is discussed formally hereinafter, in A-Cor 4.1.)

Note that '$\overset{\text{A}}{=}$' does not appear to be transitive, but '$\overset{\text{A}}{>}$' and '$\overset{\text{A}}{<}$' are.

3.3.1 Describing Communication

A-*Theorem 1.* *Communication is adequately describable for sufficiently small range of linear orthonormal variables x_0, x_1, x_2, and x_3 by a linear combination with finitely many relative amplitudes and phases of finitely many independent solutions $w_{z,k}$ of the coupled equations*

$$\left.\begin{aligned} &\nabla \times w_{z,k} + (-1)^k \frac{1}{c}\frac{\partial}{\partial x_0} w_{z,1-k} = 0 \\ &\nabla \cdot w_{z,k} = 0 \end{aligned}\right\} k = 0, 1,$$

where '$\nabla \times$' and '$\nabla \cdot$' respectively denote curl and divergence with respect to x_1, x_2, x_3, and z is a label of a finite set.

1. From Postulate 2, information is communicable.
2. For communication, a source and a receiver must exist.
3. From A-Def 1, existence requires 1 bit of information.
4. From 1–3, existence of communication requires 2 bits of information.
5. From 4, mathematical functions describing communication are adequately describable in terms of functions of four independent variables.
6. From Postulate 3, and 5, for each of these four variables the accessible range is a finite multiple of the smallest interval which can be defined.
7. From Postulate 3, and 5 and 6, for sufficiently small range of said four variables, each is adequately describable as linear.
8. From 5 and 7, for sufficiently small range of said four variables, all mathematical functions describing communication are adequately expressable as functions of four orthonormal linear variables: designate said four orthonormal linear variables by x_0, x_1, x_2, and x_3.
9. From Postulate 3, and 8, for small enough range of x_0, x_1, x_2, and x_3, all mathematical functions describing availability of information must be linear in x_0, x_1, x_2, x_3. (location of discontinuity would represent infinite information; if region does not contain information, then functions therein may be regarded as continuations from neighboring regions and fully determined thereby)
10. From Postulate 1, and 9, the mathematical function describing communication is not monotonic. (not any information is left behind)
11. From Postulate 3, and 10, each basis function of a mathematical function space in which the mathematical function describing communication can be adequately expressed represents finite information. (e.g., not (Dirac's) δ functions)
12. From Postulate 3, and 9–11, said basis functions are both positive- and negative-going, and can be combined with relative amplitudes and phases.
13. From Postulate 3, and 12, for only finitely many such basis functions is amplitude different from zero possible; only finitely many amplitudes are possible; and only finitely many relative phases and amplitudes are possible.
14. From Postulates 1–3, A-Def 1, and 9–13, for communication, coding of information in a representation for which availability of information is adequately describable as in steps 9–13 must occur.
15. From 3 and 14, existence of such coding requires 1 bit of information.

16. From 11–15, the function space in step 11 can be expressed as the product of finitely many 2-spaces, for each of which two orthonormal basis functions can be chosen with identical dependence on x_0, x_1, x_2, x_3. (isolating dependence on step 15 existence of coding bit)
17. From 8, 9, 11, 12, 13, and 16, by Fourier expansion, for small enough range of x_0, x_1, x_2, x_3, with suitable choice of x_0, the mathematical function describing communication may be adequately expressed as a linear combination with finitely many relative amplitudes and phases of finitely many independent solutions $w_{z,k}$ of the coupled equations

$$\left.\begin{array}{l} \nabla \times w_{z,k} + (-1)^k \dfrac{1}{c} \dfrac{\partial}{\partial x_0} w_{z,1-k} = 0 \\ \nabla \cdot w_{z,k} = 0 \end{array}\right\} k = 0, 1,$$

where '$\nabla \times$' and '$\nabla \cdot$' respectively denote the conventional curl and divergence operations with respect to x_1, x_2, x_3, with z as a label of a finite set.

●

The constant c arises solely from the requirement, from finite accessibility of information, of local linearity, and may be regarded as a 'linear coefficient of local propagative communication'. Each pair of functions $w_{z,k}$, $k=0,1$, spans one of the 2-spaces associated with the existence of coding bit, as noted in steps 14–16.

Note that, although c appears in description of local communication, and thus in communication 'pieced together' from local communication, this does not appear to require that c appear in adequate description of communication, if any, not involving adequately-describable-as-continuous traversal of successive local regions ('local propagation').

For convenience of reference,

A-Definition 3. *Let 'w', 'w_j', 'w_z', '$w_{j,k}$', '$w_{j,z}$', '$w_{z,k}$', where $k=0,1$, j and/or z is any finite set or element thereof, denote a solution or solutions of the 'wave equations' in* A-*Theorem* 1, *and let 'WE' denote such equations.*

3.3.2 3-Space and Apparently Unidirectional Time

A-Definition 4. *Call that which is capable of transcribing information among all modes of representation with which it can exchange information and of transporting information among different values of x_0 an 'observer'.*

Having for specificity designated the (unique) 'independent' variable in mathematical descriptions of availability of information by x_0, we can write a 'local' definition of observation, choosing therewith a sign convention for x_0:

A-*Definition 5.* *'Observation' is said to occur for an observer O participating in communication if for* $x_0' \hat{<} x_0$ *there is not available in O information of the occurrence of an 'event' and for* $x_0' \hat{>} x_0$ *there is available in O such information.*

With A-Definitions 4 and 5, noting that observation requires the existence of an event (1 bit) and an observer (1 bit) with communication therebetween of information, we obtain from A-Theorem 1:

A-*Corollary 1.1.* *For sufficiently small ranges of* x_0, x_1, x_2, x_3, *an observer can adequately describe observed events by mathematical functions of* x_0, x_1, x_2, *and* x_3 *as in* A-*Theorem* 1.

For example, an observer can systematize 'local' observed events in terms of a flat 4-space.

A-*Definition 6.* *Define the 'universe' ('U') of an observer O to be all about which O can have information.*

Postulate 1 is interpreted in terms of the unique one of the four variables appearing in description of availability of information, locally designated x_0 in A-Theorem 1. With this interpretation, we obtain

A-*Corollary 1.2.* *Observer O can adequately describe O's universe in terms of a* 4-*space, one variable of which (locally* x_0*) appears homogeneous and strictly unidirectional, which* 4-*space as factored by said variable appears homogeneous and isotropic, and O does not see any coordinate transformation as interchanging said variable with any variable in said factor* (3-)*space. (in conformity with the statement of this corollary, every* x_0 *value and every relation among* x_0 *values stated in the following derivation is stated as adequate description)*

1. From Postulates 1, 2, and 3, A-Theorem 1, A-Cor 1.1, and A-Def 4 and 5, O at x_0, in describing all events E of the occurrence of which O has

any information of observation for respective x_0 value x_{0E}, by designating all events with $x_{0E} < x_0$ as observed and all events with $x_{0E} > x_0$ as caused (or vice versa) can adequately describe all such events in terms of unidirectional x_0.

2. From 1, with such description as in step 1, O describes O as transporting information from smaller to larger values of x_0 (or vice versa).
3. Suppose O as in steps 1 and 2 were to receive information I from O' as in '( )' in steps 1 and 2 at x_0.
4. From 1–3, in case as in step 3, information I would be carried by O' from $x_0' > x_0$ to x_0, and by O from x_0 to $x_0' > x_0$.
5. From 4, in case as in step 3, information I would exist only for $x_0' > x_0$.
6. Postulate 1 is interpreted herein in terms of the unique one of the four variables appearing in description of availability of information.
7. From Postulate 1, and 3, 5, and 6, case as in step 3 is not observable.
8. From 3–7, similarly, the case that O' receive information from O is not observable.
9. From A-Def 6, and 7 and 8, each (if any) of O's and O''s (respective) universe(s) does not contain any such communication between any such observers.
10. Suppose both and only O and O' have either observational or causal communication with event E at x_{0E}.
11. From Postulates 1 and 2, and 1, 2, and 6, E is (a node in) one way communication between O and O'.
12. From Postulate 2, and 1, 2, 10, and 11, (factoring such communication node) all observers with the same direction of x_0 as O or O' and so communicating with E can be grouped respectively with O or O'.
13. From 10 and 12, E as in step 10 can be generalized to any event with any set of observers including at least one observer for each direction of x_0.
14. From 9–11, and 13, each (if any) of O's and O''s (respective) universe(s) does not contain any such event E as in steps 10 and/or 13.
15. Suppose O and O' were to communicate in a way not prohibited with Postulate 1.
16. From 3–15, communication as in step 15 cannot result in net transfer of information between O and O'.
17. From 10–13, 15, and 16, for communication as in step 15 to provide to O or O' information of the other's direction of x_0, such communication would have to include transfer between O and O' of information of an event as in steps 10–13.
18. From 14–17, by communication as in steps 15 and 16, O and/or O' does not obtain information of the other's direction of x_0.

19. From 15, 16, and 18, each of O and O' could adequately describe any communication as in steps 15 and 16 as (at most) transformation of representation of conserved information transported by each respectively both to and from x_0.
20. From 9, 14, and 19, observer(s) with the same direction of x_0 can treat such observer(s) universe(s) as disjoint from universe(s) of any observer(s) with opposite direction of x_0.
21. From A-Def 6, and 9, 14, 19, and 20, O can adequately describe O's universe in terms of unidirectional x_0.
22. From 20 and 21, O does not see any coordinate transformation as reversing direction of x_0.
23. From Postulates 1–3 and A-Theorem 1, any observer at any set of values of the four variables in A-Theorem 1 (step 5) could have applied Postulates 1–3 and constructed A-Theorem 1 and the 4-space for description of observed events stated in A-Theorem 1, as at any other set of values for such four variables.
24. From 23, said 4-space is homogeneous.
25. From Postulates 1–3, A-Theorem 1, A-Cor 1.1, and 1–24, there is not any dependence on orientation in the 3-space obtained by factoring said 4-space by the unique one of the four variables (locally x_0) appearing in description of availability of information. (e.g., see WE)
26. From A-Theorem 1 and A-Cor 1.1, and 25, said 3-space is a formal construct for description of communication (including observed events).
27. From 25 and 26, said absence of dependence on orientation in said 3-space is formally equivalent to isotropy of said 3-space.
28. From 27, if there were any coordinate transformation in O's universe that could transform between x_0 and a linear combination of x_1, x_2, x_3 spanning said 3-space, the steps of such transformation, rotation allowed in isotropic 3-space, and such transformation back to x_0, could together comprise a transformation reversing direction of x_0.
29. From 22 and 28, O does not see any coordinate transformation as interchanging x_0 with any linear combination of x_1, x_2, x_3 spanning said 3-space.

●

From A-Corollary 1.2, designation of the unique one of the four variables in description of availability of information appears to O to apply for all of O's universe, and to uniquely factor O's 4-space into that variable and an isotropic homogeneous 3-space. This appears to correspond formally and conceptually with notions of time and space, including apparent unidirectionality of time, prior to and relatively near this writing.

A-Definition 7. *Call the unique one of the four variables appearing in description of availability of information in the 4-space of A-Theorem 1 and A-Corollary 1.1 'time' (denoted by 'x_0' or 't'); call the 3-space obtained by factoring said 4-space by time '3-space' or 'space' (denoted by $\vec{x}=(x_1,x_2,x_3)$); and call said 4-space 'spacetime' (denoted by $x=(x_0,x_1,x_2,x_3)=(t,\vec{x})$).*

With A-Definition 7, we can restate A-Corollary 1.2 as

A-Corollary 1.3. *Observer O can adequately describe O's universe in terms of spacetime; time appears homogeneous and unidirectional to O; space appears homogeneous and isotropic to O; and O does not see any coordinate transformation as interchanging space and time.*

With A-Definition 7, A-Corollaries 1.2 and 1.3 show that, with Postulates 1, 2, and 3, an observer O would not have any information requiring O to regard time as other than unidirectional. O's 'physics' includes only descriptions necessary for adequate description of O's observations. Therefore, O's physics does not include time as other than unidirectional.

The formal purpose herein of mathematical description of (observed) events (or other (received) communication) is to represent information thereof. A-Theorem 1 and A-Corollaries 1.1–1.3 indicate that such mathematically descriptive functions can be adequately written as functions of a locally flat 4-space. Referring to 'coordinates' for all of an observer O's universe U as 'global', one can regard any nonflat global coordinates as themselves representing information.

From Postulate 3, mathematical functions describing such nonflat coordinates in terms of a global flat space would represent a finite amount of information (e.g., no singularities or divergences, in keeping also with time never being seen as interchanged with space in U). Considering the appearance of Taylor series restricted to finite information and the nature of the (local) 'wave equations' of A-Theorem 1, it appears that one formal statement for such nonflat coordinates to not represent any information in a region is that the mathematical functions describing them in terms of flat 4-space be analytic and pole-free continuations therein of such functions in neighboring regions. Therefore, if such coordinates do not represent information anywhere in U, and not, by definition of U, from any region outside of U, then they can adequately be described as flat everywhere in U.

There appears to be a choice as to whether information is represented by use of nonflat coordinates, or by descriptions in terms of global flat space(time), or by combinations thereof: from Postulates 1 and 3, it is the same information. By choosing a formalism in which said 4-space is globally flat, one in effect attributes all information to information representations adequately describable by mathematical functions of such flat 4-space, and none to the coordinates themselves.

Much of the formal development of concepts explored herein is concerned with describing amounts and representations of information ('bookkeeping'), an activity which would seem not to be made easier by choosing global coordinates which themselves represent information. For this reason, I have used for this formalism globally flat coordinates, on which may be modeled nonflat coordinates as (a) description of various physical phenomena (e.g., Theorems 12 and 13) for regions small compared to a universe, in which regions phenomena on a much smaller scale may for convenience be described in terms of such nonflat coordinates.

A-Definition 8. *Absent specification otherwise, all mathematical descriptions herein are written in terms of global extensions of local coordinates used in* A-*Theorem* 1, A-*Corollaries* 1.1, 1.2, *and* 1.3, *and* A-*Definition* 7, *adequately describable as flat everywhere in universe U.*

Isotropy and homogeneity of 3-space (A-Def 7 and A-Cor 1.3), finite accessibility of information (Postulate 3), and choice of globally flat coordinates (A-Def 8) allow a single number to be used for scaling size of a universe:

A-Corollary 1.4. 3-*space sizes in universe U may be scaled by assigning linear size R_U to U.*

1. From Postulate 3, A-Theorem 1, A-Cor 1.1–1.3, and A-Def 7, each x_i, $i=1,2,3$, has a range $[x_i]$ which is a finite multiple of the smallest distinguishable interval thereof. (i.e., in which 1 bit could be represented; this is like A-Theorem 1 step 6, with flatness and orthonormality as in A-Def 8 (square brackets used in this corollary for range, not as in Section 2))

2. From A-Cor 1.2 and 1.3, 3-space is isotropic.
3. From 1 and 2, $[x_1]=[x_2]=[x_3]$.
4. From 1 and 3, the range in any direction of 3-space is a finite multiple of the smallest distinguishable linear extent (i.e., length) in any direction.
5. From 2–4, range of 3-space is spherically symmetric, with radius $R_U = [x_i]/2$, $i=1,2,3$, measured in terms of local units of length in any direction in 3-space, thus defining such local units in terms of range $[x_i]$.
6. From A-Def 8, and 5, measurement of any size in U in terms of such local units so defined corresponds to scaling such size in terms of R_U.

●

A-Corollary 1.4 may give you some indication of a difficulty I have had in finding any meaning in theories of an 'expanding universe' : compared to *what*? : and of part of why I have sought a conceptual basis for size in an information picture (e.g., see discussion of 'red shift' in Cor 10.2 and 10.3, Section 2).

A-Definition 9. *Call R_U of A-Corollary 1.4 the 'linear size' of universe U.*

This may be thought of as calibrating length measuring units. What appears in calculations depending on length, of properties of a system S with linear size R in U, is R/R_U, which, with A-Definition 9, may be written in terms of a 'size' of S or scale $s=R/R_U$:

A-Definition 10. *Call $s=R/R_U$ the 'scale' of a system in U with linear size R.*

From Postulate 3, A-Theorem 1, and A-Definitions 1, 3, 6, 7, and 8, noting that, from linearity of WE, any linear combination of solutions of WE is a solution of WE,

A-Corollary 1.5. *In U, there is a finite maximum spatial period possible for any possible solution(s) of WE, and any linear combination of such solutions, having amplitude* >0.

●

For convenience of reference,

A-*Definition 11.* *Call the maximum spatial period in* A-*Corollary* 1.5 'λ_1'.

Scale s of a system S in U might appear in terms of maximum w wavelength λ' possible in S, as $s=\lambda'/\lambda_1$. (This is discussed more fully hereinafter, near after A-Theorem 2 in relation to its step 8, in A-Theorems 3 and 4, and in A-Theorem 6 steps 18–33.)

3.3.3 Separability of Internal and External Information

Next, we test sufficiency of Postulates 1, 2, and 3 for more detailed description of representation, separability, and amount of information. For this, we proceed by considering some aspects of information representations adequately describable by ws; then discuss B_0s and separability of internal and external information; and, then, with formal methods based on such separability, discuss quantification.

A-*Lemma 2.1.* *In U; for $w_{z,k}$; amplitude $A'_{z,k}$ of $w_{z,k}$; least amount $I_z>0$ of information representable by irad $w_{z,k}$: every possible amplitude of $w_{z,k}$ is given by $A'_{z,k}(lI_z)=l^{1/2}A_z(I_z)$, where l is a finite integer $\geqslant 0$ and A_z is least possible amplitude >0 of w_z, adequately describing representation of amount lI_z of information.*

1. From Postulate 3, for every possible $w_{z,k}$, only finitely many amplitudes are possible.
2. From 1, there is a least amplitude $A_{z,k}>0$ of $w_{z,k}$.
3. From Postulates 1 and 2, and A-Theorem 1, $w_{z,k}$ with zero amplitude does not describe representation of any information.
4. From given, with amplitude $A_{z,k}$, $w_{z,k}$ adequately describes representation of amount $I_{z,k}$ of information.
5. Starting with irad zero amplitude $w_{z,k}$, add amount $I_{z,k}$ of information to irad $w_{z,k}$.
6. From Postulate 1, and 4 and 5, procedure as in step 5 conserves information, representation of which information is subsequently adequately describable by $w_{z,k}$ with amplitude $A_{z,k}$.
7. From 5 and 6, noting that $w_{z,k}$ with zero amplitude does not have phase, adding amount $I_{z,k}$ of information to be so represented corresponds to adding amplitude $A_{z,k}$ without phase to $w_{z,k}$.
8. Starting with irad zero amplitude $w_{z,k}$, add $I_{z,k}$ l times to irad $w_{z,k}$.
9. From Postulate 1, and 8, procedure as in step 8 conserves information, which amount $lI_{z,k}$ of information is subsequently so represented.

10. From 7 and 8, procedure as in step 8 corresponds to adding amplitude $A_{z,k}$ of $w_{z,k}$ l times without phase. (specifying phase could describe representation of additional information)
11. From A-Theorem 1, and 8 and 10, procedure as in step 8 results in total $w_{z,k}$ amplitude $l^{1/2} A_{z,k}(I_{z,k})$.
12. From 8, 9, and 11, such representation of amount $l I_{z,k}$ of information is adequately describable by $w_{z,k}$ with amplitude $A'_{z,k}(l I_{z,k}) = l^{1/2} A_{z,k}(I_{z,k})$.
13. From Postulate 3, and 8, l is a finite integer $\geqslant 0$.
14. From Postulate 2, and 3–5, amount $I_{z,k}$ of information can be exchanged with irad $w_{z,k}$.
15. Starting with irad $w_{z,k}$ with any amplitude $\geqslant 0$, extract information $I_{z,k}$ (repeatedly) until $w_{z,k}$ amplitude $< A_{z,k}$.
16. From Postulates 1 and 2, and 7, 14, and 15, (as in step 10) procedure as in step 15 corresponds to (repeatedly) subtracting amplitude $A_{z,k}$ without phase (information) until $w_{z,k}$ amplitude $< A_{z,k}$.
17. From 2, 15, and 16, after procedure as in step 15, $w_{z,k}$ amplitude$=0$.
18. From 3, 12, 13, 15, and 17, every possible $w_{z,k}$ amplitude is given by $A'_{z,k}(l I_{z,k}) = l^{1/2} A_{z,k}(I_{z,k})$, where l is a finite integer $\geqslant 0$.
19. From A-Theorem 1 steps 14–16, A-Cor 1.3, A-Def 8, and 3, 12, and 18, noting the origin of subscript k as relating to the bit required by existence of coding, every possible amplitude of $w_{z,k}$ is given by $A'_{z,k}(l I_z) = l^{1/2} A_z(I_z)$, where l is a finite integer $\geqslant 0$ and A_z is least possible amplitude >0 of w_z, adequately describing representation of amount $l I_z$ of information.

●

That is, information is exchanged with representation by irad $w_{z,k}$ in 'lumps' I_z, independent of k.

For the following, A-Definitions 7 and 8 provide a formal basis for usual definitions in prior theory of frequency (e.g., $c/\lambda, 1/x_0, 1/t$), velocity (e.g., $\vec{v} = (\partial x_1/\partial x_0, \partial x_2/\partial x_0, \partial x_3/\partial x_0) = d\vec{x}/dt$), and acceleration (e.g., $\vec{a} = d\vec{v}/dt = (\partial^2 x_1/\partial x_0^2, \partial^2 x_2/\partial x_0^2, \partial^2 x_3/\partial x_0^2)$), in terms of which parts of the following are expressed, in each case relying on this formal basis.

A-Definition 12. *In observer O_U's universe U, let 'B_0' designate two ($i = 1,2$) oppositely directed w_{z_i}, each respectively with least amplitude $A_{z_i} > 0$, wavelength λ_{z_i}, and frequency f_{z_i}; with the following names (in ' ') for various properties as seen by O_U, for $f_{z_1} \geqslant f_{z_2}$: '$w_+$'$= w_{z_1}$, '$\lambda_+$'$= \lambda_{z_1}$, '$f_+$'$= f_{z_1}$,*

'w_-' = w_{z_2}, 'λ_-' = λ_{z_2}, 'f_-' = f_{z_2}, 'f_d' = $f_+ - f_-$, spatial period of f_d called 'λ_b'; call velocity v at which an observer O' moving substantially parallel to direction of propagation of w_+ ('with' B_0) would see w_+ and w_- as both having the same frequency 'f_0' with which O' associates at least one wavelength 'λ_0', the 'velocity of B_0'; all operations on B_0 not creating or destroying B_0 are defined to be those for which O' could adequately describe f_0 and/or λ_0 as unchanged; scale 's_0' of B_0 defined by $s_0 = (\lambda_0$ as seen by O_U of B_0 with least v possible in $U)/\lambda_1$.

("at least one wavelength 'λ_0'" avoids using result that in some nonzero velocity domain λ_0 could be unique, and relation '$\geqslant$' in '$f_{z_1} \geqslant f_{z_2}$' avoids using result $f_+ \neq f_-$, before they are properly obtained.)

A-Definition 13. *Define 'transitions among allowed states of B_0' to be transitions of w_+ and w_- among allowed states w_{z_i}, $i = 1,2$, for which f_+ and f_- appear to O' of A-Definition 12 as f_0 adequately describable as unchanged.*

Without yet specifying in detail what states of w or B_0 are allowed in U, we define in a restricted way 'gently accelerating' and 'gently scaling' in terms of states in a 'box' B_1 as compared to states in U:

A-Definition 14. *For irad box B_1, containing one w with least amplitude >0 or one B_0, in universe U; call substantially linearly accelerating such information representation so as to perform a process adequately describable by substantially one by one transitions of said w or B_0 among adjacent possible states thereof in U. 'gentle acceleration 1' or 'GA1'; and call such performance adequately describable by one such transition, one 'step' of GA1.*

A-Definition 15. *For irad box B_1, containing one w with least amplitude >0 or one B_0, in universe U; call a process on such information representation adequately describable by substantially linearly and proportionately changing scale of B_1 in U so as to perform substantially one by one transitions of said w or B_0 among adjacent possible states thereof in U, 'gentle scaling 1' or 'GS1'; and call such performance adequately describable by one such transition, one 'step' of GS1.*

A-Definition 16. *Call information represented by system S_1 adequately describable by w('s) and/or B_0('s) in box B_1 'internal information' I_1 of B_1 (including existence bit for S_1 if coding so described), information represented by S_1 in B_1 in box B_2 internal information I_2 of B_2, and $I'_{1,2} = I_2 - I_1$ 'external information' of B_1 (containing S_1) in B_2.*

Unlike in Theorem 8 and its lemmas and corollaries, we are not starting from a local nonrelativistic theory and extending it by an information argument: here, we are starting from global statements about information, and the derivation is more concise:

A-Theorem 2. *In observer O_U's universe U, for system S with scale s, moving with velocity v as seen by O_U in U, and containing only either irad one w with least amplitude >0 and longest wavelength possible in S, or irad two w's with different k values in such way that S is by containing such irad w's itself in effect irad one B_0, (both cases designated 'IR 1') representing information I_1 in S: for all v which can be reached from least v possible to S containing IR 1 in U by N_1 steps of GA 1 and/or N_2 steps of GS 1, if $\sum_{i=1}^{N_1+N_2} s_i^2 < 1$, where s_i is the scale of S before each step of GA 1 or GS 1, both amount and representation of internal and external information of S containing IR 1 in U are adequately describable as separable.*

1. From A-Theorem 1, A-Cor 1.3, and A-Def 6, 8, and 12, states of w or B_0 can be decomposed parallel to v.
2. From Postulates 1 and 2, and A-Lemma 2.1, a process transcribes information between information representations adequately describable by amplitudes of states in an amount adequately describable as the amount of information in the state(s) times the square of the amplitude coupling such states. (square of 'matrix element'; with suitable normalization and noting possibility of transcription in more than one direction)
3. From A-Theorem 1, A-Cor 1.3, and A-Lemma 2.1, w of irad w with longest wavelength possible in S and least amplitude >0 does not specify any direction in, and/or subregion of, S. (note that 'relative phase' is not defined within such w with such amplitude, itself)
4. From Postulates 1 and 2, A-Def 1, and 3, such w does not by spatial function describe representation of any information in S.
5. From A-Theorem 1, A-Def 1 and 3, and 3 and 4, each such w adequately describes representation of only 1 bit of information in S. ($\log_2(\text{card}\{k\}) = 1$)
6. From A-Def 12, and 3 and 4, noting contents specified for S, each w of irad w in S or S as B_0 is a w as in step 3.
7. From A-Theorem 1, A-Cor 1.3, and A-Def 3, 14, and 15, amount of information transcribed by GA1 and/or GS1 of S between irads $w_{z,k}$s with different k values is adequately describable as zero.
8. From Postulates 1–3, A-Cor 1.3, A-Def 10, 12, 14, and 15, and 2, (noting isotropy and homogeneity of 3-space and absence of

'boundary conditions' on U) for one step of GA1 or GS1 of S containing one IR1, amplitude for transcription of information between irads amplitudes of w states in S coupled by GA1 or GS1 is adequately describable as s^2.

9. From A-Theorem 1 and A-Def 12, IR1 can be chosen such that as a B_0 S's IR1 ws would have different k values, and is here if as B_0 so chosen. (as 'source' states)
10. From A-Def 14 and 15, and 7 and 9, (noting initial contents of S) all irads w states in S to which GA1 or GS1 might transcribe information ('target states') from initial irads w states ('source states') are adequately describable by zero amplitude ws ('empty').
11. From 7, 9, and 10, S may contain more than one sets of source and target states, such that the amount of information transcribed between such sets by GA1 and/or GS1 is adequately describable as zero. ('disjoint' sets)
12. From A-Theorem 1, A-Cor 1.3 and 1.4, A-Def 3, 6, 8, 10, and 12–15, and 1, 2, and 5–11, one step of GA1 or GS1 of S with scale s in U transcribes information between representations adequately describable by such states of w in S in an amount adequately describable for each, under GA1 and/or GS1 disjoint, set of source and target states as $1s^2$.
13. From A-Def 15, s may be changed by steps of GS1 (if any).
14. From 12 and 13, s in step 12 is to be taken before each step of GA1 or GS1.
15. From Postulates 1–3, and A-Def 1, there is not any virtual information.
16. From A-Def 2, and 3–15, for $N_1 \geqslant 0$ steps of GA1 and $N_2 \geqslant 0$ steps of GS1, with $\Sigma_{i=1}^{N_1+N_2} s_i^2 < 1$, where s_i is taken before each step, amount of information transcribed thereby among IR1s is adequately describable as zero.
17. From A-Def 1, 2, and 16, and 16, under such process, both amount and representation of S's internal information are adequately describable as unchanged.
18. From 16 and 17, under process as in step 16, for only one IR1 (in S), all laws and constants of physics, including derivation herein from A-Definition 1 through this step, would be adequately describable as adequately describable in S as unchanged. (including c in equations in A-Theorem 1; 'adequately describable as adequately describable' avoids assuming presence of observer)
19. From A-Cor 1.2, and 18, as in A-Corollary 1.2, every observable process in S is adequately describable as a transformation of repre-

sentation of I_1. (events viewed as nodes in conserved ('flow' of) information)

20. From 16–19, for only one IR1 (in S), for N_1 steps of GA1 and N_2 steps of GS1, if $\Sigma_{i=1}^{N_1+N_2} s_i^2 < 1$, where s_i is taken before each step of GA1 or GS1, amount and representation of S's internal and external information are adequately describable as separable.

●

(See also Section 6 question 5 in relation to step 18.)

(w_+ and w_- having different k values is discussed further herein near after A-Theorem 6 and A-Cor 6.1.)

w spectrum in U, and the concept of scale of S in terms of w spectra in U and S, are in some sense present in step 8. One might think of this in terms of isotropy and homogeneity of 3-space and absence of 'boundary conditions' on U, as seeming to allow only a 'linear' w spectrum in U, the only direction defined along which to 'decompose' the 'wave equations' of A-Theorem 1 being the 'direction of propagation' of a 'wave'. (This is discussed more fully hereinafter, in A-Theorems 3 and 4, A-Cor 4.1 and 4.3, and text near after A-Cor 4.3.)

A-Theorem 2 has in this axiomatic form a role similar to PODIC; information conservation is a postulate (1) in this approach, and what A-Theorem 2 provides is a formal statement of adequate description as separable of amount and representation of internal and external information.

For this reason, for convenience of reference,

A-Definition 17. *Call 'adequately describable as separability' (of amount and representation of information) 'adseparability' (adjective form: 'adseparable').*

(Formal treatment of systems containing irad combinations of B_0s and/or w amplitude greater than least amplitude >0 and/or more than one ws is presented in A-Theorems 5 and 6.)

A-Theorem 2 step 18 is explicitly, for one IR1, a derived statement in this axiomatic form of the postulates of Einstein's special theory of relativity (SR), applicable within the domain derived in steps 1–17. Therefore, the formalism of SR may be used in that domain, and provides a formal method of including adseparability in calculations; and, from adseparability (as so expressed), in such domain, λ_0 (A-Def 12) is unique.

For convenience in using adseparability in further consideration of irad *ws* and/or B_0s, some results of calculations relating to B_0's, using computational methods of SR to formally include effects of adseparability, are collected in A-Corollary 2.1:

A-Corollary 2.1. *In observer O_U's universe U, in domain of adseparability, for a B_0 moving with velocity v in U (using nomenclature as in* A-*Def* 12); *observer O′ moving substantially with B_0;* (*a*) *length L parallel to v, L as seen by O_U; where λ_+, λ_-, and λ_b fit into L n_+, n_-, and n_b times, respectively*; *with f, n, and n_0 as defined herein*; *the following are adequate descriptions*:

$$(a)\ f_{\pm} = f_0 \frac{\sqrt{1-v^2/c^2}}{1 \mp v/c} = \frac{f \pm \sqrt{f^2 - 4f_0^2}}{2} = \frac{\sqrt{f_d^2 - 4f_0^2} \pm f_d}{2}$$

where upper and lower signs are taken together

$$(b)\ f = f_+ + f_- = \frac{2f_0}{\sqrt{1-v^2/c^2}}$$

$$(c)\ n_{\pm} = n_0 \frac{\sqrt{1-v^2/c^2}}{1 \mp v/c} = \frac{n \pm \sqrt{n^2 - 4n_0^2}}{2} = \frac{\sqrt{n_b^2 + 4n_0^2} \pm n_b}{2}$$

where upper and lower signs are taken together

$$(d)\ n = n_+ + n_- = \frac{2n_0}{\sqrt{1-v^2/c^2}}$$

$$(e)\ \frac{v}{c} = \frac{f_+ - f_-}{f_+ + f_-} = \frac{f_+ - f_-}{f} = \frac{1}{\sqrt{1 + 4\left(\frac{f_0}{f_d}\right)^2}}$$

$$= \frac{n_+ - n_-}{n_+ + n_-} = \frac{n_+ - n_-}{n} = \frac{1}{\sqrt{1 + 4\left(\frac{n_0}{n_b}\right)^2}}$$

$$(f)\ f_+ f_- = f_0^2$$

(g) $n_+ n_- = n_0^2$

(h) $$\lambda_b = \frac{c}{f_+ - f_-} = \frac{c}{f_d} = \frac{c^2}{(f_+ + f_-)v} = \frac{c^2}{fv} = \frac{c^2\sqrt{1-v^2/c^2}}{2f_0 v}$$

$$= \frac{L}{n_+ - n_-} = \frac{L}{n_b} = \frac{Lc}{(n_+ + n_-)v} = \frac{Lc}{nv} = \frac{Lc\sqrt{1-v^2/c^2}}{2n_0 v}$$

(i) $$n_+ - n_- = n_b = (n_+ + n_-)\frac{v}{c} = n\frac{v}{c} = \frac{2n_0 v}{c\sqrt{1-v^2/c^2}}$$

(j) $$(2f_0)^2 = f^2 - \left(\frac{c}{\lambda_b}\right)^2 = f^2 - f_d^2 = 4f_+ f_-$$

(k) $$(2n_0)^2 = n^2 - n_b^2 = 4n_+ n_-$$

(l) $$\sqrt{1-v^2/c^2} = \frac{2\sqrt{f_+ f_-}}{f_+ + f_-} = \frac{2f_0}{f_+ + f_-} = \frac{2\sqrt{f_+ f_-}}{f} = \frac{2f_0}{f}$$

$$= \frac{2\sqrt{n_+ n_-}}{n_+ + n_-} = \frac{2n_0}{n_+ + n_-} = \frac{2\sqrt{n_+ n_-}}{n} = \frac{2n_0}{n}$$

(m) $$f_{\substack{+\\-\\0\\d}} = \frac{c}{\lambda_{\substack{+\\-\\0\\b}}} = n_{\substack{+\\-\\0\\b}}\frac{c}{L}$$ *where corresponding levels of subscripts are taken together*

(n) $$f = n\frac{c}{L}.$$

1. From A-Def 12, f_+ and f_- appear to O' as f_0.
2. From A-Theorem 1 and adseparability (computationally SR),

$$f_\pm = f_0 \frac{\sqrt{1-v^2/c^2}}{1 \mp v/c}.$$

3. Define f as $f = f_+ + f_-$.

4. From 2 and 3,

$$f = f_+ + f_- = f_0\left(\frac{\sqrt{1-v^2/c^2}}{1-v/c} + \frac{\sqrt{1-v^2/c^2}}{1+v/c}\right)$$

$$= f_0\left(\frac{(1+v/c)\sqrt{1-v^2/c^2} + (1-v/c)\sqrt{1-v^2/c^2}}{1-v^2/c^2}\right)$$

$$= f_0\left(\frac{(1+v/c+1-v/c)\sqrt{1-v^2/c^2}}{1-v^2/c^2}\right)$$

$$= \frac{2f_0}{\sqrt{1-v^2/c^2}}.$$

5. From A-Theorem 1, A-Def 8 and 12, and given definition of n_+ and n_-,

$$n_\pm = \frac{L}{\lambda_\pm} = f_\pm \frac{L}{c}.$$

6. Define n_0 as $n_0 = f_0 L/c$. (a construct, not specified here to be an integer)
7. Substituting 5 and 6 in 2, and cancelling,

$$n_\pm = n_0 \frac{\sqrt{1-v^2/c^2}}{1 \mp v/c}.$$

8. Define n as $n = n_+ + n_-$.
9. Substituting 7 in 8, as in step 4,

$$n = n_+ + n_- = \frac{2n_0}{\sqrt{1-v^2/c^2}}.$$

10. From 2,

$$\frac{f_+ - f_-}{f_+ + f_-} = \frac{f_0 \dfrac{\sqrt{1 - v^2/c^2}}{1 - v/c} - f_0 \dfrac{\sqrt{1 - v^2/c^2}}{1 + v/c}}{f_0 \dfrac{\sqrt{1 - v^2/c^2}}{1 - v/c} + f_0 \dfrac{\sqrt{1 - v^2/c^2}}{1 + v/c}}$$

$$= \frac{\dfrac{1}{1 - v/c} - \dfrac{1}{1 + v/c}}{\dfrac{1}{1 - v/c} + \dfrac{1}{1 + v/c}}$$

$$= \frac{\left(\dfrac{1 + v/c - 1 + v/c}{1 - v^2/c^2} \right)}{\left(\dfrac{1 + v/c + 1 - v/c}{1 - v^2/c^2} \right)}$$

$$= \frac{2v/c}{2}$$

$$= \frac{v}{c}.$$

11. Substituting from 3 in 10,

$$\frac{v}{c} = \frac{f_+ - f_-}{f}.$$

12. Substituting from 5 in 10, and cancelling,

$$\frac{n_+ - n_-}{n_+ + n_-} = \frac{v}{c}.$$

13. Substituting from 8 in 12,

$$\frac{v}{c} = \frac{n_+ - n_-}{n}.$$

14. From 2,

$$f_+ f_- = \left(f_0 \frac{\sqrt{1-v^2/c^2}}{1-v/c} \right)\left(f_0 \frac{\sqrt{1-v^2/c^2}}{1+v/c} \right) = f_0^2 \frac{1-v^2/c^2}{1-v^2/c^2} = f_0^2.$$

15. Substituting from 5 and 6 in 14, and cancelling, $n_+ n_- = n_0^2$.
16. From A-Theorem 1 and A-Def 8 and 12,

$$\lambda_b = \frac{c}{f_+ - f_-} = \frac{c}{f_d}.$$

17. From 10 and 16,

$$\lambda_b = \frac{c^2}{(f_+ + f_-)v}.$$

18. From 3 and 17,

$$\lambda_b = \frac{c^2}{fv}.$$

19. From 4 and 18,

$$\lambda_b = \frac{c^2\sqrt{1-v^2/c^2}}{2f_0 v}.$$

20. From 5 and 16,

$$\lambda_b = \frac{L}{n_+ - n_-}.$$

21. From definition in given,

$$\lambda_b = \frac{L}{n_b}.$$

22. From 12 and 20,

$$\lambda_b = \frac{Lc}{(n_+ + n_-)v}.$$

23. From 8 and 22,

$$\lambda_b = \frac{Lc}{nv}.$$

24. From 9 and 23,

$$\lambda_b = \frac{Lc\sqrt{1 - v^2/c^2}}{2n_0 v}.$$

25. From 20–24, inverting, setting all equal to each other through equality to λ_b^{-1}, eliminating λ_b, and cancelling L,

$$n_+ - n_- = n_b = (n_+ + n_-)\frac{v}{c} = n\frac{v}{c} = \frac{2n_0 v}{c\sqrt{1 - v^2/c^2}}.$$

26. From 3, squaring, and from 16, solving for c/λ_b, squaring, and subtracting,

$$f^2 - \left(\frac{c}{\lambda_b}\right)^2 = (f_+ + f_-)^2 - (f_+ - f_-)^2$$

$$= f_+^2 + 2f_+ f_- + f_-^2 - f_+^2 + 2f_+ f_- - f_-^2$$

$$= 4f_+ f_-.$$

27. From 14, 16, and 26,

$$f^2 - \left(\frac{c}{\lambda_b}\right)^2 = f^2 - f_d^2 = 4f_+ f_- = (2f_0)^2.$$

28. From 8 and 25, as in 26,

$$n^2 - n_b^2 = (n_+ + n_-)^2 - (n_+ - n_-)^2$$

$$= n_+^2 + 2n_+ n_- + n_-^2 - n_+^2 + 2n_+ n_- - n_-^2$$

$$= 4n_+ n_-.$$

29. From 15 and 28,

$$n^2 - n_b^2 = 4n_+ n_- = (2n_0)^2.$$

30. From 3 and 14, solving 14 for $f_\mp = f_0^2/f_\pm$ and substituting in 3, from 3,

$$f = f_+ + f_-;$$

substituting,

$$f=\frac{f_0^2}{f_\pm}+f_\pm;$$

solving for $f_\pm$,

$$f_\pm^2-ff_\pm+f_0^2=0,$$

$$f_\pm=\frac{f\pm\sqrt{f^2-4f_0^2}}{2}.$$

31. From 8 and 15, as in 30,

$$n_\pm=\frac{n\pm\sqrt{n^2-4n_0^2}}{2}.$$

32. From 10,

$$\sqrt{1-v^2/c^2}=\sqrt{1-\frac{(f_+-f_-)^2}{(f_++f_-)^2}}$$

$$=\frac{\left((f_++f_-)^2-(f_+-f_-)^2\right)^{1/2}}{f_++f_-}$$

$$=\frac{(f_+^2+2f_+f_-+f_-^2-f_+^2+2f_+f_--f_-^2)^{1/2}}{f_++f_-}$$

$$=\frac{2\sqrt{f_+f_-}}{f_++f_-}.$$

33. From 3, 5, 14, and 32, substituting and cancelling (solving 14 for $f_0=\sqrt{f_+f_-}$),

$$\sqrt{1-v^2/c^2}=\frac{2\sqrt{f_+f_-}}{f_++f_-}=\frac{2\sqrt{f_+f_-}}{f}=\frac{2f_0}{f_++f_-}=\frac{2f_0}{f}.$$

34. From 5, 6, 8, and 33, variously substituting and cancelling,

$$\sqrt{1-v^2/c^2}=\frac{2\sqrt{n_+n_-}}{n_++n_-}=\frac{2\sqrt{n_+n_-}}{n}=\frac{2n_0}{n_++n_-}=\frac{2n_0}{n}.$$

35. From 15 and 25, solving 15 for n_-, substituting, and solving for n_+,

$$n_b = n_+ - n_-$$

$$n_b = n_+ - \frac{n_0^2}{n_+}$$

$$n_+^2 - n_b n_+ - n_0^2 = 0$$

solving for n_+, and excluding $-\sqrt{\ }$ by $n_+ \geqslant 0$,

$$n_+ = \frac{n_b + \sqrt{n_b^2 + 4n_0^2}}{2}.$$

36. From 15 and 25, solving 15 for n_+, substituting, and solving for n_-,

$$n_b = n_+ - n_-$$

$$n_b = \frac{n_0^2}{n_-} - n_-$$

$$n_-^2 + n_b n_- - n_0^2 = 0$$

solving for n_-, and excluding $-\sqrt{\ }$ by $n_- \geqslant 0$,

$$n_- = \frac{-n_b + \sqrt{n_b^2 + 4n_0^2}}{2}.$$

37. From 35 and 36,

$$n_\pm = \frac{\sqrt{n_b^2 + 4n_0^2} \pm n_b}{2}.$$

38. From 5, 6, 16, 21, and 25,

$$f_{\substack{+\\-\\0\\d}} = \frac{c}{\lambda_{\substack{+\\-\\0\\b}}} = n_{\substack{+\\-\\0\\b}} \frac{c}{L},$$

where corresponding levels of subscript are taken together.

39. From 3, 8, and 38,

$$f = n\frac{c}{L}.$$

40. From 37 and 38, substituting, factoring, and cancelling c/L terms,

$$f_{\pm} = \frac{\sqrt{f_d^2 + 4f_0^2} \pm f_d}{2}.$$

41. Substituting from 37 in 12,

$$\frac{v}{c} = \frac{n_+ - n_-}{n_+ + n_-} = \frac{\dfrac{\sqrt{n_b^2 + 4n_0^2} + n_b}{2} - \dfrac{\sqrt{n_b^2 + 4n_0^2} - n_b}{2}}{\dfrac{\sqrt{n_b^2 + 4n_0^2} + n_b}{2} + \dfrac{\sqrt{n_b^2 + 4n_0^2} - n_b}{2}}$$

$$= \frac{n_b}{\sqrt{n_b^2 + 4n_0^2}}$$

$$= \frac{1}{\sqrt{1 + 4\left(\dfrac{n_0}{n_b}\right)^2}}.$$

42. From 38, 39, and 41, substituting and cancelling,

$$\frac{v}{c} = \frac{1}{\sqrt{1 + 4\left(\dfrac{f_0}{f_d}\right)^2}}.$$

●

(Note that L is a spatial period within which various of these terms are defined, and, if one were to compare this with classical EMT analyses of 'cavities' as if with boundary conditions, L might not correspond to one traverse of such a 'cavity', depending on 'boundary conditions'.)

Next, using relations appearing in A-Corollary 2.1, we examine part of the domain of applicability of adseparability:

A-Corollary 2.2. *In observer O_U's universe U, for any information representation adequately describable by one w with least amplitude >0 and longest wavelength possible in box B_1 in U with allowed velocities as a B_0*

with two such w's in U, or by one B_0 treated as having w's with different k values for purposes of A-*Theorem* 2: *adseparability applies for all possible external states of B_1 or B_0 in U with velocity v less than maximum possible velocity v_m thereof in U*; *and relations appearing in* A-*Corollary* 2.1 *for B_0 are for $v < v_m$ adequate descriptions.*

1. Starting with B_1 with scale $s = s_1$, containing irad one w with longest wavelength possible in B_1 and least amplitude >0, shrink B_1 by steps of GS1 until $s = s_2$.
2. From A-Theorem 1, A-Lemma 2.1, A-Def 15, and 1, any transition of w in description of process as in step 1 would be to w states fitting into B_1 a smaller number of times.
3. From 1 and 2, for irad w with least amplitude >0 and longest wavelength possible in B_1, process as in step 1 does not perform any information transcription describable by transitions among possible w states in B_1.
4. From 1 and 3, amount and representation of information represented by irad such w are adequately describable as not changed by process as in step 1.
5. From A-Def 12, each w in description of B_0 may be regarded as having least amplitude >0 and longest wavelength possible in 'box' B_0 with scale s_0.
6. From 1, 4, and 5, internal information of irad one B_0 is not changed by process as in step 1.
7. From A-Cor 1.3, A-Def 10, and 1, any scales possible for B_1 or B_0 in U can be related by process as in step 1.
8. Starting with irad B_0 substantially at rest in U as seen by O_U, by monotonic colinear steps of GA1 change irad B_0's velocity in a way adequately describable as putting B_0 in any allowed B_0 state with velocity $v < v_m$ in U.
9. From A-Theorem 1, A-Lemma 2.1, and A-Def 14, GA1 is adequately describable as performing more transcription adequately describable by transition among allowed states of w in B_1 or B_0 for larger step size Δv.
10. From A-Cor 2.1(e), A-Def 12–14, and 9, such GA1 would be adequately describable as performing the most such transcription for the 'coarsest' spacing of possible w states in U.
11. From Postulate 3, A-Theorem 1, A-Cor 1.5, and A-Def 11, the 'coarsest' that w states in U could be is for w to have $\lambda_j = \lambda_1 / j$, where j is a finite integer >0.

12. From A-Def 12, and 11, a B_0 defined over such spectrum of possible *ws* has $\lambda_{\pm} = \lambda_1 / n_{\pm}$, where n_+ and n_- are finite integers >0.
13. From A-Def 12, B_0 is a linear combination of two solutions of WE.
14. From A-Theorem 1, and 13, B_0 can be a solution of WE.
15. From A-Theorem 1 (step 15) and A-Def 12 and 16, each of B_0's two *ws* describes representation internally of 1 bit. (also, noting A-Def 12, 'internally' as of B_0 'made of' such two *ws*)
16. From Postulate 2 and A-Def 2, 12, and 16, (as in A-Theorem 1 step 15) for B_0 to describe existence of coding in irad B_0, B_0 must describe representation internally of 1 bit.
17. From A-Theorem 2, A-Def 17, and 15 and 16, in domain of applicability of adseparability, one of said two bits described by B_0's two *ws* as represented internally must be represented internally as said existence of coding bit by said irad B_0.
18. From Postulates 1 and 2, A-Theorem 1, and 15–17, such B_0's two *ws* do not as parts of such B_0 describe information representations having separate existences.
19. From 13, 14, and 16–18, such B_0 must itself be a solution of WE adequately describing coding in an information representation.
20. From A-Theorem 1, A-Cor 1.5, A-Def 11 and 12, and 19, such B_0 has $\lambda_b \leqslant \lambda_1$.
21. From A-Cor 2.1(h, i), A-Def 11 and 12, and 11, 12, 19, and 20, for such spectrum of *w* states, such B_0 has $n_b = n_+ - n_- \geqslant 1$.
22. From A-Cor 2.1(e), A-Def 12 and 13, and 21, for such spectrum of *w* states, least velocity v_1 of irad such B_0 is for $n_+ - n_- = 1$.
23. From A-Theorem 1, A-Cor 1.3 and 1.5, A-Def 11 and 12, and 22, with such spectrum of *w* states, such B_0 of irad such B_0 with v_1 has the same relative spatial distribution of amplitude as that of *w* with longest wavelength λ_1 possible in *U*.
24. From A-Theorem 1 and A-Cor 1.3 and 1.5, *w* with wavelength λ_1 does not describe direction in *U*.
25. From 19 and 21–24, v_1 of irad such B_0 does not have direction in *U*.
26. From A-Theorem 1, A-Cor 1.5 and 2.1(e), A-Def 11–13, and 11 and 12, for such spectrum of *w* states, v of irad such B_0 is largest for $n_- = 1$.
27. From A-Cor 2.1(g), and 26, for such spectrum of *w* states and irad such B_0, for v largest, $n_+ = n_0^2 / n_- = n_0^2$.
28. From 27, for such spectrum of *w* states and such B_0, for $v < v_m$, $n_+ < n_0^2$.
29. From A-Cor 2.1(e, h), A-Def 12 and 13, and 8, 11, 12, 22, 25, 26, and 28, for such spectrum of *w* states and irad such B_0, adequate descrip-

tion of process as in step 8 involves at most

$$n' = n_+ - n_- + 2 - 1 = n_+ - n_- + 1 \leqslant n_+ - 1 + 1 = n_+ < n_0^2$$

steps of GA1. ('2' from possibility of n_+ for v_1 being n_- for desired velocity direction; first '-1' from number of steps between states being one less than number of states; second '-1' from least $n_- = 1$)

30. From A-Cor 2.1, A-Def 11 and 12, and 11, 12, and 22, for such spectrum of w states and irad such B_0, s_0 is adequately describable by $s_0 = (\lambda_0$ as seen by O_U of B_0 with $n_+ - n_- = 1)/\lambda_1$.
31. From A-Theorem 2 and A-Def 12 and 17, in domain of applicability of adseparability, O_U would see irad B_0 as shorter parallel to the direction of its velocity v, and as having the same width perpendicular to the direction of v, compared with what O' moving substantially with irad B_0 would see.
32. From 22, 25, 30, and 31, for such spectrum of w states and irad such B_0, noting that 'velocity' v_1 associated with $n_+ - n_- = 1$ does not have direction, s_0 is adequately describable by $s_0 = \lambda_0/\lambda_1$.
33. From A-Cor 2.1, A-Def 12, and 11, 12, and 32, for such spectrum of w states and irad such B_0, s_0 is adequately describable by $s_0 = \lambda_0/\lambda_1 = 1/n_0$.
34. From A-Theorems 1, and 2 (steps 2–16), A-Lemma 2.1, A-Def 12 and 14, and 15, 16, and 33, as in A-Theorem 2 steps 2–13, for such spectrum of w states and irad such B_0, one step of GA1 would transcribe information among internal representation in irad B_0 in an amount adequately describable by $1s_0^2 = 1/n_0^2$. (first '1' from amount of 'source state' information, in bit(s))
35. From 8, 10, 11, 12, 29, and 34, process as in step 8 would transcribe information among internal representations in irad B_0 at most in an amount adequately describable by $n's_0^2 = n'/n_0^2 < 1$.
36. From Postulates 1–3, A-Def 2, and 35, transcription of information among representations in irad B_0 by process as in step 8 is adequately describable as zero. (there is not any 'virtual' information)
37. From 8 and 36, under process as in step 8 to any possible $v < v_m$ of irad B_0 in U, amount and representation of internal information of irad B_0 in U are adequately describable as not changed.
38. From 5, 8, and 37, amount and representation of information represented by irad one w with least amplitude >0 and longest wavelength possible in B_1 with scale s_0 with allowed velocities as a B_0 with two such ws in U are adequately describable as not changed by process as in step 8 to any such allowed $v < v_m$ thereof in U.

39. From Postulates 1 and 2, and 5–8, 37, and 38, for all possible scales and/or velocities $v < v_m$, amount and representation of internal and external information represented by irad such w in B_1 or irad B_0, in U, are adequately describable as separable.
40. From A-Def 17, and 39, in observer O_U's universe U, adseparability applies to information represented in irad one w with least amplitude >0 and longest wavelength possible in B_1 in U with allowed velocities as of irad B_0 with two such ws in U, or by irad B_0, in any possible external state with $v < v_m$ in U.
41. From A-Cor 2.1, and 40, for irad B_0 in any possible external state with $v < v_m$ in U, relations appearing in A-Corollary 2.1 are adequate descriptions.

●

('external' from choice of different k values; discussed more fully herein near after A-Theorem 2 and hereinafter.)

Therefore, we can use A-Corollary 2.1 in considering applicability of adseparability to systems with larger scale containing more than one B_0 and/or w and/or w amplitudes larger than least amplitude >0, without overreaching the domain of applicability of the computational method (of SR) used therein.

Note that A-Corollary 2.2 steps 13–25 are formal in this axiomatic form, allowing formal exclusion of $v=0$ for all B_0s.

Steps 13–25 might offer some insight into the notion of 'quantum mechanical spreading', and part of a formal basis, in this axiomatic form, for various aspects of 'quantum mechanics'.

3.3.4 *w* Spectrum in *U*

Next, one might consider the spectrum of solutions w of the A-Theorem 1 'wave equations' in universe U without boundary conditions (for definitions of various terms used in A-Theorems 3 and 4, see especially A-Theorem 1, A-Def 12, and A-Cor 2.1):

A-Theorem 3. *In universe U, all possible* $w_{j,z}$ *have respectively wavelengths* $\lambda_j = \lambda_1/j$, *where j is a finite integer* >0.

1. From A-Cor 1.5 and A-Def 11, there is a 'longest' wavelength λ_1 possible for any solution of WE not constant everywhere in U.

2. Write the wavelength λ_j of each $w_{j,z}$ solution of WE as $\lambda_j=\lambda_1/j$, where j is not (yet) specified to be integer.
3. From 1 and 2, $j \geqslant 1$.
4. From A-Cor 1.3 and A-Def 8, noting absence of any boundary conditions due to U, for any two possible wavelengths λ_{j_1} and λ_{j_2}, corresponding possible w_{j_1} and w_{j_2} could be chosen to have opposite directions of propagation.
5. From A-Def 12, A-Cor 2.1(i), and 4, for any two possible wavelengths λ_{j_1} and λ_{j_2} could be considered a B_0 with $n_b=n_+-n_-=|j_1-j_2|$, where n_b, n_+, n_-, j_1, and j_2 are not (yet) specified to be integers.
6. From A-Theorem 1, A-Def 12, and 5, hypothetical B_0 as in step 5 is a solution of WE.
7. From A-Def 12, and 1, 5, and 6, if such hypothetical B_0 is not constant everywhere in U, then $\lambda_b \leqslant \lambda_1$.
8. From A-Cor 2.1(h), A-Def 12, and 7, noting $\lambda_b=\lambda_1/n_b$, if $j_1 \neq j_2$ then $|j_1-j_2|=n_b \geqslant 1$.
9. From 4 and 8, about every possible value of j there is not any other value of j in (in terms of 'real numbers') 'open interval' $j \pm 1$.
10. From A-Theorem 1 (WE) and A-Cor 1.3, (noting absence of boundary conditions due to U) all values of j not specifically excluded are possible.
11. From 9 and 10, all possible values of j have integer spacings.
12. From 3 and 11, all possible values of j are integers.
13. From Postulate 3, j is finite.
14. From 2, 12, and 13, all $w_{j,z}$ possible in U have wavelengths $\lambda_j=\lambda_1/j$, where j is a finite integer >0.

●

3.3.5 Describing Information Representation by *ws*

Having obtained the spectrum of λ_js of WE solutions $w_{j,z}$ in U, one might next consider least amount I_j of information which can be exchanged with irad $w_{j,z}$:

A-Theorem 4. *In observer O_U's universe U, least amount $I_j>0$ which can be exchanged with irad $w_{j,z}$, with j as in A-Theorem 3, is j bit(s).*

1. From A-Theorem 1, A-Cor 1.3, A-Lemma 2.1, A-Theorem 3, and A-Def 8, I_j depends only on λ_j.

2. From A-Cor 1.3 and 2.1, A-Theorem 3, and A-Def 12, for each pair of successive values j and $j+1$ a hypothetical observer O' for which adseparability applies can be considered as if for a hypothetical B_0 ('hB_0') with $n_-=j$, $n_+=j+1$, and $n_b=n_+-n_-=1$. (e.g., if O' has scale smaller than s_0 of hB_0, adseparability would (always) apply to O')
3. From A-Theorem 1, A-Def 12, and 2, hB_0's w_+ and w_- have frequencies as seen by O_U respectively $f_\pm=(c/\lambda_1)n_\pm$.
4. From A-Cor 2.1(f) and 2.2, and 2, solving for f_0, O' moving substantially with hB_0 at velocity v would see hB_0's w_+ and w_- as having the same frequency $f_0=(f_+f_-)^{1/2}$.
5. From A-Theorem 2, A-Cor 2.1 and 2.2, and 3 and 4, O' would see f_0 as associated with wavelength

$$\lambda_0=\frac{c}{f_0}=\frac{c}{(f_+f_-)^{1/2}}=\frac{\lambda_1}{(n_+n_-)^{1/2}}.$$

6. From A-Theorems 1 and 3, A-Def 12, and 1 and 5, O' would see w_+ and w_- as hypothetical waves of equal wavelength each adequately describing representation of the same amount of information as the other.
7. From 4 and 6, O' would see adequate description of representation of a total amount of information equal to the sum of the amounts of information representation of which is adequately describable by each such hypothetical wave separately and the amount of information representation of which is adequately describable by relative phase between such two hypothetical waves of equal frequency (f_0).
8. From Postulates 1 and 2, and 2, 4, and 6, the number of possible relative phases between such two hypothetical waves is the same as seen by O' and by O_U.
9. From A-Theorem 2, A-Cor 2.1(l) and 2.2, and 5, O_U would see O''s said wavelength of said hypothetical waves as

$$\lambda_h=\sqrt{1-v^2/c^2}\,\frac{\lambda_1}{(n_+n_-)^{1/2}}=\frac{2(n_+n_-)^{1/2}}{n_++n_-}\,\frac{\lambda_1}{(n_+n_-)^{1/2}}=\frac{2\lambda_1}{n_++n_-}.$$

(note that hB_0 least $v>0$ would seem able to have direction, unlike least $v>0$ of an existent B_0)
10. From 2 and 9,

$$\lambda_h=\frac{2\lambda_1}{j+j+1}=\frac{\lambda_1}{j+1/2}.$$

11. From A-Theorem 3, and 10, such two hypothetical waves would be seen by O_U as having λ_h half integral in λ_1.
12. From 11, O_U would see such two hypothetical waves as each repeating in two traverses of λ_1.
13. From A-Def 12, and 2, 6, and 12, noting that B_0 *ws* have least possible amplitude >0, O_U would see such two hypothetical waves as having exactly two possible relative phases.
14. From 8 and 13, O' would see such two hypothetical waves as having exactly two possible relative phases.
15. From A-Def 1, and 14, O' would see such two hypothetical waves as adequately describing representation of 1 bit of information by relative phase.
16. From 6, 7, and 15, O' would see said two hypothetical waves as together adequately describing representation of 1 bit more information than twice the amount of information representation of which is adequately describable by one such hypothetical wave.
17. From Postulates 1 and 2, both O_U and O' would see representation of the same total amount of information as being adequately describable by w_+ and w_- or said two hypothetical waves.
18. From 1, 2, 6, 16, and 17, O_U would see, for least amount >0 I_j or I_{j+1} respectively in irad w_j or w_{j+1},

$$|2I_j-(I_j+I_{j+1})|=1=|2I_{j+1}-(I_j+I_{j+1})|.$$

19. From 18,

$$I_{j+1}=\begin{Bmatrix} I_j+1 \\ \text{or} \\ I_j-1 \end{Bmatrix} \text{ bit (s).}$$

20. From A-Theorem 1, A-Cor 1.3 and 1.5, and A-Def 11, w_1 with least possible amplitude >0 does not specify any direction in, and/or subregion of, U.
21. From Postulates 1 and 2, A-Def 1, and 20, such w_1 does not by spatial function describe representation of any information in U.
22. From A-Theorem 1 (step 15), every w_j with amplitude >0 describes representation of at least 1 bit. ('occurrence of coding', or 'existence', bit)
23. From A-Theorem 1, and 21 and 22, $I_1=1$ bit.

24. From 19 and 23, $I_{j+1}=I_j+1$ bits.
25. From 23 and 24, by repetition, $I_j=j$ bit(s).
26. From A-Lemma 2.1, and 25, least amount >0 of information exchangeable with irad w_j is $I_j=j$ bit(s).

●

Intuitively, A-Theorem 4 might be thought of as 'working' by comparison of two equivalent representations, one in which $w_{j,z}$ and $w_{j+1,z'}$ have two different frequencies, and, therefore, do not have a set of possible constant relative phases, and the other in which two (hypothetical) waves are treated as having equal frequency, and, therefore, a set of constant relative phases, the cardinality of which set is calculated.

A-Theorem 4 provides a formal basis for 'adequately', as in A-Definition 2, being sufficient for all observable quantities:

A-Corollary 4.1. *In observer O_U's universe U, 'adequate' description does not lead to any result observably, representably, and/or communicably different from such 'result' as would be obtained from 'exact' description.*

1. From A-Def 2, (an) adequate description has error less than 1 bit of information for description and every observable dependence thereon.
2. From A-Theorems 1, 3, and 4, for all modes by which information may be represented and/or communicated, information is exchangeable with such modes only in amounts of integer numbers of bits. (for $v\approx 0$, representation might be regarded as in some respects (also noting existence bit of coding for local propagative transport) like communication nearly parallel to x_0)
3. From 2, amount of information less than 1 bit cannot be represented and/or communicated in U.
4. From A-Def 5 and 6, and 1 and 3, any 'error' associated with an adequate description cannot be represented, communicated, and/or observed, in U.
5. From 4, in U, (an) 'adequate' description does not lead to any result observably, representably, and/or communicably different from such 'result' as would be obtained from (an) 'exact' description.

●

Thus, there is not any problem of having to use description which to be exact might require an infinite amount of information: 'adequate' appears always good enough.

'$\overset{A}{=}$' does not appear to be transitive.

In view of A-Corollary 4.1, it appears formally consistent to write correspondences with observation using adequate description as the strongest statements ever necessary; and to define 'x', where 'x' is '$=$', '$>$', '$<$', '$\geqslant$', or '$\leqslant$', as the (prior) mathematical 'x' with respect to all calculational manipulations, and as '$\overset{A}{x}$' with respect to satisfaction of such expression in (a) (finite) universe U with discrete quantities.

In deriving adseparability, and algebraic expressions incorporating adseparability appearing in A-Corollary 2.1 and used in A-Theorems 3 and 4, a formalism based on B_0s (A-Def 12) and transitions among allowed B_0 states (A-Def 13) was used. With A-Theorems 3 and 4, a B_0's n_+ and n_- are integers. However, observer O' moving substantially with such B_0 would 'see' $n_0=(n_+n_-)^{1/2}$ (A-Cor 2.1(g), solving for n_0), and would adequately describe n_0 as unchanged for changes $|\Delta n_0|<1$. (see also Section 6 question 49)

Thus, n_+ and n_- can be integers within A-Definitions 12 and 13, and

A-Corollary 4.2. *A-Definitions 12 and 13, A-Theorems 2, 3, and 4, and A-Corollaries 2.1, 2.2, and 4.1, are consistent with respect to properties of B_0's.*

●

Ranges of n_+ and/or n_- consistent with $|\Delta n_0|<1$ might be considered in relation to 'phase space' arguments about 'cross sections', 'rates', and the like. (see also Section 6 question 49) Note that for $n_-\sim 1$ and $n_+\gg 1$, $|\Delta n_+|\sim n_+^{1/2}$, so such ranges do not eliminate the 'fast B_0' 'graininess' considered in Corollaries 8.6–8.16 and Theorem 15 and its corollaries (Section 2).

From A-Theorem 4, noting step 3,

A-Corollary 4.3. *For all w_j frequencies f_j possible in universe U, least amount $I(f_j)>0$ of information exchangeable with irad having frequency f_j is $I(f_j)=(\lambda_1/c)f_j=I_j=j$ bit(s), j finite integer >0.*

●

Note that generalizing this to frequency $f=0$ would yield $I(f=0)=0$, consistent with the notion that w with 'infinite wavelength', which would 'reach outside of' observer O_U's universe U, not describe communication or representation of information, that is, accessibility of information between O_U and any region(s) 'outside' U.

This proportionality of I and f suggests that 'hard' 'boundary conditions' for U might be disposed of in the following general way: In observer O_U's universe U, while conserving information, O_U can exchange finite amounts of information with only representations whose exchange 'lump' sizes are with such (finite) amounts of information exactly commensurable with such sizes used by O_U, and, similarly, (from A-Def 4) all of O_U's 'lump' sizes must be commensurable with each other. Thus, from finite accessibility of information (Postulate 3), and that O_U's universe contains only representations with which O_U can exchange information (A-Def 5 and 6), U contains only representations with commensurable information exchange 'lump' sizes. Frequencies previously assumed in EMT and/or QM for a typical small 'simple' enclosure with 'hard' boundary conditions (e.g., a rectangular box much smaller than U, with 'perfectly conducting' walls) are generally not all commensurable. If one were to generalize the proportionality of $I(f)$ and f to describe information exchange 'lump' sizes for such incommensurable frequencies, it would seem that such 'hard enclosure' spectra would not properly describe the possible spectrum in any one universe.

Thus, formal consideration of the spectrum of frequencies possible in U might be partly expressed by the 'shorthand' characterization of the 'boundary conditions' of U as 'soft', although such formal considerations explicitly avoid using any boundary conditions due to U.

The longest length for λ_1 defined in a flat isotropic homogeneous 3-space of radius R_U is $\sim 2\pi R_U$. This is not the same as an enclosure with 'hard' 'walls', in that the available path length is not required to depend on frequency. For this reason, and with regard to 'whispering mode' properties for $j \gg 1$, it would seem appropriate to choose 'linear size' R_U to be defined in terms of the spectrum of ws possible in U, by

A-*Definition 18.* $R_U=\lambda_1/2\pi$.

Note that every observer might appear to be at the 'center' of such observer's universe; Postulate 3 does not presume that entirely the *same* information be accessible to every observer. (See also Section 6 question 47.)

For systems in which information can be represented in more than one co-possible modes y, where the presence of representation in mode y

would represent at least amount I_y of information, from conservation of information (Postulate 1), information proportional to square of amplitude (A-Lemma 2.1), and commensurability of information representations (A-Theorems 3 and 4),

A-*Corollary* 4.4. *In universe U, amplitudes A_y in adequate description of information representation by different co-possible modes y by which information might be represented, each if present representing at least amount I_y of information, are proportional to $1/I_y^{1/2}$.*

●

Note that co-possibility is a kind of 'coherence' condition. This is like Corollary 6.4 (Section 2), but here obtained by an axiomatic derivation. In view of A-Lemma 2.1 and A-Theorems 3 and 4, the following definition appears useful:

A-*Definition* 19. *For an information representation by which information is representable in integer multiples only of some amount $I>0$ of information, call the amount of information so represented divided by I the 'occupation number' ('l') of such representation.*

From Postulate 3, A-Lemma 2.1, and A-Theorems 3 and 4, with A-Definition 19,

A-*Corollary* 4.5. *Amount I of information in irad set S of w's is adequately describable in terms of occupation number(s) l_j of all $w_{j,z}$ having wavelength $\lambda_j=\lambda_1/j$, by $I=\Sigma_j l_j j$ for all $w_{j,z}$ possible in S, where l_j and j are finite integers, $l_j \geqslant 0, j>0$.*

●

From Postulates 1 and 2, A-Definition 12, and A-Corollary 4.5 (noting information conservation and communicability (e.g., from past to future)):

A-*Corollary* 4.6. *Amount I of information represented by irad set S of noninteracting B_0's is adequately describable in terms of occupation number(s) l_z of B_0's having $\lambda_{\pm,z}=\lambda_1/n_{\pm,z}$ by $I=\Sigma_z l_z(n_{+,z}+n_{-,z})$ all z*

labeling B_0's in S, where l_z, $n_{+,z}$, and $n_{-,z}$ are finite integers, $l_z \geqslant 0$, $n_{\pm,z} > 0$.

●

From A-Theorem 1 and A-Corollary 1.3, the following condition is chosen for A-Corollaries 4.7 and 4.8:

Condition 1. *All $w_{j,z}$ are orthonormal in U, $w_{j_1,z_1} \cdot w_{j_2,z_2} = \delta_{j_1,j_2}\delta_{z_1,z_2}$, where*

$$\delta_{l,n} = \begin{Bmatrix} 1 & \text{if} & l = n \\ 0 & \text{if} & l \neq n \end{Bmatrix},$$

'•' denotes a suitable function space scalar product, '$|x|$' denotes $(x \cdot x)^{1/2}$, index z represents all values of all labels other than j (including k and any designation of direction of propagation), and A-Definition 12 is used for B_0 nomenclature

("Kronecker's δ")

Using such orthonormal set, one can define a 'wave' W:

A-*Corollary 4.7.* C1; *irad set of $w_{j,z}$'s is adequately describable by 'wave'*

$$W = \sum_{j,z} l_{j,z}^{1/2} j^{1/2} w_{j,z}$$

adequately describing representation of amount $I = |W|^2 = \Sigma_{j,z} l_{j,z} j$ bit(s) of information, where z denotes finitely many labels, and $l_{j,z}$ and j are finite integers, $l_{j,z} \geqslant 0, j > 0$.

1. From A-Lemma 2.1, where I_j is least amount >0 of information representable by irad $w_{j,z}$, $A_{j,z}(lI_j) = l^{1/2}A_j(I_j)$, where l is a finite integer $\geqslant 0$.
2. From A-Theorem 4, $I_j = j$ bits.
3. From 1 and 2, amount of information represented in irad $A_{j,z}(lI_j)w_{j,z}$ is lj bits.
4. From A-Lemma 2.1, A-Theorems 3 and 4, and 1 and 3, for information to be proportional to amplitude squared, as in A-Lemma 2.1, $A_{j,z}(lI_j) = l^{1/2}j^{1/2}A_1(1)$, where $A_1(1)$ is least amplitude >0 of $w_{1,z}$ with wave-

length $=\lambda_1$ as in A-Theorem 3, adequately describing representation of 1 bit.

5. From C1 orthonormality of $w_{j,z}$s, and 4, for

$$W=\sum_{j,z} l_{j,z}^{1/2} j^{1/2} w_{j,z}, \qquad W\cdot W=\sum_{j,z} l_{j,z} j.$$

6. From A-Cor 4.5, and 4 and 5, $I=W\cdot W=|W|^2=\sum_{j,z} l_{j,z} j$ bit(s).
7. From A-Theorem 1 and A-Cor 4.5, noting linearity of WE, such W adequately describes all information represented in irad such set of $w_{j,z}$s with respective occupation numbers $l_{j,z}$.
8. From Postulate 3, A-Theorem 3, and 5–7, z denotes finitely many elements (labels), and $l_{j,z}$ and j are finite integers, $l_{j,z}\geqslant 0, j>0$.

●

Note that, from Postulates 1 and 2 and A-Theorem 1, use of w phase in describing representation of information is an alternative to using only occupation numbers and w states. From Postulate 1, the total amount of information would be the same. One can see how such use of phase might appear in this formulation, by noting that phase coherent addition or subtraction of amplitude ΔA_z from w state z, which, to define relative phase, must (already) have $A_z>0$, would change A_z by more than adding or subtracting ΔA_z without relative phase, such larger change in A_z corresponding to exchange, with irad $A_z w_z$, of both amount of information adequately describable by such ΔA_z without phase and amount of information adequately describable in terms of such relative phase for such ΔA_z and A_z. Note that, for $A_z>0$ and $\Delta A_z>0$, for addition or subtraction in phase, $|(A_z\pm\Delta A_z)^2-(A_z^2\pm(\Delta A_z)^2)|>0$, from which it may be seen that the amount of information corresponding to such relative phase of A_z and ΔA_z depends on both A_z and ΔA_z. From this, it appears that the amount of information corresponding to relative phase is not more than that specifiable by comparison of two 'oscillatory' quantities each corresponding to a finite amount of information. For phase coherent addition, writing ϕ for relative phase and I_ϕ for amount of phase information, $I_\phi(A_z,\Delta A_z,\phi)=I(A_z+\Delta A_z)-I(A_z)-I(\Delta A_z)$. Thus, it would seem that, in adequate description of such process of addition, the number of distinguishable relative phases would be finite, with, for a specific form of phase description, card $\{\phi\}=2^{I_\phi}$. (This concept and formal structure might be extended to specifically include (coherent) combination of noncolinear 'waves', and distinguishable relative (3- and/or 4-space) directions in such process(es).)

A-*Corollary 4.8.* C1; *in universe U, irad set of noninteracting B_0's is adequately describable by 'wave function'* (*writing* $(-1)^{1/2}$ *as i*)

$$\psi=\sum_z l_z^{1/2}\left(n_{+_z}^{1/2}+in_{-_z}^{1/2}\right)w_{n_{+_z},z}w_{n_{-_z},z}$$

adequately describing representation of amount $I=|\psi|^2=\Sigma_z l_z(n_{+_z}+n_{-_z})$ *bits of information, where z denotes finitely many elements* (*labels*), *and* l_z, n_{+_z}, *and* n_{-_z} *are finite integers,* $l_z\geqslant 0$, $n_{+_z}>0$, $n_{-_z}>0$.

1. From A-Theorem 3 and A-Def 12, each B_0 is by definition equivalent to two $w_{j,z}$s, each with least amplitude >0, with wavelengths $\lambda_\pm=\lambda_1/n_\pm$.
2. From A-Theorem 4, and 1, each B_0 adequately describes representation of n_++n_- bits in U.
3. From A-Cor 4.7, C1 orthonormality of $w_{j,z}$s, and 2, (noting complex conjugation in scalar product) for l_z noninteracting B_0s, each respectively with n_{+_z} and n_{-_z},

$$\psi=\sum_z l_z^{1/2}\left(n_{+_z}^{1/2}+in_{-_z}^{1/2}\right)w_{n_{+_z},z}w_{n_{-_z},z},$$

$$|\psi|^2=\psi\cdot\psi=\sum_z l_z(n_{+_z}+n_{-_z})=I.$$

4. From Postulates 1 and 2, A-Theorem 1, and 3, (noting linearity of WE) such ψ adequately describes all information represented by irad such set of B_0s with respective occupation numbers l_z.
5. From Postulate 1, A-Theorem 3, A-Def 12, and 3 and 4, z denotes finitely many elements (labels), and l_z, n_{+_z}, and n_{-_z} are finite integers, $l_z\geqslant 0$, $n_{+_z}>0$, $n_{-_z}>0$.

●

With complex normalization $(n_+^{1/2}+in_-^{1/2})$ and orthonormality of $w_{j,z}$s, the norm square of the amplitude 'comes apart' into in effect two sums over n_+s and n_-s, each like sum in A-Corollary 4.7.

Note that, from Postulates 1 and 2, use of B_0 phase in describing information transcription is an alternative to using only occupation numbers and B_0 states, as noted more fully with respect to *ws* following A-Corollary 4.7.

Description of information representation by *ws* is considered further in the next subsection, in relation to adseparability of internal and external information.

3.3.6 Adseparability for Systems

As was used in A-Theorem 1 steps 4 and 5 (for $I=2$) and steps 15 and 16 (for $I=1$), mathematical functions adequately describing a system containing I bits of information may be written as functions of 2^I independent variables. These independent variables do not have to be linear. Thus,

A-Theorem 5. *For internal information (amount I_0) and external information (amount I_1) of irad system S, representing in universe U amount $I=I_0+I_1$ of information, adseparable, S is adequately describable by: one element of a set with cardinality 2^I; a pair of elements one each from two sets with cardinality 2^{I_0} and 2^{I_1} respectively; mathematical functions of 2^{I_j} identical variables, each of which variables is a mathematical function adequately describing representation of amount I_{1-j} of information, where $j=0,1$; two mathematical functions, each of which adequately describes representation of amount I_0 or I_1 of information, respectively; or one mathematical function adequately describing representation of amount I of information.*

1. From A-Def 1, amount of information is $\log_2$ of the cardinality of the set of possibilities from which one is designated.
2. From 1, irad S representing amount I of information is adequately describable by one element of a set with cardinality 2^I.
3. From A-Def 17, in domain of adseparability, I_0 and I_1 have representations each adequately describable independently of the other.
4. From 1 and 3, I_j is adequately describable by one element of a set with cardinality 2^{I_j}, $j=0,1$.
5. From 1, 2, and 4, noting that $2^I=2^{I_0}\,2^{I_1}$, irad S is adequately describable by a pair of elements one each from two sets with cardinality 2^{I_0} and 2^{I_1} respectively.
6. From Postulates 1 and 2, A-Theorem 1, and 3, representation of amount I_j of information is adequately describable by a mathematical function independent of representation of amount I_{1-j} of information, $j=0,1$.
7. From 4 and 6, irad S is adequately describable by designation of one element of a set with cardinality 2^{I_j}, all elements of which are the same

mathematical function adequately describing representation of amount I_{1-j} of information, $j=0,1$.

8. From 7, noting that such designation can be adequately described by a mathematical function defined over such set, irad S is adequately describable by a mathematical function of 2^{I_j} identical variables each of which is a mathematical function adequately describing representation of amount I_{1-j} of information, $j=0,1$.
9. From 3 and 6, irad S is adequately describable by two mathematical functions each of which adequately describes representation of amount I_0 or I_1 of information, respectively.
10. From Postulates 1 and 2, and A-Theorem 1, irad S is adequately describable by a mathematical function adequately describing representation of amount I of information.

●

A-Theorem 5 provides a formal basis for factoring 'wave functions', or otherwise separately describing internal and external information in convenient ways.

In particular,

A-Corollary 5.1. *In domain of applicability of adseparability, irad S, representing internally amount I_0 of information, in U, is adequately describable by a mathematical function of 2^{I_0} identical variables, each of which variables is a ψ as in* A-*Corollary* 4.8.

1. From A-Theorem 5, in domain of applicability of separability, irad S is adequately describable by a mathematical function of 2^{I_0} identical variables each of which is a mathematical function adequately describing representation of amount I_1 of information.
2. From A-Cor 4.8, representation of amount I_1 of external information by B_0 is adequately describable by ψ as in A-Corollary 4.8.
3. From 1 and 2, irad S is adequately describable by a mathematical function of 2^{I_0} identical variables, each of which variables is a ψ as in A-Corollary 4.8.

●

Mathematical descriptions as in A-Corollary 5.1 of relatively simple systems appear particularly convenient for describing transformation of representation of conserved amount of internal information when representations of internal and external information are not adseparable, and for describing appearance of representation of conserved internal information as seen by different observers.

As examples out of sequence and not as steps in formal derivation:

1. irad a stable B_0 is one for which coupling of w_+ and w_- has occurred, for which, as from A-Corollary 2.1 (e) and A-Corollary 2.2 step 22, or in Corollary 14.1, $v \neq 0$, and the 'occurrence of coupling' bit can appear in the use of two ψ ('wave function')s as in A-Corollary 4.8, in this case, for example, in a mathematical function defined over two oppositely directed ψs required to describe as a 'standing wave' an irad B_0 which lasts longer than the time it would take to 'cross' universe U.
2. Description of the 'electromagnetic field' as defined over two components provides a means for explicitly describing the way it appears to observers with different velocities ('electric' and 'magnetic' 'fields').
3. The 3 bits of internal information of irad B_0 as in Theorem 14 and its corollaries can appear in description of such irad by a mathematical function defined over eight ψs, the 'occurrence of coupling' bit appearing in such description as a 'standing wave' of two sets, of four ψs each, propagating in opposite directions; with each such set of four ψs used in describing the four combinations spanned by relative polarization of B_0's two ws (1 bit, statistically 'spin $\frac{1}{2}$') and relative phase of B_0's relatively more and less coupled modes (1 bit, charge and matter sign); which description as four components provides a means for explicitly describing the way such irad B_0 appears to observers with different velocities (covariance of Dirac's equation with 4–component 'spinors', for electron and positron with 'spin $\frac{1}{2}$').
4. When representations of internal and external information are not adseparable but amount of internal information is conserved, as, for example, in 'precession' of 'spin' of an electron, time dependent functions over the subspace of two ψ's spanning 'spin $\frac{1}{2}$' can be used to describe such process (e.g., Dirac's 'spinors' after Foldy and Wouthuysen's transformation).
5. Adseparability also might offer insight into the adequacy of describing diffraction of an atom passing through slits by a 'wave function' having wavelength as if the atom were a single particle, instead of various longer wavelengths of each of its 'component' particles.

Using the spectrum of w states described in A-Theorem 3, properties of B_0s with such spectrum, and description as in A-Corollary 5.1 of external information of irad system S in terms of a B_0 ψ as in A-Corollary 4.8, we can now treat formally adseparability of information represented by S in universe U for states of S in U changed by acceleration and/or scale change:

A-Theorem 6. *In observer O_U's universe U, for irad system S of w's and/or B_0's in 'box' B_1 with scale s and velocity v as seen by O_U, information represented by irad S in B_1 in U is adseparable iff*

$$v < c\frac{1-s^2}{1+s^2}.$$

1. From A-Theorem 5, in domain of adseparability, representation of amount I of information by irad S in U is adequately describable, outside of irad S moving with velocity v in U, as would be described representation of amount I of information by one B_0, 'B_{0S}', with velocity v in U.
2. Suppose irad S contained irad one B_0 in state z_1 in B_1. ('irad $\mathfrak{A}$', 'case $\mathfrak{A}$')
3. From A-Cor 1.3, A-Theorem 2, and 2, starting with velocity $v' \approx 0$, adseparability would apply in case $\mathfrak{A}$ for some number x of unidirectional colinear steps of GA1 to higher velocity, where in at least one case $x > 0$. (e.g., for B_0 coincident with B_1 and $s \ll 1$, at least one such step would be possible)
4. Suppose irad S contained irad l_1 B_0s in state z_1 in B_1. (occupation number l_1; 'irad $\mathfrak{B}$', 'case $\mathfrak{B}$')
5. From A-Theorem 1, A-Cor 1.3, A-Theorem 3, A-Def 12 and 13, and 1, 2, and 4, in domain of adseparability, velocity change for each least step to higher velocity of B_{0S} in U in case $\mathfrak{B}$ would be $1/l_1$ times as large as for each least step to higher velocity of B_{0S} in U in case $\mathfrak{A}$, for each v possible for B_{0S} in U in case $\mathfrak{A}$.
6. From A-Lemma 2.1, and 5, in domain of adseparability, each such step to higher velocity in case $\mathfrak{B}$ would perform on irad S a process adequately describable as performing on each B_0 in state z_1 $1/l_1^2$ times as much transition as in case $\mathfrak{A}$.
7. From 4 and 6, in domain of adseparability, each such step to higher velocity in case $\mathfrak{B}$ would perform on irad $\mathfrak{B}$ a process adequately describable as performing $l_1/l_1^2 = 1/l_1$ times as much transition of B_0 among states in B_1 as would such step to higher velocity in case $\mathfrak{A}$.
8. From 5 and 7, in domain of adseparability, so stepping from $v' \approx 0$ to $v' \approx v$ would perform on irad $\mathfrak{A}$ or irad $\mathfrak{B}$ a process adequately describable as performing the same amount of transition of B_0 in B_1 in both cases $\mathfrak{A}$ and $\mathfrak{B}$.

9. From 8, applicability of adseparability to irad S in B_1 in processes involving change by unidirectional colinear least steps of velocity of B_1 in U is independent of l_1.
10. Suppose irad S contained irad one w in state z_2 in B_1. ('irad $\mathcal{C}$', 'case $\mathcal{C}$')
11. From A-Theorem 2, and 10, starting with $v' \approx 0$, adseparability applies in case $\mathcal{C}$ for some number y of steps of GS1 for which in at least one case $y > 0$. (e.g., w with longest wavelength possible in B_1 and least amplitude > 0)
12. Suppose irad S contained l_2 ws in state z_2 in B_1. (occupation number l_2; 'irad $\mathcal{D}$', 'case $\mathcal{D}$')
13. From A-Theorems 3, 4, and 5, A-Def 12 and 15, and 1, 10, and 12, in domain of adseparability, least change of scale of B_1 in U observable by O_U outside B_1 in case $\mathcal{D}$ would be $1/l_2$ times as large as in case $\mathcal{C}$.
14. From A-Lemma 2.1, and 13, in domain of adseparability, each such least scale change in case $\mathcal{D}$ would perform on irad $\mathcal{D}$ a process adequately describable as performing on each w in state z_2 $1/l_2^2$ times as much transition as in case $\mathcal{C}$.
15. From 12 and 14, in domain of adseparability, each such least change of scale in case $\mathcal{D}$ would perform on irad $\mathcal{D}$ a process adequately describable as performing $l_2/l_2^2 = 1/l_2$ times as much transition of w among states in B_1 as would such least scale change in case $\mathcal{C}$.
16. From 13 and 15, in domain of adseparability, changing scale of irad S in B_1 in U from s_1 to s_1', by sequence of fewest sufficient for scale change from s_1 to s_1' of such least scale changes, would perform on irad $\mathcal{C}$ or irad $\mathcal{D}$ a process adequately describable as performing the same amount of transition of w in B_1 in both cases $\mathcal{C}$ and $\mathcal{D}$.
17. From 16, applicability of adseparability to information represented by irad S in B_1 in processes involving sequences of fewest such least scale change(s) of B_1 in U is independent of l_2.
18. From A-Theorem 1, for amount and representation of information represented by irad S in B_1 to be independent of amount and representation of information represented in U outside B_1, there must not be any communication of information between information representation(s) inside B_1 and outside B_1. (this appears as a concept of 'contained')
19. From A-Theorem 1, A-Cor 1.3, A-Theorems 3 and 4, A-Cor 4.3, and A-Def 2 and 5, if adequate description of a process involves a w with frequency > 0 and amplitude > 0 in two 3-space regions, such process involves communication of information between such regions.
20. From A-Def 17, and 19, information represented by irad S in B_1 in U

is adseparable iff irad S is adequately describable by a w or ws not any of which has both frequency >0 and amplitude observably >0 both inside and outside B_1.

21. From A-Theorem 1, A-Cor 1.3, A-Def 12, and 20, information represented by irad S in B_1 is adseparable iff irad S is adequately describable by ws (solutions of A-Theorem 1 'wave equations') with zero amplitude at B_1's 'walls'. (' ' on 'walls', because not assumed to be 'hard boundary conditions')
22. From A-Theorems 3 and 4, A-Cor 4.3, A-Def 6, and 21, to be in U, all such ws possible in B_1 must also be possible in U.
23. From A-Theorem 3, all $w_{j,z}$ possible in U have wavelength $\lambda_j = \lambda_1/j$.
24. From A-Theorem 1, A-Cor 1.3, A-Theorems 2, 3, and 4, A-Cor 4.8, and 22 and 23, all $w_{j',z'}$ possible in B_1 with zero amplitude at B_1's 'walls' have wavelength $\lambda_{j'} = \lambda_{B_1}/j'$, where λ_{B_1} is the longest w wavelength possible in B_1.
25. From 21 and 24, information represented by irad S in B_1 is adseparable iff irad S is adequately describable by $w_{j',z'}$s as in step 24.
26. From A-Theorems 3 and 4, A-Cor 4.3, and 24, all such $w_{j',z'}$s adequately describe representation of amounts of information adequately describable as integer multiples of least amount I_{B_1} of information representable by irad $w_{j',z'}$ in B_1.
27. From A-Theorems 1, 3, and 4, and 24, least amount of information representable by irad such $w_{j',z'}$ is that represented by irad one w with least amplitude >0 and longest wavelength possible in B_1.
28. From A-Theorem 5, A-Def 12, and 1, in domain of adseparability, choice of representation for information represented by irad S in B_1 is independent of representation of information outside B_1.
29. From A-Theorem 1, and 26–28, irad S can be chosen adequately describable by occupation number l in longest wavelength w(s) possible in B_1.
30. From 9, 17, 28, and 29, the domain in which adseparability applies to information represented by irad S in B_1 in U is the same as the domain in which adseparability applies to information represented by irad two ws in B_1 with different k values, least amplitude >0, and longest wavelength possible in B_1.
31. From A-Cor 2.2, A-Def 12, and 30, adseparability applies to irad such ws in B_1 for all possible velocities v less than the maximum possible velocity v_m of B_1 containing only such two ws in U.
32. From A-Def 12, and 1, 29, 30, and 31, such velocities are equal to the velocities possible for irad B_0 in U representing the same amount of information as represented by irad such ws in B_1 in U.

33. From A-Theorems 3 and 4, longest w wavelength possible in B_1 with scale s is $s\lambda_1$.
34. From A-Cor 2.1(d) and 2.2, A-Theorem 4, A-Cor 4.6, A-Def 12, and 30, (as in A-Cor 2.2 steps 32 and 33) irad B_0, representing for least v possible in U the same amount of information as represented by irad such ws in B_1 in U, would have n_0 adequately describable by $n_0 = 1/s$.
35. From A-Cor 2.2, and 31, 32, and 34, adseparability applies to irad S in B_1 in U for v less than maximum velocity v_m possible for irad B_0 with $n_0 \stackrel{\text{A}}{=} 1/s$ in U.
36. From A-Cor 2.1(e), A-Theorems 3 and 4, and A-Def 12, for given n_0, irad B_0 maximum velocity is for $n_- = 1$.
37. From A-Cor 2.1(g), solving for n_+, for $n_- = 1$, $n_+ \stackrel{\text{A}}{=} n_0^2$.
38. From A-Cor 2.1(e), and 36 and 37, for given n_0, irad B_0 maximum velocity is

$$\frac{v}{c} \stackrel{\text{A}}{=} \frac{n_0^2 - 1}{n_0^2 + 1}.$$

39. From 35 and 38, adseparability applies to information represented by irad S in B_1 in U for

$$v < c\frac{\dfrac{1}{s^2} - 1}{\dfrac{1}{s^2} + 1} = c\frac{1 - s^2}{1 + s^2}.$$

40. From Postulates 1 and 2, A-Theorem 1, A-Theorem 2 step 2, and 1–39, for $v \geqslant c(1 - s^2)/(1 + s^2)$, information is transcribable in amount adequately describable as >0 among representation(s) in S in a way adequately describable by transitions among states of components of S in B_1.
41. From 40, amount and representation of internal and external information of irad S in B_1 in U are not adequately describable as separable for $v \geqslant c(1 - s^2)/(1 + s^2)$.
42. From A-Def 17, and 41, adseparability does not apply to irad S in B_1 in U for $v \geqslant c(1 - s^2)/(1 + s^2)$.
43. From 39 and 42, adseparability applies to information represented by

irad S in B_1 moving with velocity v in U iff

$$v < c\frac{1-s^2}{1+s^2}.$$

●

From A-Theorems 2, 5, and 6, choice as in step 30 of w_+ and w_- as having different k values would seem not to restrict this result. (See also discussion herein near after A-Cor 6.1.)

The notion here of 'box' might be taken as not limited to 'hard walled enclosure', but rather as denoting inclusion in a system or description thereof.

As discussed hereinbefore in relation to A-Theorem 2, where adseparability applies, all laws and constants of physics appear independent of the velocity of the observer (this subsumes Einstein's special theory of relativity in such domain of applicability).

From A-Theorem 6, for $v \ll c$, adseparability breaks down with size at $s \approx 1$. (scale of $U = 1$)

For $s \ll 1$, neglecting terms of order s^4, adseparability breaks down with velocity at $v \approx c(1-2s^2)$.

Substituting $n_0 = 1/s$ (from step 34) and $n = n_0^2 + 1$ (from step 37) in A-Corollary 2.1(1), for such v,

$$\sqrt{1-v^2/c^2} = \frac{2\dfrac{1}{s}}{\dfrac{1}{s^2}+1} = \frac{2s}{1+s^2};$$

or, equivalently, from A-Theorem 6, solving for v/c, and substituting, adseparability breaks down with velocity for

$$\sqrt{1-v^2/c^2} = \left(1-\left(\frac{1-s^2}{1+s^2}\right)^2\right)^{1/2} = \frac{\left((1+s^2)^2-(1-s^2)^2\right)^{1/2}}{1+s^2}$$

$$= \frac{(1+2s^2+s^4-1+2s^2-s^4)^{1/2}}{1+s^2}$$

$$= \frac{(4s^2)^{1/2}}{1+s^2}$$

$$= \frac{2s}{1+s^2}.$$

For $s \ll 1$, neglecting terms of order s^2 in comparison to 1, $\sqrt{1-v^2/c^2} \approx 2s$. From A-Definition 10, calling B_1's linear size R_1, $s = R_1/R_U$; neglecting terms of order s^2 in comparison to 1, adseparability would break down at about that velocity at which an observer moving with B_1 would see U as having linear size parallel to $v \approx 2sR_U = 2R_1$, about twice as long as B_1.

This size and velocity relation for breakdown with velocity of adseparability, and with it special relativity, agrees with that obtained in Corollary 8.16 (Section 2) by considering at what velocity an observer would begin to be able to see graininess in behavior of irad substantially co-moving B_0s. In this axiomatic derivation, it is shown that such *observer's internal history would not necessarily contain any information that such observer was moving* until such breakdown velocity. The derivation of Corollary 8.16 is based on considering at what velocity an observer would begin to be able to *by current observation* observe effects providing to such observer information about such observer's velocity. That these two different approaches give consistent result(s) appears to illustrate conformity of these derivations with the concept that it be sufficient for an observer's physics to include only information accessible to such observer. (This comparison is discussed in more detail in Comment 4.1.)

In relation to amount of information accessible to an observer in an adseparable system,

A-Corollary 6.1. *In observer O_U's universe U, in domain of adseparability, for irad system S including observer O_S, irad S representing internally information I_S; one bit of I_S as measured by O_U coding for existence of irad S in U is not accessible to O_S.*

1. For O_S to be in irad S, irad S must exist.
2. To O_U, for irad S in U, irad S can exist or not.
3. From A-Theorem 6, and 1 and 2, for I_S adseparable, nonexistence of irad S is not a possibility observable by O_S in irad S.
4. From A-Def 1 and 16, and 2 and 3, for I_S adseparable in U, the 1 bit of I_S as measured by O_U coding existence of irad S is not accessible to O_S. $(\log_2 1 = 0)$

●

With A-Corollary 6.1, one might think of choice of w_+ and w_- with different k values, as in A-Theorem 2 and A-Theorem 6 step 30, as conforming to availability inside adseparable irad S of 1 bit less of irad S's internal information than available outside irad S.

3.3.7 Coding

For represented information to be accessible, coding of such information must be 'decodable', for which decoding a 'code book', itself information, must be suitably represented. As in Section 2 Definition 2,

A-*Definition 20.* *Two information representations which can together decode each other so that all information in both is available for representation in any other way are said to be 'complementary', and the codes therein are said to be 'complementary'.*

In transcription of information into or from representation by a pair of complementary information representations without using any other 'code book', there is not any information represented by distinction of one such representation as being a 'code book'. From Postulate 1, information is conserved in such transcription. From Postulate 2, for all information so represented, such transcription can occur. Therefore, by symmetry,

A-*Theorem* 7. *Complementary information representations represent equal amounts of information.*

●

From Postulates 1, 2, and 3, and A-Theorems 3, 4, and 7, noting (as in A-Cor 4.1) that only finite integer numbers of bits of information are possible in U,

A-*Corollary 7.1.* *All information represented by irad system S can be coded in complementary representations iff amount in bits thereof is even.*

●

From Postulates 1, 2, and 3, A-Definition 20, and A-Corollary 7.1,

A-*Corollary 7.2.* *All information represented by irad system S can be coded fully transcribable to be represented in some other way without any other*

information (without other 'code book') iff amount thereof is an even number of bits.

●

Following reasoning as in A-Corollaries 6.1 and 7.2, from A-Corollaries 6.1 and 7.2,

A-*Corollary 7.3.* *In observer O_U's universe U, in domain of adseparability, irad system S, representing internally information I_S: all of that part of I_S accessible in S or if S is required to exist, can be coded completely transformable by transcription alone (without other 'code book') thereof into different coding iff amount of I_S as measurable by O_U is an odd number of bits.*

●

This 'oddness' or 'evenness' might offer some insight into 'parity' 'conservation' and/or 'nonconservation' for 'particles', systems, and/or a universe.

3.3.8 Quantized Electromagnetic Theory and Relativistic Quantum Mechanics

We can choose a set of constructs, in terms of which various results obtained so far in this axiomatic derivation may be expressed in forms in which they may be used as a basis for the nonaxiomatic derivation presented in Section·2 hereinbefore, to facilitate completing the test of sufficiency of the postulates used. Various of these constructs and forms appear as formally derived results for which I was not previously aware of a derivational basis.

A-*Definition 21.* *Where I is in bits:*

(*a*) (*Noting* A-*Cor* 2.1, A-*Theorem* 4, *and* A-*Cor* 4.3) *call* $E=(\hbar c/R_U)I$ *'energy'*;

(*b*) (*Noting* A-*Cor* 2.1 (*h*)) *call* $m=(\hbar/R_U c)I$ *'mass'*;

(*c*) (*Noting* A-*Cor* 2.1 (*j and k*)) *call* $p=(\hbar/R_U c)vI$, $v \leqslant c$,*'momentum'*.

Note that Planck's constant h (in $\hbar = h/2\pi$) appears solely as a calibration factor, and not in the conceptual basis of these definitions.

For convenience, A-Definition 21 may be written in terms of the 'energy' E_1 corresponding to representation in U of one bit of information:

A-Definition 22. *Where I is in bits, let $E_1 = \hbar c / R_U$;*

call (*a*) $E = E_1 I$ *'energy'*;
(*b*) $m = (E_1/c^2)I$ *'mass'*;
(*c*) $p = (E_1/c^2)vI$, $v \leqslant c$, *'momentum'*.

In domain of adseparability, one can obtain from A-Corollary 2.1 and A-Definition 21 or 22 (by inspection and combination) expressions for I, v, E, m, p, and $\lambda = \lambda_b$, respectively $I = n_+ + n_-$, $v = c(n_+ - n_-)/(n_+ + n_-)$, $E = E_1(n_+ + n_-)$, $m = (E_1/c^2)(n_+ + n_-)$, $p = (E_1/c)(n_+ - n_-)$, and $\lambda = 2\pi R_U/(n_+ - n_-)$; and, noting A-Corollary 2.1 (d and g), for rest frame energy E_0 and mass m_0, respectively, $E_0 = E_1 2(n_+ n_-)^{1/2}$ and $m_0 = (E_1/c^2)2(n_+ n_-)^{1/2}$.

If one were to write a 'wave function' or set of 'wave functions' ψ for a B_0, for example, as in A-Corollary 4.8, A-Theorem 5, and A-Corollary 5.1, and to define operators $\mathbf{n}_+$ and $\mathbf{n}_-$ such that $\mathbf{n}_\pm \psi = n_\pm \psi$, then, substituting such operators, these expressions would serve as definitions for operators, respectively, for I, v, E, m, p, λ, E_0, and m_0. Note that relations among representations of the two internal bits of the two w's of a B_0 as in Theorem 14 and its corollaries yield a 4-component ψ, as discussed also following A-Corollary 5.1.

Note that these derived relations have as consequences

$$E_0^2 = E^2 - (pc)^2,\ \ E = mc^2,\ \ p = mv,\ \text{and}\ \lambda = h/p.$$

With these correspondences, formal structure in this Section 3, especially A-Theorem 1, A-Lemma 2.1, and A-Corollaries 4.3 and 4.7, appears to subsume a consistent *quantized electromagnetic theory,* with A-Theorem 1 'wave equations' as *source-free Maxwell's equations,* A-Lemma 2.1 as *energy proportional to field amplitude squared,* and A-Corollary 4.3 as $E = hf$; and various features of mechanics of B_0s and systems, especially as presented in A-Corollaries 2.1 and 4.8, A-Theorem 5, A-Corollary 5.1, and A-Theorem 6, with use of squared amplitude from A-Lemma 2.1, appear to subsume a consistent *relativistic quantum mechanics,* including as special cases *nonrelativistic quantum mechanics, Einstein's special relativistic classical mechanics,* and *Newton's (nonrelativistic) (classical) mechanics.* Information conservation appears sufficient for unitarity of the time development operator of the resulting forms, with Hermiticity of the resulting Hamiltonian form; and A-Corollary 4.8, A-Theorem 5, A-Corollary 5.1, and A-Theorem 6 appear to include 'wave packets' defined over B_0s, 'internal variables' of 'particles' and systems, and the like.

3.3.9 Sufficiency of Postulates

With the constructs introduced in A-Definition 21 (or 22) (and using $R_B = R_U$), the axiomatic structure presented hereinbefore in this Section 3 appears to provide a basis for all concepts, including, without limitation, those for space, time, mass, energy, momentum, quantized electromagnetic theory, and quantum mechanics, used in Section 2 hereinbefore. The development appearing in Section 2 may therefore be taken in sequence, following this part of Section 3, using this axiomatic treatment to replace Theorem 1 through Corollary 8.4 inclusive and Principles 1 and 2, continuing with the examples following PODIC (p. 68) (see also map in Appendix 3).

Thus,

A-Theorem 8. *Postulates* 1, 2, *and* 3 *are sufficient for all conceptual and formal structure presented hereinbefore in Section* 2 *and in this Section* 3, *with reference in Section* 2 *to*

(*a*) '*electromagnetic theory*' ('*EMT*') '*nonrelativistic quantum mechanics*' ('*NRQM*'), *and/or* '*quantum mechanics*' ('*QM*'), *absent more specificity, deemed reference to as sufficient* ('*drtas*') *all conceptual and formal structure hereinbefore in this Section* 3;

(*b*) '*photon*' *drtas* '(*irad*) *w with least amplitude* > 0';

(*c*) *Definition drtas* A-*Definition*

1	*1*
2	*20*
3, 4	*16*
5, 6	*12*
7	*12, and 21 and/or 22*
8	*21 and/or 22*
9	*12, 16, and 21 and/or 22*
10	*14} and generalization to systems*
11	*15} as in* A-*Theorem 6*;

(*d*) *Lemma* ('*L*'), *Theorem* ('*T*'), *Corollary* ('*C*') *drtas Postulate* ('*P*'), A-*Lemma* ('A-*L*'), A-*Theorem* ('A-*T*'), A-*Corollary* ('A-*C*'), A-*Definition* ('A-*D*')

Section 2	*Section 3*
	A-(*except for P*)
T 1, C 1.1	*P 1*
C 1.2	*T 7*
T 2	*C 4.5*
T 3, C 3.1	*P 1, P 2, T 4, and D 21 and/or 22*
L 4.1-C 4.5, T 4′	*C 2.1, and D 21 and/or 22*
T 5-C 5.3	*P 1, P 2, and D 21 and/or 22*
T 6-C 6.3	*D 21 and/or 22*
C 6.4	*C 4.4*
T 7, C 7.1	*as if reference to EMT and QM*
L 8.1-C 8.4	*P 1 and T 2, noting consideration of 'effects' of 'enclosure'*
C 8.5	*C 2.2*
C 8.16	*itself, and/or P 1 and T 6, noting conceptual and derivational distinction*;

(*e*) 'PODIC' *drtas* '*Postulate* 1 *and adseparability* (A-*Theorem* 2, A-*Cor 2.2*, A-*Theorem 6*)'.

●

3.4 AXIOMATIC FORM

With sufficiency of the stated set of postulates, axiomatic treatment hereinbefore in this Section 3 is hereby merged with treatment in Section 2, following definition of PODIC, at the location noted in Section 2, obtaining thereby an axiomatic form for the theory presented herein.

This axiomatic form uses three postulates,

1. *Information is conserved,*
2. *Information is communicable,*
3. *Information is finitely accessible,*

and three numerical quantities,

I_U = *amount in bits of information in an observer's universe U,*
R_U = *linear size of U,*
c = *linear coefficient of local propagative communication.*

Of these three numbers, I_U seems more basic than R_U and c: from A-Theorems 3 and 4 and A-Corollary 4.1, it might seem that the bit is the natural unit of amount of information. R_U appears as calibration for 3-space length measurement units, and, noting from A-Corollary 1.3 that an observer never sees time interchanged with any space coordinate, c appears as calibration for time interval units in terms of space length units: both R_U and c may be set equal to any finite number >0, for example, 1.

For convenience in using instruments calibrated in energy units, one can add a fourth number, E_1 (A-Def 22), the amount of 'energy' per bit of information represented in U; like R_U and c, E_1 may be set equal to any finite number >0, for example, 1.

From A-Theorem 5, with I_U, the maximum number of possible *alternatives* of which only one is observably present, noting 1 bit for existence of observer and 1 bit for existence of the alternative present, might be $\sim 2^{I_U-2}$. Noting requirement of 1 bit for existence of each, the maximum number of *elements* observable together might be $\sim I_U - 1$.

Noting A-Definition 6 and A-Corollary 6.1, in relation to I_U and application of these concepts in adseparable systems in U,

Hereinafter in this axiomatic form, for other than postulate 3, *'finite' may occasionally be taken as meaning 'consistent with amount of accessible information'.*

This particularizes also range of j in A-Theorems 3 and 4. (In some sense, 'infinitely many' might be thought of as 'more than (an) observer (in (an) adseparable system) can count'.) (See also Section 6 question 5.)

From Table 2 (Section 2), we can take numerical values for I_U, R_U, and E_1, and, with c from Cohen and Taylor (ref. 4), obtain

$$I_U \approx 3.64344 \times 10^{122} \text{ bits,}$$

$$R_U \approx 3.084568 \times 10^{26} \text{ meters,}$$

$$c \approx 299792458 \text{ meters/second,}$$

$$E_1 \approx 1.024966 \times 10^{-52} \text{ joule } (\approx 6.39728 \times 10^{-34} \text{ electron volt).}$$

Please see discussion following Corollaries 14.15 and 14.16, and Table 2, in relation to propagation of 'error' in these approximate values for I_U, R_U, and E_1.

Various other numerical quantities may be expressed in terms of these quantities, such as

$$\hbar \approx \frac{E_1 R_U}{c} \qquad \text{(A-Def 22)},$$

$$M_U \approx \frac{I_U E_1}{c^2} \qquad \text{(A-Def 22)},$$

$$H \approx \frac{c}{R_U} \qquad \text{(Cor 10.2–10.4)},$$

$$G \approx \frac{R_U c^4}{I_U E_1} \qquad \text{(Theorem 13)}.$$

Alternatively, one could use the set of numerical quantities for $G, H, c,$ and h, of which H appears ('locally') at the time of this writing to be the least accurately measured; or, as is done in the next subsection hereof, G, c, $\hbar$, and m_e. (See also Section 6 questions 47, 52, and 66.)

Thus, concepts in this physical picture appear to be expressed in this axiomatic form formally and without any arbitrary parameter.

3.5 PARTIAL TEST OF AXIOMATIC FORM

By this stage of the development, we have at our disposal an axiomatic form of the theory presented herein, and values for all numerical constants appearing therein are approximately known: for G, c, and $\hbar$, from observation by others; and for R_U, from Corollaries 10.3 and 10.4 ($R_U \approx c/H$), and 14.15 and 14.16, obtaining an approximate value for H from G, c, $\hbar$, and m_e, as was done following Corollary 14.15.

One of the aspects of this physical picture which appear different from prior physical theory of which I am at the time of this writing aware is that, in this picture, it appears that structure and properties of an entity can depend on representing internally a few bits of information, even though the amount of 'energy' which would be required to represent that amount of information directly by photons might be substantially less than the 'mass energy' of the entity. For example, the structure and properties derived hereinbefore for an electron or positron, with 'rest mass energy' $\approx$511003.4 eV, were obtained by considering representation internally of

3 bits of information, which amount of information would, if represented directly by independent photons, require total 'energy' of $\approx 1.919\times10^{-33}$ eV, or only about 3.756×10^{-39} of the 'mass energy' of the electron.

Also, finite 'size' of accessible universe U appears directly and quantitatively in such calculations, both as a limit of integration in calculation as in Corollaries 14.3 and 14.8 (F), and through E_1 in consideration of the lowest few w states in U. (see also Section 6 question 52)

Although one might argue that prior theory has treated small multiplicities for such descriptors as 'spin', 'charge' sign, and other 'internal quantum numbers', I am not aware of any prior theory appearing to me to have treated the possibility that the amount of energy which can form a particle might in some sense be determined by representation of information in an amount corresponding to an amount of 'energy' a factor of roughly 3.756×10^{-39} smaller.

For this reason, such (approximate) calculations of amounts of internal information and masses of particles seem, near and before this writing, to be addressed more effectively by the physical picture presented herein and formalism based on it than by any prior theory of which I am at the time of this writing aware. For this reason, one of the remarkable features of this theory might be partially tested by calculating systematically the amounts of internal information of various particles and approximate values for the masses of several.

Because the following systematic calculation uses a chain of reasoning in which observations are used in various steps, it does not have the formal purity of a theorem, and I have labeled it 'Chain 17', including 14 'Links', 17.1–17.14, following in numerical sequence after Theorem 16 of the axiomatic form produced by merging Sections 2 and 3; for the same reason, the 'therefore' of formal logic, often expressed herein by 'From (list), (statement)' in various derivations, is occasionally replaced in Chain 17 and its links by what might be expressed by 'it seems to me reasonably likely that', written as 'likely'.

Before reading the intervening material, you might wish to turn to Table 3, p. 268, and look at some of the results collected there. (report(s) of observation not otherwise specified in Chain 17 through Link 17.14 refer to reference 8 compilations and various references listed therein)

Chain 17

Conditions for Chain 17. *In observer O_U's universe U with linear size R_U, containing total amount I_U in bits of information; for $x=\gamma$, ν_e, $\bar{\nu}_e$, e^+, e^-, ν_μ, $\bar{\nu}_\mu$, μ_L^+, μ_L^-, μ_H^+, μ_H^-, π^0, π_L^+, π_L^-, π_H^+, π_H^-; with I_x = amount in bits of* *information represented internally by x,* I'_x *= amount in bits of information represented by x in U,* I''_x *= amount in bits of information represented by x in U and* m_x *= 'mass' of x both as would be observed by an observer moving substantially with x,* E_x *= 'energy' (including any 'mass') of x as would be observed by* O_U*; let 'b' designate coding for 1 bit of information describing 'existence of coding' coded consistent with coding 1 bit of information by relative polarization of a* B_0*'s associated two photons; let 'd' designate code for contribution to Theorem 13 interaction among, associated with coding for 1 bit of information by relative phase of, an entity's associated photons; let* τ_x *denote time for 1/2 of an assembly of substantially isolated x's to decay.*

Link 17.1. $E_1 \approx \hbar H \approx \hbar c/R_U$*; there are exactly two photon states in U with* $E_1 = E_1$*, exactly six with* $E_2 = 2E_1$*, and at least six with* $E_3 = 3E_1$*.*

1. From A-Def 22, $E_1 = \hbar c/R_U$.
2. From Cor 10.3, $H \approx c/R_U$.
3. From 1 and 2,

$$E = \frac{\hbar c}{R_U} \approx \hbar H.$$

4. From A-Theorems 1, 3, and 4, and 1, there are exactly two photon states in U with $E_1 = E_1$. (first 'E_1' designates 'energy' of 'state'; corresponding to two 'polarizations' or 'helicities', $k = 0, 1$, of (longest wavelength) $w_{1,k}$ in U)
5. From A-Theorem 1, A-Cor 1.3, A-Theorems 3 and 4, and A-Def 8, noting separability of 'wave equations' in A-Theorem 1 in 3-space of A-Corollary 1.3 and A-Definition 8, there are exactly six photon states in U with $E_2 = 2E_1$. (corresponding to $k = 0, 1$, for $w_{2,k}$ in each of three mutually orthogonal directions ('separability' of WE in 3-space))
6. From A-Theorem 1, A-Cor 1.3, A-Theorems 3 and 4, and A-Def 8, as in step 5, there are at least six photon states in U with $E_3 = 3E_1$.

●

Step 4 here is like step 24 of A-Corollary 2.2, in that w with longest wavelength possible in U and least amplitude >0 does not by its spatial distribution define direction in U. Although appearing as a 'link', Link 17.1 appears formal.

Note that, with Corollary 14.2, phase coherent addition of one more photon to each of the two E_1 states as in step 4 would result in total energy $4E_1$ in each such state, so so adding the second such photon would require $3E_1$ of 'energy', and such use of the two E_1 w states might be counted in the set of $E_3 = 3E_1$ states, and similarly for higher 'energies'.

State definitions appearing in Link 17.1 appear variously in some subsequent links of Chain 17.

3.5.1 Photon

Link 17.2. *$x = \gamma$ (photon): $I_\gamma = 1$, $m_\gamma \overset{\mathrm{A}}{=} 0$, $I_\gamma / I'_\gamma = E_1 / E_\gamma$; I_γ coded as 'linear polarization' or 'helicity'.*

1. From A-Theorem 1 steps 14–16, $I_\gamma = 1$.
2. From A-Theorem 1 steps 16 and 17, coding of I_γ is adequately describable as linear polarization.
3. From A-Theorem 1 step 16, by choosing linear combinations with equal amplitude and relative phase $\pm\pi/2$ of linear polarization basis functions spanning each factor 2-space as in A-Theorem 1 step 16, coding of I_γ is adequately describable as 'helicity'.
4. From A-Theorem 1, γ locally propagates at velocity c.
5. From A-Cor 1.2, and 4, a 'rest state' for γ in U is not observable.
6. From A-Def 6, and 5, U does not contain m_γ.
7. From A-Cor 2.1 and A-Def 21 and 22, if A-Corollary 2.1 and A-Definition 21 or 22 were to apply for velocity $= c$, if m_γ were > 0, I'_γ would be infinite.
8. From Postulate 3, and 6 and 7, m_γ is adequately describable as zero.
9. From A-Theorems 3 and 4, and A-Def 22, $I'_\gamma = E_\gamma / E_1$.
10. From 1 and 9, $I_\gamma / I'_\gamma = E_1 / E_\gamma$.

Although designated as a 'link', Link 17.2 appears formal. $m_\gamma \overset{\mathrm{A}}{=} 0$ might be thought of in terms of apparent unidirectionality of time: it might seem that, if a photon had an accessible 3-space 'rest frame', transformation into and out of that 'frame of reference' from $v < c$ might interchange time with a 3-space coordinate; so transformation into such 'rest frame', rotation through π radians about an axis perpendicular to the space coordinate imaging 'time', and transformation back to the $v<c$ system, might reverse the 'direction' of the time coordinate in the $v<c$ system. This was treated more formally hereinbefore, in A-Corollaries 1.2 and 1.3.

3.5.2 Electron Neutrino and Antineutrino

Link 17.3. *$x=\nu_e, \bar{\nu}_e$ ('electron' neutrino and antineutrino): if computational method be approximately applicable, likely: $I_x=2$, $I_x''\approx 2(E_x/E_1)^{1/2}\approx 2I_x'^{1/2}$, $I_x/I_x'=2E_1/E_x$; I_x 1 bit coded as 'spin $\frac{1}{2}$', 1 bit coded for ν_e as $b=1$, for $\bar{\nu}_e$ as $b=-1$, none of I_x coded as 'charge' sign; ν_e approximately describable as having only one possible 'helicity', and $\bar{\nu}_e$ approximately describable as having only one possible 'helicity' opposite to that of ν_e.*

1. From report(s) of observation, neutron decays to proton, electron, and electron antineutrino $\bar{\nu}_e$.
2. From Cor 14.15, 14.16, 15.11, and 15.12, and 1, likely $I_{\bar{\nu}_e}=2$.
3. From Cor 14.15, 14.16, 15.11, and 15.12, and 1, likely coding of internal information of $\bar{\nu}_e$ complements coding of the 2 bits of electron's internal information represented in a way adequately describable by relative polarization of associated two photons and b.
4. As a convention, designate b coding in an electron as $+1$.
5. From Cor 14.6, 14.10, 14.15, and 14.16, likely 1 bit of information represented internally by electron and coded in a way adequately describable by relative polarization of associated two photons corresponds through 'statistics' to 'spin $\frac{1}{2}$'.
6. From Cor 15.11 and 15.12, and 3 and 5, likely 1 bit of $I_{\bar{\nu}_e}$ is coded as 'spin $\frac{1}{2}$'.
7. From Cor 15.11 and 15.12, and 3 and 4, by complementarity of coding, likely 1 bit of $I_{\bar{\nu}_e}$ is coded as $b=-1$.
8. From Cor 14.12 and 14.16, likely electron (e^-) and positron (e^+) have complementary internal codes.
9. From report(s) of observation, ν_e is associated with production of e^+.
10. From Cor 15.11 and 15.12, and 3–5, 8, and 9, likely $I_{\nu_e}=2$, 1 bit of I_{ν_e} is coded as 'spin $\frac{1}{2}$', and 1 bit of I_{ν_e} is coded as $b=+1$.
11. From Cor 8.11, 8.16, 14.16, 15.11 and 15.12, if computational method be approximately applicable, likely $m_{\nu_e}=m_{\bar{\nu}_e}\approx 2(E_xE_j)^{1/2}/c^2$.
12. From L 17.2, and 2, likely down-shifted photon associated with ν_e or $\bar{\nu}_e$ represents only $I_x-I_\gamma=2-1=$ 1 bit of information in U, $x=\nu_e,\bar{\nu}_e$. ($I_\gamma=1$ in up-shifted photon)

13. From A-Theorems 3 and 4, A-Def 22, and 12, likely $E_j = E_1$.
14. From 11 and 13, likely $m_x \approx 2(E_x E_1)^{1/2}/c^2$, $x = \nu_e, \bar{\nu}_e$.
15. From A-Theorems 3 and 4, and A-Def 22, $I'_x = E_x/E_1$, $x = \nu_e, \bar{\nu}_e$.
16. From A-Theorem 6 and A-Def 22, $I''_x = m_x c^2/E_1$.
17. From 14–16, likely

$$I''_x \approx \frac{m_x c^2}{E_1} \approx \frac{2(E_x E_1)^{1/2}}{c^2}\,\frac{c^2}{E_1} = 2(E_x/E_1)^{1/2} = 2I'^{1/2}_x, \qquad x = \nu_e, \bar{\nu}_e.$$

18. From 2, 10, and 15, likely

$$\frac{I_x}{I'_x} = \frac{2}{E_x/E_1} = \frac{2E_1}{E_x}, \qquad x = \nu_e, \bar{\nu}_e.$$

19. From Cor 14.15, 14.16, 15.11, and 15.12, and 1 and 9, likely ν_e and $\bar{\nu}_e$ do not represent internally any information coded in a way adequately describable as relative phase of associated two photons.
20. From Cor 14.3–14.5, 14.9, and 14.14–14.16, likely 1 bit of information represented internally in a way adequately describable as relative phase of associated two photons corresponds to 'charge' sign.
21. From 19 and 20, likely ν_e and $\bar{\nu}_e$ do not represent internally any information in code for 'charge' sign.
22. From A-Cor 6.1, if adseparability be approximately applicable, only 1 bit of I_x is available internal to x, $x = \nu_e, \bar{\nu}_e$.
23. From 3–10, and 22, likely both 'particle' versus 'antiparticle' and two 'spin' states are coded by 1 bit of I_x with complementary codes in ν_e and $\bar{\nu}_e$, $x = \nu_e, \bar{\nu}_e$.
24. From A-Theorem 1, Cor 8.10 and 15.11, Def 13, and 14 and 23, likely to O_U x's two 'spin' states are approximately describable as two 'helicity' states.
25. From 23 and 24, by enumeration of complementary representations possible with exactly 1 bit of accessible information, likely ν_e is approximately describable as having only one possible 'helicity' and $\bar{\nu}_e$ is approximately describable as having only one possible 'helicity' opposite to that of ν_e.

●

(cautionary note on applicability of computational method based on adseparability due to electron neutrino and/or antineutrino velocity likely $= v$ maximum for it in U)

Note that, as indicated by use of 'likely', Link 17.3 is not 'formal'.

The concept of coding 'existence of coding' consistent with 'spin $\frac{1}{2}$' appears to correspond to availability of 2 bits of information represented internally by two associated unbound photons, as the polarization or 'helicity' of each: representing 'spin' by relative polarization of such two photons 'uses' one of these 2 bits, 'leaving' only 1 bit to describe 'existence of coding' for both photons. In this way, the 'existences' of such two photons appear to be combined, and the 1 bit describing such combined existence likely is coded in a representation consistent with use of both photons together in representing 1 bit of 'spin'. This is the consistency specified in the definition of 'b' in the statement of conditions for Chain 17. (see also A-Cor 2.2 steps 13–19)

3.5.3 Electron and Positron

Link 17.4. $x = e^+, e^-$ (*positron, electron*):

likely:

$$I_{e^\pm} = 3, \quad I_x''\left(32\pi^2(\ln(2I_x''))^2\right)^{1/3} \approx 3.6^{1/3} I_U^{1/3},$$

$$I_x / I_x'' \approx 7.5^{1/3}\left(32\pi^2(\ln(2I_x''))^2\right)^{1/3} / I_U^{1/3};$$

$I_{e^\pm}$ coded 1 bit as 'spin $\frac{1}{2}$', 1 bit as 'charge' $+$ in e^+ and $-$ in e^-, 1 bit as $b = +1$ in e^- and -1 in e^+; in $e^\pm$, $d = +1$ only.

1. From Cor 14.7, 14.15, and 14.16, likely $I_x = 3$.
2. From Cor 14.15 and 14.16, likely

$$m_x\left(32\pi^2(\ln(R_U 2m_x c/\hbar))^2\right)^{1/3} \approx \left(\frac{\hbar^2 M_U 3.6}{c^2 R_U^2}\right)^{1/3}.$$

3. From A-Def 22, $E_1 = \hbar c / R_U$.
4. From A-Cor 2.1, A-Theorems 3 and 4, A-Def 22, and 3, $I_x'' = m_x c^2 / E_1$ and $I_U = M_U c^2 / E_1$.
5. Substituting 3 and 4 in 2, and variously factoring and performing the

cube root indicated, likely

$$m_x\left(32\pi^2(\ln(R_U 2m_x c/\hbar))^2\right)^{1/3} \approx \left(\frac{\hbar^2 M_U 3.6}{c^2 R_U^2}\right)^{1/3}$$

$$\frac{m_x c^2}{E_1}\left(32\pi^2(\ln(2m_x c^2/E_1))^2\right)^{1/3} \approx \left(\frac{\hbar^2 M_U 3.6 c^6}{c^2 R_U^2 E_1^3}\right)^{1/3}$$

$$I_x''\left(32\pi^2(\ln(2I_x''))^2\right)^{1/3} \approx \left(\frac{\hbar^2 M_U 3.6 c^6 R_U^2}{c^2 R_U^2 E_1 \hbar^2 c^2}\right)^{1/3}$$

$$I_x''\left(32\pi^2(\ln(2I_x''))^2\right)^{1/3} \approx \left(\frac{M_U c^2 3.6}{E_1}\right)^{1/3}$$

$$I_x''\left(32\pi^2(\ln(2I_x''))^2\right)^{1/3} \approx 3.6^{1/3} I_U^{1/3}, \qquad x = e^+, e^-.$$

6. From Cor 14.9–14.11 and 14.14–14.16, and L 17.3 step 4, likely I_{e^-} coded 1 bit as 'spin $\frac{1}{2}$', 1 bit as 'charge' sign (by convention) $-$, and 1 bit as (by convention) $b = +1$.
7. From Cor 14.12, 14.15, and 14.16, and 6, by complementarity of coding, likely I_{e^+} coded 1 bit as 'spin $\frac{1}{2}$', 1 bit as 'charge' sign $+$, and 1 bit as $b = -1$.
8. From Cor 14.3–14.5, and 14.8, Chain 17 definition of 'd', and 1, 6, and 7, likely d does not describe coding of any of $I_{e^\pm}$.
9. From Cor 14.3–14.5, and 14.8, A-Def 1, and 8, likely there is only one possible value for d in $e^\pm$.
10. From Cor 14.15 and 14.16, Chain 17, and 9, likely in $e^\pm$ $d = +1$ only.
11. From 1 and 5, likely

$$\frac{I_x}{I_x''} \approx \frac{3\left(32\pi^2(\ln(2I_x''))^2\right)^{1/3}}{3.6^{1/3} I_U^{1/3}} = \frac{\left(32\pi^2(\ln(2I_x''))^2\right)^{1/3}}{\left(\frac{2}{15}\right)^{1/3} I_U^{1/3}}$$

$$= 7.5^{1/3}\left(32\pi^2(\ln(2I_x''))^2\right)^{1/3} / I_U^{1/3}, \qquad x = e^+, e^-,$$

3.5.4 Muons

Link 17.5. $x=\mu_L^+, \mu_L^-, \mu_H^+, \mu_H^-$ *(antimuons and muons, 'L' for 'lighter', 'H' for 'heavier'):*

likely: $I_x=7$,

$$I''_{\mu_L^\pm}\approx 11.76^{1/3}I_U^{1/3}, \qquad I''_{\mu_H^\pm}\approx 12.25^{1/3}I_U^{1/3},$$

$$\frac{I_{\mu_L^\pm}}{I''_{\mu_L^\pm}}\approx\left(\frac{175}{6}\right)^{1/3}\Big/ I_U^{1/3}, \qquad \frac{I_{\mu_H^\pm}}{I''_{\mu_H^\pm}}\approx 28^{1/3}/I_U^{1/3};$$

I_x 1 bit coded as apparent 'spin $\frac{1}{2}$', 1 bit coded as 'charge' sign + in $\mu_{L,H}^+$ and − in $\mu_{L,H}^-$, 1 bit coded as $b=+b_\mu$ in $\mu_{L,H}^-$ and $-b_\mu$ in $\mu_{L,H}^+$, 1 bit coded as $d=+1$ in $\mu_L^\pm$ and -1 in $\mu_H^\pm$;

$$\mu_L^\pm\to e^\pm 4\gamma,$$

$$\mu_L^\pm\to e^\pm\nu_e\bar{\nu}_e,$$

$$\mu_H^\pm\to e^\pm\overset{(-)}{\nu}_e\overset{(-)}{\nu}_e \quad \left(\text{'}\overset{(-)}{\nu}_e\text{' indicates '}\nu_e \text{ and/or } \bar{\nu}_e\text{'}\right);$$

$$\tau_{\mu_L^\pm}\ll\tau_{\mu_H^\pm}.$$

1. From report(s) of observation, there are $\mu^\pm$ which decay entirely or nearly entirely by $\mu^\pm\to e^\pm\overset{(-)}{\nu}\overset{(-)}{\nu}$. ('$\overset{(-)}{\nu}$' denotes some neutrino and/or antineutrino)
2. From Cor 14.15, 14.16, 15.11, and 15.12, and 1, likely there are $\mu^\pm$ such that $\mu^+\to e^+\nu_e\overset{(-)}{\nu}$ and $\mu^-\to e^-\bar{\nu}_e\overset{(-)}{\nu}$. (Note that in a formal argument this step would be unnecessary; however, this is a likeliness argument, and various steps are included on the ground that they might contribute to an understanding based on likeliness.)
3. From Cor 15.8, 15.11, and 15.12, likely the other neutrino in $\mu^\pm\to e^\pm\overset{(-)}{\nu}_e\overset{(-)}{\nu}$ represents internally at least 2 bits of information.
4. From L 17.3 and L 17.4, and 2 and 3, likely $I_{\mu^\pm}\geqslant 7$.
5. From report(s) of observation, μ rarely or never decays to 3 e.
6. From L 17.4, and 5, likely $I_{\mu^\pm}<9$. (Note that steps 5 and 6, although unnecessary in a formal argument, appear to contribute to a likelihood argument.)
7. From report(s) of observation, $\mu\to e\overset{(-)}{\nu}\overset{(-)}{\nu}\gamma$ is rarely or never seen.

8. From L 17.2–17.4, and 7, likely $I_{\mu^\pm} < 8$.
9. From 4, 6, and 8, likely $I_{\mu^\pm} = 7$
10. From report(s) of observation, $\mu^\pm$ has two states of 'spin' and two signs of 'charge', with (by convention) charge sign + for μ^+ and − for μ^-.
11. From Cor 15.11 and 15.12, L 17.3 and 17.4, and 1, 9, and 10, likely 1 bit of $I_{\mu^\pm}$ is coded as 'spin $\frac{1}{2}$' and 1 bit as 'charge' sign.
12. From L 17.3 and 17.4, and 2 and 9, likely other neutrino in $\mu^+ \to e^+ \nu_e \overset{(-)}{\nu}$ or $\mu^- \to e^- \bar{\nu}_e \overset{(-)}{\nu}$ represents internally 2 bits of information. (here, '$\overset{(-)}{\nu}$' denotes both ν and $\bar{\nu}$ as under consideration)
13. From Cor 15.11 and 15.12, L 17.3, and 12, likely such other neutrino represents internally information coded 1 bit as 'spin $\frac{1}{2}$' and 1 bit as $b = \pm 1$.
14. From L 17.3, and 12 and 13, likely electron neutrino or antineutrino and 'other' neutrino in $\mu \to e \overset{(-)}{\nu}_e \overset{(-)}{\nu}$ each represents internally exactly 2 bits of information, coded 1 bit as 'spin $\frac{1}{2}$' and 1 bit as $b = \pm 1$.
15. From 14, likely there is not any internal information or coding thereof that would distinguish such other neutrino or antineutrino from an electron neutrino or antineutrino.
16. From A-Def 1, Postulate 2, and 1, 2, 14, and 15, likely $\mu^\pm \to e^\pm \overset{(-)}{\nu}_e \overset{(-)}{\nu}_e$.
17. From report(s) of observation, $\mu^\pm$ somewhat resembles $e^\pm$, but with different mass. (e.g., with respect to 'spin', interactions, and approximate 'charge' magnitude)
18. From Cor 14.3–14.6, 14.15, and 14.16, and 1, information represented internally in extended interaction mode appears defined as 3 bits; likely if more than 3 bits decay would be too fast for approximate description of previously reported (as recognized) $\mu^\pm$.
19. From Postulate 1, Theorem 14, and Cor 14.8, relatively localized bound two photon configuration appears able to represent internally a finite amount of information $\geqslant$ 3 bits.
20. From 17–19, likely $\mu^\pm$ is approximately describable in terms of relatively localized two-photon configuration substantially as in Corollary 14.8.
21. From Cor 14.3, 14.4, 14.9, 14.14–14.16, 15.11, and 15.12, L 17.3 and 17.4, and 1, 2, 11, and 20, likely 1 bit of $I_{\mu^\pm}$ is represented in a way adequately describable by ± relative phase $\pi/2$ of two associated photons.
22. From Cor 14.5, 14.8, 14.15, and 14.16, L 17.4 steps 2–5, and Def 19, comparing the right sides of expressions in Corollary 14.8 step 15 and

L 17.4 steps 2 and 5, for such relatively localized two-photon configuration to represent internally I bits of information of which one is coded in a way adequately describable as $\pm$ relative phase $\pi/2$ of associated two photons,

$$m_0 \approx \left(\frac{Y(I)}{1 \pm 1/I^2} \right)^{1/3} I_U^{1/3} E_1/c^2,$$

with + or − sign corresponding to relative phase of parts of representation for each state of lead or lag.

23. From L 17.1, and Def 18, $Y(7) = (2 \times 1) + (5 \times 2) = 12$.
24. From 9, 21, 22, and 23, likely

$$m_{\mu^\pm} \approx \left(\frac{12}{1 \pm \frac{1}{49}} \right)^{1/3} I_U^{1/3} E_1/c^2,$$

with + or − sign in denominator corresponding to relative phase of photons for each state of lead or lag. (not to $\pm$ labeling on $\mu^\pm$)

25. From 24, likely there are two types of $\mu^\pm$, with mass approximately describable respectively with + or − sign in denominator of step 24, $\mu_L^\pm$ and $\mu_H^\pm$, $m_{\mu_L^\pm} < m_{\mu_H^\pm}$. ('$L$' for 'lighter' and '$H$' for 'heavier')
26. From Cor 14.5 steps 8 and 9, Cor 14.8 step 15, Cor 14.15 and 14.16, L 17.4, and 24 and 25, likely local configuration approximately describable with + sign in denominator of step 24 would have substantially the same relative phase of parts of representation in each state of lead or lag as electron (or positron).
27. From 11, 15, and 26, likely $\mu_L^\pm$ would have internal representation able to supply coding for all 3 bits of $e^\pm$'s internal information.
28. From L 17.2 and 17.3, and 27, likely $\mu_L^\pm$ would be able to decay by a process conserving internal information and supplying all coding needed to make $e^\pm$ by $\mu_L^\pm \to e^\pm 4\gamma$ and/or $\mu_L^\pm \to e^\pm \nu_e \bar{\nu}_e$.
29. From 24, 27, and 28, likely $\mu_L^\pm$ would decay relatively rapidly by $\mu_L^\pm \to e^\pm 4\gamma$. ($d$ coding not changed for that information remaining in a stable bound configuration after decay)
30. From 1, 25, and 29, likely $\mu^\pm$ which decays substantially entirely by $\mu^\pm \to e^\pm \overset{(-)}{\nu}_e \overset{(-)}{\nu}_e$ is approximately describable as $\mu_H^\pm$.
31. From 29 and 30, likely $\tau_{\mu_L^\pm} \ll \tau_{\mu_H^\pm}$.

32. From 24 and 25, likely

$$m_{\mu_L^\pm} \approx \left(\frac{12}{1+\frac{1}{49}} \right)^{1/3} I_U^{1/3} E_1/c^2.$$

33. From A-Def 22, and 32, noting $I_x'' = m_x c^2/E_l$, likely

$$I''_{\mu_L^\pm} \approx 11.76^{1/3} I_U^{1/3}.$$

34. From 9 and 33, likely

$$\frac{I_{\mu_L^\pm}}{I''_{\mu_L^\pm}} \approx \frac{7}{I''_{\mu_L^\pm}} \approx \left(\frac{175}{6} \right)^{1/3} \Big/ I_U^{1/3}.$$

35. From 24 and 25, likely

$$m_{\mu_H^\pm} \approx \left(\frac{12}{1-\frac{1}{49}} \right)^{1/3} I_U^{1/3} E_1/c^2.$$

36. From A-Def 22, and 35, as in step 33, likely

$$I''_{\mu_H^\pm} \approx 12.25^{1/3} I_U^{1/3}.$$

37. From 9 and 36, likely

$$\frac{I_{\mu_H^\pm}}{I''_{\mu_H^\pm}} \approx \frac{7}{I''_{\mu_H^\pm}} \approx 28^{1/3} / I_U^{1/3}.$$

38. From definition of 'b', and 9–11, likely 1 bit of $I_{\mu_{L,H}^\pm}$ coded as 'b' $= +b_\mu$ for $\mu_{L,H}^-$ and $-b_\mu$ for $\mu_{L,H}^+$. (choice of sign as stated, relation between 'b_μ' and 'b' of $e^\pm$ and ν_e, $\bar{\nu}_e$ is *not* stated in this step)
39. From definition of 'd', and 9 and 25, likely 1 bit of $I_{\mu_{L,H}^\pm}$ coded as $d = +1$ in $\mu_L^\pm$ and -1 in $\mu_H^\pm$.

It would seem that from prior theory one might expect to see $\mu^{\pm}\rightarrow e^{\pm}\gamma$ if in $\mu\rightarrow e2\nu$ the neutrinos were electron neutrino and antineutrino; with the physical picture presented herein, such process of $\mu^{\pm}\rightarrow e^{\pm}\gamma$ may be considered in terms of transcription of information between representation internally and externally. In view of the 'degree' of separability of amounts of internal and external information noted hereinbefore following Corollary 15.12, one might reasonably expect such process to be much less likely than $\mu^{\pm}\rightarrow e^{\pm}\overset{(-)}{\nu_e}\overset{(-)}{\nu_e}$, which appears to conserve amount of internal information.

Likeliness of $\tau_{\mu_L^{\pm}} \ll \tau_{\mu_H^{\pm}}$ appears traceable substantially to likeliness that d not describe coding of any of $e^{\pm}$ internal information, with likeliness of in $e^{\pm}$ $d=+1$ only: likely $\mu_L^{\pm}\rightarrow e^{\pm}4\gamma$ would preserve internal information coding approximately describable by d, and likely, if occurring, any direct $\mu_H^{\pm}\rightarrow e^{\pm}4\gamma$ would not.

I do not yet, at the time of this writing, have a clear formal understanding of what relation 'b_μ' might have to 'b' of $e^{\pm}$, ν_e, and $\bar{\nu}_e$; this is discussed further in Section 6 question 21.

3.5.5 Muon Neutrino and Antineutrino

Link 17.6. $x=\nu_{\mu(L,H)}$, $\bar{\nu}_{\mu(L,H)}$ (*'muon' neutrino and antineutrino*): *likely*: $I_x=6$, $I_x''\approx 2(5E_x/E_1)^{1/2}\approx 2\sqrt{5}\,I_x'^{1/2}$ *or* $I_x''\approx 2(10E_x/E_1)^{1/2}\approx 2\sqrt{10}\,I_x'^{1/2}$, $I_x/I_x'=6\,E_1/E_x$; 1 *bit of* I_{ν_μ} *or* $I_{\bar{\nu}_\mu}$ *coded as 'spin* $\frac{1}{2}$*',* 1 *bit as* 'b'$=+b_\mu$ *in* ν_μ *and* $-b_\mu$ *in* $\bar{\nu}_\mu$; *not any* of I_{ν_μ} or $I_{\bar{\nu}_\mu}$ coded as 'charge' sign.

1. From report(s) of observation, there are π^+ and π^- such that $\pi^+\rightarrow\mu^+\nu_\mu$ and $\pi^-\rightarrow\mu^-\bar{\nu}_\mu$ more than 99.7% fractional decay mode, $\pi^{\pm}$ has 'charge' sign ± 1, and $\pi^{\pm}$ does not represent internally any information corresponding to states distinguished by 'spin'.
2. From 1, likely $\pi^{\pm}$ is able to supply coding for 'charge' sign, and not able to supply coding for 'spin'.
3. From L 17.5, and 2, likely $\pi^{\pm}$ is not able to supply coding for 'b' and/or 'b_μ'.
4. From L 17.5, likely $\mu_H^{\pm}\rightarrow e^{\pm}\overset{(-)}{\nu_e}\overset{(-)}{\nu_e}$.
5. From 4, likely coding of all of $I_{\mu^{\pm}}$ except for 'charge' sign bit is 'traceable' to coding for 'spin $\frac{1}{2}$', b, or d coded consistent with both. (*both*; see also text near after L 17.7)
6. From 2, 3, and 5, likely $\pi^{\pm}$ supplies to $\mu^{\pm}$ coding for only 'charge' sign bit.
7. From A-Theorem 7, Cor 15.11 and 15.12, and 6, as in Cor 15.11, likely decay of $\pi^{\pm}$ to make $\mu^{\pm}$ involves making an entity complementing

coding of all information represented internally in $\mu^{\pm}$ for which $\pi^{\pm}$ does not supply coding.

8. From L 17.5, and 6 and 7, likely $I_x = I_{\mu^{\pm}} - 1 = 6, \quad x = \nu_\mu, \bar{\nu}_\mu$.
9. From L 17.5, likely 1 bit of $I_{\mu^{\pm}}$ coded as 'spin $\frac{1}{2}$' and 1 bit as $b = +b_\mu$ in ν_μ and $-b_\mu$ in $\bar{\nu}_\mu$.
10. From 1, 6, 7, and 9, likely 1 bit of I_{ν_μ} or $I_{(\bar{\nu})_\mu}$ coded as 'spin $\frac{1}{2}$' and 1 bit as $b = +b_\mu$ in ν_μ and $-b_\mu$ in $\bar{\nu}_\mu$.
11. From 6 and 7, likely not any of I_{ν_μ} or $I_{(\bar{\nu})_\mu}$ is coded as 'charge' sign.
12. From A-Theorems 3 and 4, L 17.1, A-Def 12, and 11, likely 1 bit of I_{ν_μ} or $I_{(\bar{\nu})_\mu}$ is represented by associated w_+.
13. From Cor 15.11 and 15.12, and 11, likely ν_μ and $\bar{\nu}_\mu$ are not stable bound entities.
14. From discussion preceding Cor 14.2, Cor 14.2, discussion following Cor 15.11 and 8, 12, and 13, likely representation of other 5 bits of I_{ν_μ} or $I_{(\bar{\nu})_\mu}$ is by the down-shifted photon of such entity; or, as an 'alternative', representation of 6 bits as by photons with fixed relative phase.
15. From A-Theorems 3 and 4, A-Def 22, L 17.1, and 14, likely such representation of such 5 bits involves 'energy' $5E_1$; or (for such 'alternative') $\lesssim Y(6)E_1 = ((2\times 1)+(4\times 2))E_1 = 10E_1$.
16. From 12–15, likely such representation of such 5 bits approximately corresponds to such entity's associated w_- being in the w_5 state in U, with 'energy' $5E_1$; or, to photons with total 'energy' $\lesssim Y(6)\ E_1$. (note that this does not specifically discuss localized form or interaction 'strength')
17. From Cor 8.11 and 15.12, and 14–16, likely $m_x \approx 2(E_x 5E_1)^{1/2}/c^2$; or $m_x \lesssim 2(E_x 10E_1)^{1/2}/c^2; \quad x = \nu_\mu, \bar{\nu}_\mu$.
18. From A-Def 22, and 17, noting $I_x'' = m_x c^2/E_1$, likely

$$I_x'' \approx 2(5E_x/E_1)^{1/2} \approx 2\sqrt{5}\ I_x'^{1/2}$$

or

$$I_x'' \approx 2(10E_x/E_1)^{1/2} \approx 2\sqrt{10}\ I_x'^{1/2}, \qquad x = \nu_\mu, \bar{\nu}_\mu.$$

19. From A-Def 22 (or L 17.1), and 8, likely $I_x/I_x' = 6/(E_x/E_1) = 6E_1/E_x$, $x = \nu_\mu, \bar{\nu}_\mu$.

I have not yet, at the time of this writing, understood formally in detail whether neutrinos produced with $\mu_L^{\pm}$ are the same as those produced with $\mu_H^{\pm}$. Subscripts '(L)', '(H)', and/or '(L,H)' may be used herein with 'ν_μ', for example, '$\nu_{\mu_{(L,H)}}$', to indicate this lack of understanding.

From A-Theorem 3, A-Corollaries 6.1 and 7.3, and Link 17.6, likely amount of information readily accessible in $\overset{(-)}{\nu}_{\mu_{(L,H)}}$ is an odd number of bits, and, therefore, likely such internally accessible information would not be fully transcribable into different representation (coding) without another information representation ('code book'). This might offer some insight into 'parity nonconservation'; in particular, from Link 17.3, one might suspect that the 'odd' ('leftover') bit likely readily accessible in $\overset{(-)}{\nu}_{\mu_{(L,H)}}$ might play a role like that of the only bit likely readily accessible in $\overset{(-)}{\nu}_e$.

From Links 17.3 and 17.6, it would seem that the difference in amount of information represented internally by ν_e $(\bar{\nu}_e)$ and ν_μ $(\bar{\nu}_\mu)$ might offer some insight into why they appear on observation to be different.

Formal treatment of interaction strength might involve clearer understanding and sharper discussion of m_x, $x = \nu_{\mu_{(L,H)}}, \bar{\nu}_{\mu_{(L,H)}}$, but comparison with report(s) of experimental measurement specifically of such mass prior to and near this writing does not appear substantially sensitive to distinction between factors of $\sqrt{5}$ and $\sqrt{10}$, and this relatively crude treatment appears sufficient for such use hereinafter (I have not, at the time of this writing, (yet?) clearly and formally understood this).

3.5.6 Pions

Link 17.7. $x = \pi_L^+, \pi_L^-, \pi_H^+, \pi_H^-$ (*'charged pions' (including both 'particles' and 'antiparticles'); 'L' for 'lighter', 'H' for 'heavier'*):

likely: $I_x = 13$,

$$I''_{\pi_L^{\pm}} \approx \left(\frac{557.7\sqrt{3}}{34}\right)^{1/3} I_U^{1/3},$$

$$I''_{\pi_H^{\pm}} \approx \left(\frac{1859\sqrt{3}}{112}\right)^{1/3} I_U^{1/3},$$

$$I_{\pi_L^{\pm}} / I''_{\pi_L^{\pm}} \approx \left(\frac{442}{3.3\sqrt{3}}\right)^{1/3} \Big/ I_U^{1/3},$$

$$I_{\pi_H^{\pm}} / I''_{\pi_H^{\pm}} \approx \left(\frac{1456}{11\sqrt{3}}\right)^{1/3} \Big/ I_U^{1/3};$$

1 *bit of* I_x *coded as 'charge' sign* + *in* $\pi^+_{L,H}$ *and* − *in* $\pi^-_{L,H}$, 1 *bit as* $d=+1$ *in* $\pi^\pm_L$ *and* −1 *in* $\pi^\pm_H$; *not any of* I_x *coded as* b;

$$\pi_L^+ \to \nu_{\mu(L)} + \mu_L^+ \begin{array}{l} \nearrow e^+4\gamma \\ \searrow e^+\nu_e\bar{\nu}_e \end{array}$$

$$\pi_L^- \to \bar{\nu}_{\mu(L)} + \mu_L^- \begin{array}{l} \nearrow e^-4\gamma \\ \searrow e^-\nu_e\bar{\nu}_e \end{array}$$

$$\pi_L^+ \to e^+ + \nu_e + (8\gamma \text{ or } e^+e^-2\gamma)$$

$$\pi_L^- \to e^- + \bar{\nu}_e + (8\gamma \text{ or } e^+e^-2\gamma)$$

$$\pi_L^+ \to e^+ + 2\nu_e + \bar{\nu}_e + 4\gamma$$

$$\pi_L^- \to e^- + \nu_e + 2\bar{\nu}_e + 4\gamma$$

$$\pi_L^+ \to e^+ + 3\nu_e + 2\bar{\nu}_e$$

$$\pi_L^- \to e^- + 2\nu_e + 3\bar{\nu}_e$$

$$\pi_H^+ \to \nu_{\mu(H)} + \mu_H^+ \searrow e^+ \overset{(-)}{\nu}_e \overset{(-)}{\nu}_e$$

$$\pi_H^- \to \bar{\nu}_{\mu(H)} + \mu_H^- \searrow e^- \overset{(-)}{\nu}_e \overset{(-)}{\nu}_e$$

$$\tau_{\pi_L^\pm} < \tau_{\pi_H^\pm}.$$

1. From report(s) of observation, there are $\pi^\pm$ such that $\pi^+ \to \mu^+\nu_\mu$ and $\pi^- \to \mu^-\bar{\nu}_\mu$ comprise >0.9975 of all reported as recognized $\pi^\pm$ decays.
2. From report(s) of observation, $\pi^+ \to e^+\nu_e$ and $\pi^- \to e^-\bar{\nu}_e$ comprise $\sim 1.24 \times 10^{-4}$ of all recognized $\pi^\pm$ decays.
3. From report(s) of observation, $\pi^\pm$ has charge sign + for π^+ and − for π^-.
4. From report(s) of observation, $\pi^\pm$ has 'spin 0'.
5. From L 17.5 and 17.6, and 1, likely $I_{\pi^\pm} = I_\mu + I_{\nu_\mu} = 7 + 6 = 13$.
6. From 3 and 5, likely 1 bit of $I_{\pi^\pm}$ is coded as 'charge' sign, + in π^+ and − in π^-.
7. From A-Cor 2.1 and 2.2, A-Theorem 6, and L 17.2, for irad two coupled photons substantially at rest in U, both such photons have equal frequency in rest frame of such irad. (as B_0)

8. From Cor 14.6, 14.10, 14.15, and 14.16, and 7, likely approximate 'spin $\frac{1}{2}$' (Fermi–Dirac) statistics of an assembly of such entities is approximately describable in terms of phase coherent superposition of such photons. (with (formally) equal frequency)
9. From Postulates 1 and 2, A-Theorem 5, and L 17.2, adequate description of entity not adequately describable as photon involves irad more than one photon.
10. From A-Theorems 5 and 6, for entity composed of three photons, such photons are not required to have equal frequency in such entity's rest frame.
11. From A-Lemma 2.1 and L 17.2, likely 'spin 0' (Bose–Einstein) statistics involves addition without phase coherence.
12. From A-Cor 4.8, A-Theorem 5, and A-Cor 5.1, in domain of adseparability, external behavior of a system is adequately describable in terms of B_0 ψ's independent of description of internal structure of such system.
13. From L 17.2, and 4 and 7–12, likely adequate description of 'internal structure' of $\pi^{\pm}$ does not relatively simply correspond to description as association of less than three photons.
14. From 12 and 13, likely simplest approximate (substantial) description ('sasd') of $\pi^{\pm}$ 'internal structure' is in terms of association of three photons.
15. From all formal material hereinbefore, the only 'direct' interaction among photons appearing so far in this theory as presented herein is interaction as approximately described in Theorem 13 and its lemmas and corollaries ('Theorem 13 interaction').
16. Consider three photons associated by Theorem 13 interaction as a relatively localized system S, representing internally I_S bits of information.
17. From A-Theorems 3 and 4, Theorem 13, A-Def 22, and 16, there is a least mass m_0 for which S can represent internally I_S bits of information.
18. From A-Lemma 2.1, A-Cor 2.1, A-Theorems 3 and 4, Theorem 13, A-Def 22, and 16 and 17, m_0 appears approximately largest in Theorem 13 interaction if S's associated three photons have approximately equal amplitude and are superposed approximately in phase.
19. From A-Def 22, and 18, in such case, each of S's associated three photons would have 'energy' $E \approx m_0c^2/3$.

20. From A-Theorem 1, A-Cor 2.1 and 4.8, A-Theorem 5, A-Cor 5.1, A-Def 12, and 10, 12, 16, and 19, S's associated three photons are not by presence in S's rest frame (formally) of (exactly) equal frequency. (note difference between this and Cor 14.8 step 9)
21. From report(s) of observation, $\pi^{\pm}$ is not an absolutely stable entity.
22. From 20 and 21, likely in sasd of S's associated three photons, such three photons are not superposable for geometric conditions. (e.g., for defining 'λ' and 'lobe' spacing(s); unlike Cor 14.8)
23. From A-Theorem 1, Theorem 13, and 22, noting likely m_0 such that Theorem 13 'weak field', likely in sasd of S's associated three photons each appears with approximately that wavelength with which it would appear were S's other two photons not present.
24. From A-Cor 2.1, A-Theorems 3 and 4, A-Def 22, and 19 and 23, likely each of S's associated three photons has wavelength

$$\lambda \approx \frac{c}{f} \approx \frac{c}{E/h} \approx \frac{hc}{E} \approx \frac{hc}{m_0c^2/3} \approx \frac{3h}{m_0c}.$$

25. From A-Lemma 2.1 and A-Def 22, 'mass' of each of S's associated photons appears with spatial distribution corresponding to square of amplitude of corresponding w. (thus two lumps in circumferential distance λ)
26. From 25, for all of S's associated three photons to be described approximately symmetrically, likely least linear size r of S is approximately that r such that S is approximately describable as comprised of three approximately symmetrically spaced approximately equal 'lobes' m.
27. From 24–26, likely $r \approx 1.5\lambda/2\pi$.
28. From 24 and 27, likely $r \approx 1.5 \times 3h/2\pi m_0 c$.
29. From Cor 14.5, 14.15, and 14.16, L 17.4 and 17.5, and 6, likely 'mass' appears in Theorem 13 interaction in description of S representing internally amount I_S of information as multiplied by factor $(1 \pm 1/I_S^2)^{1/2}$, with + or − sign corresponding to relative phase of parts of representation for each state of lead or lag.
30. From 18, 19, 26, and 29, likely each of the three approximately symmetrically spaced approximately equal 'lobes' in sasd of S appears in Theorem 13 interaction as having 'mass' m approximately $(1 \pm 1/I_S^2)^{1/2}$ times 1/3 of the 'mass' corresponding to superposing each of

S's associated three photons,

$$m \approx \left(1 \pm \frac{1}{I_S^2}\right)^{1/2} \frac{1}{3}\left(3\left(\frac{m_0 c^2}{3}\right)^{1/2}\right)^2 \Bigg/ c^2$$

$$\approx \left(1 \pm \frac{1}{I_S^2}\right)^{1/2} m_0.$$

31. From A-Theorems 1, 3, and 4, A-Cor 4.1, Theorem 13, A-Def 22, and 26, likely the distance between 'lobes' appearing in Theorem 13 interaction in sasd of S is $\approx \sqrt{3}\, r$.
32. From 28 and 31, likely the distance between 'lobes' appearing in Theorem 13 interaction in sasd of S is $\approx \sqrt{3} \times 4.5\hbar / m_0 c$.
33. From Theorem 13, and 26, 30, and 32, likely there are in sasd of S three pairs of 'lobes' appearing in Theorem 13 interaction, with each 'lobe' appearing as having 'mass' $\approx (1 \pm 1/I_S^2)^{1/2} m_0$, with separation distance in each such pair $\approx \sqrt{3} \times 4.5\hbar / m_0 c$.
34. From Theorem 13, A-Def 22, and 30 and 33, likely 'energy' $|\Delta E|$ corresponding to Theorem 13 interaction of S's associated three photons is given approximately by

$$|\Delta E| \approx \frac{3G\left(m_0\left(1 \pm \frac{1}{I_S^2}\right)^{1/2}\right)^2}{\left(\frac{\sqrt{3} \times 4.5\hbar}{m_0 c}\right)}$$

$$\approx \frac{\sqrt{3}\, G m_0^3 c}{4.5\hbar}\left(1 \pm \frac{1}{I_S^2}\right).$$

35. From A-Theorem 1, L 17.2, and 16, S can represent 2 bits of I_S as relative polarizations of S's associated three photons.
36. From 20 and 21, as in 22 and 23, likely in sasd of representation of information internally by S in terms of alternative representation by photons in U, such photons in such alternative representation do not have fixed relative phase.

37. From A-Theorems 3 and 4, A-Cor 4.7, A-Def 22, and 36, likely 'energy' for such alternative representation of j bits of S's internal information by photons in U is $(\approx)\, jE_1$.
38. From 34, 35, and 37, likely representation internally by S of amount I_S of information involves Theorem 13 interaction of S's said three photons such that $|\Delta E| \approx (I_S - 2)E_1$.
39. From 34 and 38, likely, for S to so represent internally amount I_S of information of which 1 bit is coded as 'charge' sign,

$$(I_S - 2)E_1 \approx \frac{\sqrt{3}\, G m_0^3 c}{4.5\hbar}\left(1 \pm \frac{1}{I_S^2}\right).$$

40. From 39, solving for m_0 as a function of I_S, likely such

$$m_0(I_S) \approx \left(\frac{(I_S - 2)E_1 4.5\hbar}{\sqrt{3}\, Gc\left(1 \pm \frac{1}{I_S^2}\right)}\right)^{1/3}$$

41. From Theorem 13, $G \approx c^2 R_U / M_U$.
42. From (A-Cor 2.1, A-Theorems 3 and 4, A-Cor 4.7, and) A-Def 22,

$$I_U = M_U c^2 / E_1.$$

43. Solving 42 for M_U, and substituting in 41,

$$G \approx \frac{c^4 R_U}{I_U E_1}.$$

44. From 43,

$$\frac{E_1 \hbar}{Gc} \approx \frac{E_1^2 I_U \hbar}{c^5 R_U}.$$

45. From A-Def 22, as in L 17.1, $E_1 = \hbar c / R_U$.
46. Expressing 45 as $E_1 / c = \hbar / R_U$, and substituting in 44,

$$\frac{E_1 \hbar}{Gc} \approx \frac{E_1^3 I_U}{c^6}.$$

47. From 39, 40, and 46, likely, for S representing internally amount I_S of information of which 1 bit is coded as 'charge' sign, such

$$m_0(I_S) \approx \left(\frac{(I_S - 2)4.5}{\sqrt{3}\left(1 \pm \frac{1}{I_S^2}\right)} \right)^{1/3} I_U^{1/3} E_1/c^2$$

$$= \left(\frac{(I_S - 2)1.5\sqrt{3}}{1 \pm \frac{1}{I_S^2}} \right)^{1/3} I_U^{1/3} E_1/c^2,$$

with + or − sign corresponding to relative phase of parts of representation for each state of lead or lag.

48. From 47, likely there are two types of S, with mass approximately describable respectively with + or − sign in denominator of step 47, S_L and S_H, with $m_{S_L} < m_{S_H}$. ('L' for 'lighter', 'H' for 'heavier')

49. From A-Def 22, and 5, 14, 47, and 48, noting $I_x'' = m_x c^2 / E_1$, likely for simplest approximate substantial description of $\pi^{\pm}$, there are two types of $\pi^{\pm}$, $\pi_L^{\pm}$ and $\pi_H^{\pm}$, with

$$I''_{\pi_L^{\pm}} \approx \left(\frac{(13-2)1.5\sqrt{3}}{1 + 1/13^2} \right)^{1/3} I_U^{1/3} = \left(\frac{557.7\sqrt{3}}{34} \right)^{1/3} I_U^{1/3}$$

and

$$I''_{\pi_H^{\pm}} \approx \left(\frac{(13-2)1.5\sqrt{3}}{1 - 1/13^2} \right)^{1/3} I_U^{1/3} = \left(\frac{1859\sqrt{3}}{112} \right)^{1/3} I_U^{1/3}.$$

50. From 5 and 49, likely

$$\frac{I_{\pi_L^+}}{I''_{\pi_L^+}} \approx \frac{I_{\pi_L^-}}{I''_{\pi_L^-}} \approx \frac{13}{\left(\frac{557.7\sqrt{3}}{34} \right)^{1/3} I_U^{1/3}} = \left(\frac{442}{3.3\sqrt{3}} \right)^{1/3} \Big/ I_U^{1/3}$$

and

$$\frac{I_{\pi_H^+}}{I''_{\pi_H^+}} \approx \frac{I_{\pi_H^-}}{I''_{\pi_H^-}} \approx \frac{13}{\left(\frac{1859\sqrt{3}}{112}\right)^{1/3} I_U^{1/3}} = \left(\frac{728}{5.5\sqrt{3}}\right)^{1/3} \Big/ I_U^{1/3}.$$

51. From definition of 'd', and 47–49, likely 1 bit of I_x coded $d = +1$ in $\pi_L^{\pm}$ and -1 in $\pi_H^{\pm}$ coded consistent with 'spin 0'.
52. From A-Theorem 7, L 17.2–17.6, A-Def 20, and 5, 6, and 51, likely $\pi_{L,H}^{\pm}$ decays preserving amount of internal information, coding of 'charge' sign, and 'closest' coding of d are approximately describable as

$$\pi_L^+ \to \nu_{\mu(L)} + \mu_L^+ \begin{cases} \nearrow e^+ 4\gamma \\ \searrow e^+ \nu_e \bar{\nu}_e \end{cases}$$

$$\pi_L^- \to \bar{\nu}_{\mu(L)} + \mu_L^- \begin{cases} \nearrow e^- 4\gamma \\ \searrow e^- \nu_e \bar{\nu}_e \end{cases}$$

$$\pi_L^+ \to e^+ + \nu_e + (8\gamma \text{ or } e^+e^-2\gamma)$$

$$\pi_L^- \to e^- + \bar{\nu}_e + (8\gamma \text{ or } e^+e^-2\gamma)$$

$$\pi_L^+ \to e^+ + \nu_e + \nu_e + \bar{\nu}_e + 4\gamma$$

$$\pi_L^- \to e^- + \bar{\nu}_e + \nu_e + \bar{\nu}_e + 4\gamma$$

$$\pi_L^+ \to e^+ + \nu_e + 2\nu_e + 2\bar{\nu}_e$$

$$\pi_L^- \to e^- + \bar{\nu}_e + 2\nu_e + 2\bar{\nu}_e$$

$$\pi_H^+ \to \nu_{\mu(H)} + \mu_H^+ \searrow e^+ \overset{(-)}{\nu}_e \overset{(-)}{\nu}_e$$

$$\pi_H^- \to \bar{\nu}_{\mu(H)} + \mu_H^- \searrow e^- \overset{(-)}{\nu}_e \overset{(-)}{\nu}_e$$

53. From L 17.5 and 17.6, and 49 and 52, likely rates of transcription of information approximately describable by $\pi_L^{\pm} \to \overset{(-)}{\nu}_{\mu(L)} + \mu_L^{\pm}$ and by $\pi_H^{\pm} \to \overset{(-)}{\nu}_{\mu(H)} + \mu_H^{\pm}$ are approximately describable as equal.
54. From 52 and 53, likely $\pi_L^{\pm}$ decays by modes other than in step 53 not substantially present in $\pi_H^{\pm}$ decay.
55. From 52–54, likely $\tau_{\pi_L^{\pm}} < \tau_{\pi_H^{\pm}}$.
56. From definition of 'b', and 7–52, likely not any of I_x is coded as b.

Step 31 might seem like Corollary 14.8 with respect to amount of information available appearing as not enough for description dividing 3-space more finely than for description of interacting 'parts' as 'lumps'. (See also Comment 4.8, p. 309.)

Note that in Link 17.7 treatment of phase relations is substantially different from that in Corollary 14.8. This also seems likely relevant for Link 17.6 steps 5 and 6, in that a three-photon (incoherent) entity would seem likely not able to formally supply d coding for a two-photon (coherent) entity, although 'more easily' able to supply same d value than opposite d value (" 'closest' coding of d" in step 52).

Likeliness of $\tau_{\pi_L^\pm} < \tau_{\pi_H^\pm}$ appears traceable substantially to likeliness of in $e^\pm$ $d = +1$ only: likely $\pi_L^\pm \to e^\pm +$ (other information representations not coding or complementing internal information coding approximately describable by d ('d coding')) would approximately preserve d coding, and $\pi_H^\pm \to e^\pm +$ (such other information representations) would not.

We now have several different classes of entities. For convenience in reference (definition numbered in sequence following merged sequence including Section 2),

Definition 21. *For information representations representing internally* $\leqslant 1$ *bit of information by relative phase of associated photons, with* I_x, I_x' *and* I_x'' *as defined in conditions for Chain* 17,

I_x = *amount (in bits) of internal information, finite integer* > 0;

n = *number of photons for approximate substantial description of entity, finite integer* >0;

b = $\pm b$ *or* $\pm b_\mu$, *if not any of* I_x *coded as* b, $b=0$, *may be written as* $+, -, 0$ *or as* $+b, -b, +b_\mu, -b_\mu, 0$;

d = *contribution from relative phase coding to Theorem* 13 *interaction in entity,* $d = +1, -1, 0$, *may be written as* '+', '−', '0';

f = *relative phase coding, by convention* '±' *for* ± *'charge' sign, or* '0' *if not any of* I_x *coded as relative phase,* $f = +, -, 0$;

$d=0$ *iff* $f=0$;

Let:

(*a*)	W I_x, I_x'	$I_x \geqslant 1$	*designate 'wave' as in* A-*Theorems* 1, 3, *and* 4, A-*Corollary* 4.7, *and Link* 17.2;
(*b*)	N n, I_x, b, I_x'	$I_x \geqslant n$	*designate, for* $n=2$, *'noncoupled' ('fast') form as in Corollaries* 8.6–8.15, *Theorem* 15–*Corollary* 15.12, *and Links* 17.3 *and* 17.6;

(*c*) E n, I_x, b, d, f $I_x > n$ *designate, for* $n=2$, *'extended' interaction coupled form as in Corollaries* 14.3, 14.5, 14.15, *and* 14.16, *and Link* 17.4;

(*d*) C n, I_x, b, d, f $I_x > n$ *designate relatively localized substantially 'circular' form; for* $n=2$, *as in Corollary* 14.8 *and Link* 17.5; *for* $n=3$, *as in Link* 17.7;

where only $d=0$ *may be written for* $d=f=0$;

with (*from Corollary* 14.8 *and Link* 17.5)

$$I''_{C2,I_x,b,d,f} \approx \left(\frac{Y(I_x)}{1+d/I_x^2} \right)^{1/3} I_U^{1/3},$$

and (*from Link* 17.7)

$$I''_{C3,I_x,b,d,f} \approx \left(\frac{(I_x-2)1.5\sqrt{3}}{1+d/I_x^2} \right)^{1/3} I_U^{1/3};$$

in which Definition 21 '0' might be thought of as 'not any information so coded', and restrictions in range of I_x for each form are traceable to Postulate 3 (finite accessibility of I_x) and Theorem 14 (that I_x of a 'bound' entity be at least 1 bit more than the sum of the internal information of the photons so coupled). From A-Definition 22, corresponding 'rest frame' 'energy' $E_x = I''_x E_1$ and 'mass' $m_x = I''_x E_1/c^2$.

For convenience in the next link, noting from (A-Cor 2.1, A-Theorems 3 and 4, and) A-Definition 22 that $I''_x = m_x c^2/E_1$, and noting cancellation in numerator and denominator of d terms in Definition 21,

Definition 22. *Let* $\delta 3,2,I$ *be defined by*

$$\delta 3,2,I = 1 - \frac{I_{C3,I,b,d,f}}{I_{C2,I,b,d,f}} \approx 1 - \frac{m_{C3,I,b,d,f}}{m_{C2,I,b,d,f}}$$

$$\approx 1 - \left(\frac{(I-2)1.5\sqrt{3}}{Y(I)} \right)^{1/3}.$$

For $\delta 3,2,I_x > 0$, $I''_{C3,I_x,b,d,f} < I''_{C2,I_x,b,d,f}$: this might be regarded as a 'well' or 'local minimum' in which a 'particle' (x) might exist based on a specific (3 'lobed') form of 'photon' interaction. (see also A-Theorem 5, A-Cor 5.1, Theorem 14, and Cor 14.1)

Link 17.8. *Likely an entity approximately describable as* C3,I_x,b,d,f *with* $I_x < 12$ *would decay much more rapidly than* $\pi^{\pm}_{L,H}$.

1. From A-Theorems 3 and 4, in U, I_x is an integer >0.
2. From L 17.1, Def 22, and 1, iff $I_x < 12$, $\delta 3,2,I_x < 0$.
3. From Postulate 2, A-Theorem 1, A-Def 22, and 2, likely decay of an entity approximately describable as C3,I_x,b,d,f with $\delta 3,2,I_x < 0$ into two decay products approximately describable if substantially localized as C2,I_x,b,d,f or through an intermediate state approximately describable as C2,I_x,b,d,f would be much more rapid than for an entity approximately describable as C3, I_x, b, d, f with $\delta 3, 2, 13 \approx 1 - (16.5\sqrt{3}\,/29)^{1/3}$.
4. From L 17.7, and 2 and 3, likely an entity approximately describable as C3,I_x,b,d,f with $I_x < 12$ would decay much more rapidly than $\pi^{\pm}_{L,H}$.

●;

so rapidly as to seem substantially consistent with not having seen them, prior to and near this writing, as 'pion'-like particles. This might offer some insight into muons (likely $I_{\mu^{\pm}_{L,H}} = 7 < 12$) seeming more like electrons than like charged pions.

Link 17.9. $x = \pi^0$ (*'neutral pion'*):
likely: $I_x = 12$,

$$I''_{\pi^0} \approx I''_{C3,\,12,0,0,0} \approx (15\sqrt{3}\,)^{1/3} I_U^{1/3},$$

$$\frac{I_{\pi^0}}{I''_{\pi^0}} \approx \left(\frac{192\sqrt{3}}{5}\right)^{1/3} \Big/ I_U^{1/3};$$

not any of I_x *coded as 'charge' sign or* b.

1. From report(s) of observation, π^0 does not represent internally any information coded by distinction among states distinguished by 'spin'.
2. From report(s) of observation, π^0 does not have 'charge'. (prior conventional 'charge')
3. From 1, as in Link 17.7 steps 7–13, likely simplest approximate (substantial) description ('sasd') of π^0 'internal structure' is in terms of three photons.
4. From 2, likely not any of I_x is coded as 'charge' sign.
5. From L 17.4, 17.5, and 17.7, and 4, likely $I_{\pi^0} \leqslant I_{\pi^{\pm}_{L,H}} - 1 = 13 - 1 = 12$.

6. From report(s) of substantial observability of π^0, L 17.8, and 3, likely $I_{\pi^0} \geqslant 12$.
7. From 5 and 6, likely $I_{\pi^0} = 12$.
8. From Def 21, and 3, 4, and 7, as in Link 17.7 steps 14–28, and, noting that likely not any of I_{π^0} is coded as 'charge' sign, Link 17.7 steps 30–47, likely sasd of π^0 is approximately as C3, 12, b, 0, 0.
9. From Def 21, and 8, likely

$$I''_{\pi^0} \approx \left((12-2)1.5\sqrt{3}\right)^{1/3} I_U^{1/3} = \left(15\sqrt{3}\right)^{1/3} I_U^{1/3}.$$

10. From 7 and 9, likely

$$\frac{I_{\pi^0}}{I''_{\pi^0}} \approx \frac{12}{I''_{\pi^0}} \approx \frac{12}{\left(15\sqrt{3}\right)^{1/3} I_U^{1/3}} = \left(\frac{192\sqrt{3}}{5}\right)^{1/3} \Big/ I_U^{1/3}.$$

11. From definition of 'b', and 1 and 4, as in Link 17.7 step 56, likely not any of I_{π^0} is coded as b.

•

From Corollaries 14.3, 14.4, 14.8, 14.9, 14.13–14.16, 15.11, and 15.12, and Links 17.3–17.6, coding of 1 bit of internal information as 'charge' sign appears related to appearance of photon coupling in description of entities with 'internal structure' approximately describable in terms of two photons. As a substantially three-fold symmetric entity, π^0 would not have the two-fold spatial symmetry which might seem characteristic of transcription into complementary representations; but, not having any internal information coded as 'charge' sign, as a substantially two-fold symmetric system approximately describable as (possibly 'fragmented') C2, 12, 0, 0, 0, a π^0 would seem to be (merely) substantially two uncoupled photons. This might offer some insight into observation that π^0 appears to decay by $\pi^0 \rightarrow 2\gamma$ ~0.9883 of all π^0 decays, unlike $\pi^{\pm}_{L,H}$.

In view of the principal decay mode of π^0 being $\pi^0 \rightarrow 2\gamma$, not conserving internal information, and π^0 'lifetime' seeming related to decay via A-Theorem 1 ('electromagnetic') 'coupling', it is, at the time of this writing, not clear to me whether or not π^0 may be approximately describable as nearly enough adseparable for A-Corollaries 6.1 and 7.3 to be useful in approximate description of the amount of information accessible in π^0, and whether or not such description might be relevant in considering 'parity' in transcription of information (to be) represented as π^0s.

Because coding for 'charge' sign appears to correspond to internal representation of information in terms of states distinguished in systems whose photons are coupled by Theorem 13 interaction, it would seem that 'charge' sign might be representable only in systems whose associated photons are coupled. Thus, conservation of 'charge' sign might be a formal consequence of coding complementarity, rather than an approximation depending on separability of amounts of internal and external information, and presence of 1 bit of information coded as 'charge' sign thus formally sufficient for stability against 'decay' of the least massive entity coding such 'charge' sign bit.

Note that conservation of coding of 1 bit of internal information as 'charge' sign in various processes, such as $\pi^{\pm}_{L,H} \rightarrow \mu^{\pm}_{L,H}\ \overset{(-)}{\nu}_{\mu(L,H)}$, $\pi^{\pm}_{L} \rightarrow e^{\pm} +$ (others), or $\mu^{\pm}_{H} \rightarrow e^{\pm}\ \overset{(-)}{\nu}_e\ \overset{(-)}{\nu}_e$, does not appear to imply that 'electromagnetic' or 'electrostatic' interaction 'strength' associated with presence in such entities of 1 bit of internal information coded as 'charge' sign would be the same for all such entities with 1 bit so coded.

If a system S containing two entities coding for opposite 'charge' sign were observed from a distance large enough for its internal details to be formally unobservable, then its internal codings could be adseparable from its external information, and each such charge coding might be regarded as a property of S as a whole; thus, S might be adequately describable as not having 'net' 'charge' sign, a statement about coding. Because such 'neutrality' of such distant S appears adequately describable in terms of internal coding, such neutrality of pairs of 'charged' entities together does not appear to involve cancellation of equal and opposite effects, and, thus, does not appear to formally imply that components of such 'neutral at a distance' pairs have equal interaction 'strengths' ("'charge' magnitudes") with opposite sign.

In particular, it would seem that the pairs $e^{\pm}e^{\mp}$, $e^{\pm}p^{\mp}$, $\mu^{\pm}_{L,H}p^{\mp}$, and $\pi^{\pm}_{L,H}p^{\mp}$ might all be adequately describable as 'electrically neutral' at sufficient distance, without such 'neutrality at a distance' (formally) implying that each particle of such pairs have separately the same interaction 'strength' with 'electric' and/or 'magnetic' 'field(s)', and/or that 'electrical' interaction within such pairs be the same function of 'separation' between 'particles' comprising each such pair.

3.5.7 L 17.10 Interaction

Link 17.10. *In domain of adseparability, likely Theorem* 13 *interaction between x and relatively localized entity y representing amount* I_y *of information internally and amount* I_y'' *of information in U,* $I_i'' \gg I_i$ *for* $i = x, y$, *x and y* *substantially at rest as seen by O_U, can subsume I_x iff I_y'' is approximately describable by*

$$I_y'' \geqslant \sim \frac{12 I_U}{I_x''^2} \times \left\{ \begin{array}{ll} \textit{for } x \sim \mathrm{E2}, I_x, b, d, f & Y(I_x) \\ \textit{for } x \sim \mathrm{C2}, I_x, b, d, f & Y(I_x) \\ \textit{for } x \sim \mathrm{C3}, I_x, b, d, f & (I_x - 2) \end{array} \right\},$$

where '$\sim$' is used in brackets to mean 'approximately describable by'.

1. From A-Theorem 6, in domain of adseparability, adequate description of external information representation by x and/or y may be chosen independent of description of internal information representation by x and/or y.
2. From A-Theorem 1, A-Cor 1.3, 2.1, and 2.2, A-Theorem 5, A-Cor 5.1, A-Def 7, 8, and 12, and 1, noting A-Cor 2.2 step 22 (least B_0 velocity >0), in domain of adseparability, adequate description of external information representation by x or y substantially localized and substantially at rest in 3-space as seen by O_U in U involves description of x or y as substantially at rest separably with respect to three independent 3-space basis variables, by means of linear combinations of descriptions which with respect to each such basis variable describe representation with velocity >0.
3. From A-Cor 4.8, A-Theorem 5, A-Cor 5.1, A-Theorem 6, and 2, such description of x or y involves three pairs of single B_0 ψs with oppositely directed velocities in each pair, each such single B_0 ψ comprising two ws.
4. From A-Cor 2.1, and 3, such description of x or y substantially at rest as seen by O_U in U involves 12 ws, each adequately describing representation in U of approximately the same amount of information.
5. From A-Theorems 1 and 4, A-Cor 4.3, and 4, for adequate description of representation of an amount >0 of information, each such w has approximately the same frequency $f>0$ and wavelength $\lambda = c/f$.
6. From A-Theorem 1, and 4 and 5, for adequate description of external information representation by x or y substantially at rest as seen by O_U in U, each such w must adequately describe nonpropagating information representation.
7. From A-Theorem 1, A-Cor 1.3, and A-Def 8, nonpropagating solutions of A-Theorem 1 WE with $f>0$ have amplitude which depends exponentially on distance.
8. From Postulate 1, A-Theorems 3 and 4, A-Def 11, and 4 and 5, for $I_i \ll I_i''$, $i = x,y$, each such w has $\lambda_i \approx 12\lambda_1 / I_i''$.

9. From A-Theorem 1, A-Cor 1.3, A-Def 8, and 5–8, noting isotropy and homogeneity, such external information representation by x or y is approximately describable by w amplitude, depending on approximate 3-space distance r from approximate location of x or y, $A_i(\vec{r}) \approx C_{1,i} r e^{-2\pi r/\lambda_i}$, $i = x, y$, where $C_{1,i}$ includes suitable 'normalization'.
10. From A-Lemma 2.1, and 9, spatial distribution of information, represented as approximately described by $A_i(r)$, is approximately describable by

$$\rho_i(\vec{r}) \approx C_{2,i}(A_i(\vec{r}))^2 \approx C_{3,i} r^2 e^{-4\pi r/\lambda_i},$$

$i = x, y$, where $C_{2,i}$ and $C_{3,i}$ include suitable 'normalization(s)'.
11. From Postulate 1, and 10, for $I_i'' \gg I_i$,

$$\int_U d\vec{r} \rho_i(\vec{r}) \approx I_i'', \qquad i = x, y.$$

12. From 10 and 11,

$$\int_0^{R_U} dr\, 4\pi C_{3,i} r^2 e^{-4\pi r/\lambda_i} \approx I_i'', \qquad i = x, y.$$

13. From 8, for $I_x'' \ll I_y''$, $\lambda_x \gg \lambda_y$.
14. From Theorem 13, and 13, in 'weak field' case, for $\lambda_x \gg \lambda_y$, Theorem 13 interaction between x and y is approximately describable in terms of x spatial distribution without explicit regard for spatial extent of y.
15. From Theorem 13, A-Def 22, and 10, 12, 13, and 14, largest 'energy' $\Delta E = |\Delta E|$ 'released' by Theorem 13 interaction between x and y would occur for x and y distributed about substantially the same position, and have magnitude approximately describable by

$$\Delta E \sim G' I_x'' I_y'' \frac{\int_0^{R_U} dr\, 4\pi r C_{3,x} e^{-4\pi r/\lambda_x}}{\int_0^{R_U} dr\, 4\pi r^2 C_{3,x} e^{-4\pi r/\lambda_x}}$$

$$= G' I_x'' I_y'' \frac{\int_0^{R_U} dr\, r e^{-4\pi r/\lambda_x}}{\int_0^{R_U} dr\, r^2 e^{-4\pi r/\lambda_x}},$$

where 'G''' is 'G' of Theorem 13 in appropriate units. (the numerator integrals include a factor of $1/r$ from Theorem 13 interaction, the denominator integrals provide 'normalization')

16. From 15, for $\lambda_x \ll R_U$, integrals in step 15 may be treated approximately as definite integrals conventionally written with hypothetical 'infinite' limit of integration as

$$\int_0^{`\infty'} dr\, r^n e^{-ar} = \frac{n!}{a^{n+1}}, \qquad a>0,\ n \text{ positive integer.}$$

(e.g., see item 661, p. A165, ref. 9; for I_x not too large and $\lambda \ll R_U$, $\int_{R_U}^{`\infty'} dr\, 4\pi r^2 \rho(r) \triangleq 0$, as turns out likely here)

17. From 15 and 16,

$$|\Delta E| \sim G' I_x'' I_y'' \frac{(\lambda_x/4\pi)^2}{2(\lambda_x/4\pi)^3} = \frac{G' I_x'' I_y'' 2\pi}{\lambda_x}.$$

18. From 8 and 17,

$$\Delta E \sim \frac{G' I_x'' I_y'' 2\pi}{12\lambda_1/I_x''} = \frac{G' I_x''^2 I_y'' 2\pi}{12\lambda_1}.$$

19. From Theorem 13, Cor 14.5, and L 17.4, 17.5, 17.7, and 17.9, for $I_x'' \ll I_y''$, likely Theorem 13 interaction between x and y can subsume I_x iff ΔE is partially approximately describable as related to 'energy' ΔE_x 'released' in Theorem 13 interaction coupling of *ws* in x by $\Delta E \geqslant \Delta E_x$. (this might be thought of as a condition for 'pieces' of x to be bound more strongly to y than to each other)

20. From Theorem 13, A-Def 22, and 15,

$$G' \approx c^2 \frac{R_U E_1^2}{M_U c^4} = c^2 \frac{R_U E_1^2}{I_U E_1 c^2} = \frac{E_1 R_U}{I_U}.$$

21. From 18–20, likely I_x would be so subsumed iff

$$\frac{E_1 R_U I_x''^2 I_y'' 2\pi}{I_U 12\lambda_1} \geqslant \sim \Delta E_x.$$

22. From A-Def 18, and 21, substituting for λ_1 and solving for I_y'', likely I_x

would be so subsumed iff

$$\frac{E_1 R_U I_x''^2 I_y'' 2\pi}{I_U 12(2\pi R_U)} \geqslant \sim \Delta E_x,$$

cancelling $2\pi R_U$s, and transposing,

$$I_y'' \geqslant \sim \frac{12 I_U \Delta E_x}{I_x''^2 E_1}.$$

23. From Cor 14.2, 14.3, 14.5, 14.8, 14.15, and 14.16, L 17.4, A-Def 22, and 19, likely for x's *ws* coupled by Cor 14.5 interaction, $\Delta E_x \approx Y(I_x) E_1$.
24. From L 17.5, A-Def 22, and Def 21, likely for x approximately describable by C2, I_x, b, d, f, $\Delta E_x \approx Y(I_x) E_1$.
25. From L 17.7 and 17.9, A-Def 22, and Def 21, likely for x approximately describable by C3, I_x, b, d, f, $\Delta E_x \approx (I_x - 2) E_1$.
26. From Def 21, and 14, 16, and 22–25, cancelling E_1s, for x and y substantially at rest as seen by O_U in U, $I_x'' \ll I_y''$, and $\lambda_x \ll R_U$, likely I_x would be so subsumed iff

$$I_y'' \geqslant \sim \frac{12 I_U}{I_x''^2} \times \left\{ \begin{matrix} \text{for } x \sim \text{E } 2, I_x, b, d, f & Y(I_x) \\ \text{for } x \sim \text{C } 2, I_x, b, d, f & Y(I_x) \\ \text{for } x \sim \text{C } 3, I_x, b, d, f & (I_x - 2) \end{matrix} \right\},$$

where '$\sim$' is used in brackets to mean 'approximately describable by'.

●

Note that, in approximate description as in Link 17.10, there appears to be enough available information for description dividing 3-space more finely than to characterize x as only a (single) 'lump', as distinct from treatment in Corollary 14.8 and Links 17.5, 17.7, and 17.9. (See also Comment 4.8, p. 309.)

Note also that adseparability is 'used' for choosing description of x before subsumption of I_x, and also in conserving I_x in such subsumption.

Likely subsuming I_x by interaction approximately describable as in Link 17.10 ('L 17.10 interaction') would involve loss of x's separate identity. Unlike mutual 'annihilation' of complementary information representations with transcription of all information represented thereby into representations which might not preserve any internal code from the 'annihilated' representations, L 17.10 interaction appears to be a 'mixing' interaction, which could involve formal conservation of various internal codes, such as 'charge' sign.

From Links 17.2, 17.3, and 17.6, γ and likely ν_e, $\bar{\nu}_e$, $\nu_{\mu(L,H)}$, and $\bar{\nu}_{\mu(L,H)}$ do not appear substantially at rest to O_U, as stated for L 17.10. From Links 17.4, 17.5, 17.7, and 17.9, likely e^+, e^-, μ_L^+, μ_L^-, μ_H^+, μ_H^-, π^0, π_L^+, π_L^-, π_H^+, and π_H^- can appear substantially at rest to O_U.

Link 17.11. *Likely L 17.10 interaction between substantially localized entity y and entity* $x = e^+, e^-, \mu_L^+, \mu_L^-, \mu_H^+, \mu_H^-, \pi^0, \pi_L^+, \pi_L^-, \pi_H^+$, *or* π_H^-, *in U, substantially at rest as seen by* O_U, *iff for each such x* I_y'' *is approximately describable by*

x	$I_y'' \geqslant \sim$
e^+, e^-	$\frac{48}{3.6^{2/3}}(32\pi^2(\ln(2I_{e\pm}''))^2)^{2/3}I_U^{1/3}$
μ_L^+, μ_L^-	$\frac{144}{11.76^{2/3}}I_U^{1/3}$
μ_H^+, μ_H^-	$\frac{144}{12.25^{2/3}}I_U^{1/3}$
π^0	$\frac{120}{(15\sqrt{3})^{2/3}}I_U^{1/3}$
π_L^+, π_L^-	$132\left(\frac{34}{557.7\sqrt{3}}\right)^{2/3}I_U^{1/3}$
π_H^+, π_H^-	$132\left(\frac{112}{1859\sqrt{3}}\right)^{2/3}I_U^{1/3}$

1. From L 17.1 and 17.4, likely $Y(I_{e\pm}) = 4$.
2. From L 17.4, likely $I_{e\pm}'' \approx 3.6^{1/3}F^{-1}I_U^{1/3}$, where $F = (32\pi^2(\ln(2I_{e\pm}''))^2)^{1/3}$.
3. From Cor 14.5, 14.15, and 14.16, L 17.4 and 17.10, and 1 and 2, likely L 17.10 interaction between e^+ or e^- and y iff

$$I_y'' \geqslant \sim \frac{12I_U4}{3.6^{2/3}F^{-2}I_U^{2/3}} = \frac{48}{3.6^{2/3}}F^2I_U^{1/3}$$

$$= \frac{48}{3.6^{2/3}}\left(32\pi^2(\ln(2I_{e\pm}''))^2\right)^{2/3}I_U^{1/3}.$$

4. From L 17.5 and Def 21, likely $\mu^{\pm}_{L,H}$ are approximately describable by C2, 7, $\pm$, $\pm$, $\pm$.
5. From L 17.1 and 17.5, likely $Y(I_{\mu^{\pm}_{L,H}})=12$.
6. From L 17.5 and 17.10, and 4 and 5, likely L 17.10 interaction between μ_L^+ or μ_L^- and y iff

$$I_y'' \geqslant \sim \frac{12 I_U 12}{11.76^{2/3} I_U^{2/3}} = \frac{144}{11.76^{2/3}} I_U^{1/3}.$$

7. From L 17.5 and 17.10, and 4 and 5, likely L 17.10 interaction between μ_H^+ or μ_H^- and y iff

$$I_y'' \geqslant \sim \frac{12 I_U 12}{12.25^{2/3} I_U^{2/3}} = \frac{144}{12.25^{2/3}} I_U^{1/3}.$$

8. From L 17.9, likely π^0 is approximately describable by C3, 12, 0, 0, 0.
9. From L 17.9 and 17.10, and 8, likely L 17.10 interaction between π^0 and y iff

$$I_y'' \geqslant \sim \frac{12 I_U 10}{(15\sqrt{3})^{2/3} I_U^{2/3}} = \frac{120}{(15\sqrt{3})^{2/3}} I_U^{1/3}.$$

10. From L 17.7 and Def 21, likely $\pi^{\pm}_{L,H}$ are approximately describable by C3, 13, 0, $\pm$, $\pm$.
11. From L 17.7 and 17.10, and 10, likely L 17.10 interaction between π_L^+ or π_L^- and y iff

$$I_y'' \geqslant \sim \frac{12 I_U 11}{\left(\frac{557.7\sqrt{3}}{34}\right)^{2/3} I_U^{2/3}} = 132\left(\frac{34}{557.7\sqrt{3}}\right)^{2/3} I_U^{1/3}.$$

12. From L 17.7 and 17.10, and 10, likely L 17.10 interaction between π_H^+ or π_H^- and y iff

$$I_y'' \geqslant \sim \frac{12 I_U 11}{\left(\frac{1859\sqrt{3}}{112}\right)^{2/3} I_U^{2/3}} = 132\left(\frac{112}{1859\sqrt{3}}\right)^{2/3} I_U^{1/3}.$$

13. From L 17.4, 17.5, 17.7, and 17.9, and 3, 6, 7, 9, 11, and 12, likely I_y'' substantially larger than I_x'' (and I_y' substantially larger than I_x').

3.5.8. Comparisons With Reports of Observations

We can compare various of these results with reports of observations. For these comparisons, I have used the value of $F(\approx 137.0387173)$ approximately calculated, as presented hereinbefore, from Cohen and Taylor (CT, ref. 4)'s recommended values for G, $\hbar$, c, and m_e as listed in Table 1, page 150.

As noted hereinbefore, following Corollary 14.15, the least precisely stated input quantity in such approximate calculation of F was G. In view of CT's use of a value for e/h in obtaining their recommended values for m_e and α^{-1}, and the ~ 19.54 ppm difference between their recommended value for α^{-1} and F so approximately calculated, there might seem to be some ground for questioning the consistency of various sets of input values which might be used for these approximate calculations for comparisons with reports of observations. In view of the relatively small partial derivatives of F with respect to G, $\hbar$, m_e, and c, and of such questions, such value for F, so obtained from such values for G, $\hbar$, m_e, and c, has been used for these comparisons.

Reported values for particle properties have been taken for these comparisons from N. Barash-Schmidt+ (ref. 8) except as otherwise noted, including recommended value (from CT) for m_ec^2 ($\approx$511003.37 eV); however, note that their report stated (page 60, ref. 8), in their table of reports of values for m_μ, that they used for m_e 511004.1 eV/c^2 in obtaining m_μ from at least one prior report.

For these approximate calculations, the approximate expression for m_ec^2 was used to obtain, from $F \approx 137.0387173$ and $m_ec^2 \approx 511003.3733$ eV, $I_U^{1/3}E_1 \approx 45.69123378$ MeV (last two digits may contain computational error) (1 MeV$=10^6$ eV), and all effects of variation of input values for G, $\hbar$, c, and m_e on approximate calculated values for 'mass energies' of $\mu^\pm_{L,H}$, π^0, and $\pi^\pm_{L,H}$ used in these comparisons with reports of observations are through $I_U^{1/3}E_1$. Note that $I_U^{1/3}E_1$ contains, through F, propagated 'error' from G of ~4.5 ppm/σ, but might contain (a) larger nonstatistical error traceable through m_e (see also discussion of 'error' and error, pp. 147–149). Values depending on E_1, such as approximate values for 'rest mass energies' of ν_e, $\bar{\nu}_e$, $\nu_{\mu(L,H)}$, and $\bar{\nu}_{\mu(L,H)}$, depend through E_1 on G, $\hbar$, c, and m_e.

See also Tables 1 (p. 150) and 2 (p. 155) for numerical values and propagated 'error'.

Links 17.3 – 17.11 are stated in terms of information. From A-Definition 22, to convert to 'energy' units, one may merely multiply by E_1, or, to convert to 'mass' units, by E_1/c^2.

For convenience, I have tabulated various results obtained hereinbefore, approximate calculated values, and reports, as Table 3. In Table 3, rows are numbered at the left for reference; and columns are numbered, and include the following:

1. x, particle or item;
2. likely I_x of particle;
3. likely approximate description (see Def 21);
4. reference to details presented hereinbefore;
5. likely whether or not I_x includes code for 'b', and, if so, what code;
6. L, report(s) of whether or not particle 'has' 'lepton number', and if so, what;
7. likely approximate expression for rest frame I_x'' in bits of particle, or for value of other item, except formal $m_\gamma \triangleq 0$;
8. likely approximate calculated value for item, unless otherwise noted 'rest mass energy' $I_x'' E_1$ in MeV, except formal $m_\gamma \triangleq 0$;
9. reported value for item, unless otherwise noted 'rest mass energy' in MeV;
10. approximate fractional difference (⑧–⑨)/⑧ between approximate calculated value ⑧ and reported value ⑨, or statement of substantial consistency;
11. likely approximate expression for least I_y'' in bits of particle (y) with which particle x could interact substantially at rest by L 17.10 interaction;
12. likely approximate calculated value for such least 'mass energy' $I_y'' E_1$, in MeV;
13. likely whether or not such approximate calculated least 'mass energy' $I_y'' E_1$ is less than rest mass energies $m_p c^2$ of proton ('p') or $m_n c^2$ of neutron ('n'), for which comparisons were used values $m_p c^2 \approx 938.2796$ MeV and $m_n c^2 \approx 939.5731$ MeV (ref. 8), thus whether ('YES') or not ('NO') L 17.10 interaction likely between x and p or n;
14. Report(s) as to whether ('YES') or not ('NO') particle x interacts by 'strong interaction' with p or n;
15. likely approximate expression for (A) I_x/I_x' or (B) I_x/I_x''; and

16. likely approximate calculated value for (A) I_x/I'_x or (B) I_x/I''_x.

Note (from A-Def 22 (and conditions for Chain 17)) that $I'_x = E_x/E_1$; and that, in any particular line of Table 3, E_x may be written unambiguously as E for x of that line, without subscript 'x'.

To recalculate numerical values appearing in columns 8, 10, 12, and 16 with new input values for G, c, $\hbar$, and m_e, one may first solve as in item 6 for an approximate value for F (e.g., numerically), and then obtain an approximate value for $I_U^{1/3}E_1$ from item 5 with m_e as an input; and find E_1 from $E_1 \approx Gm_e^3F^3c/3.6\hbar$ and $I_U^{1/3}$ from $I_U^{1/3}E_1/E_1$. Ten digit numerical values (last two digits arithmetically suspect) for various coefficients of $I_U^{1/3}$ have been included for your convenience. Note that $I_U^{1/3}E_1$ does not appear in ratios of 'rest mass (energy)' of particles which can exist substantially at rest relative to observer O_U for mean times long compared to one cycle of their associated photons.

Although portions of Chain 17–Link 17.11 involve at various steps 'likeliness' arguments, and (thus) do not exhibit the formal logical purity sought in development in preceding portions hereof of formal structure based on the physical picture presented herein, relatively less formal reasoning seems useful for some explorations of the relation between such formal theory and observation.

With this cautionary note, we next consider some of the relations appearing in Table 3. (Please see Table 3 on next three pages.)

3.5.8.1 Table 3, Items 13–20, and 22 (Pions). The value recommended by N. Barash-Schmidt+ (ref. 8, University of California Lawrence Berkeley Laboratory Particle Physics Group, 'UCLBL', 1976) for '$m_{\pi^\pm}$' appears to be based on a set of reported values based variously on interpretations of spectra of π-mesic atoms (refs. 10–13), range in emulsions (ref. 14), and $\pi^- + p^+ \rightarrow n + \gamma$, $\gamma \rightarrow e^+e^-$, $e^\pm$ paths curvature(s) in magnetic field (ref. 15). Such spectra and range interpretations would seem to depend on assumptions about 'electrostatic' and/or 'electromagnetic' interaction 'strength(s)' of π^+s and π^-s. As noted following Link 17.9, conservation of charge coding and neutrality of distant systems of oppositely charged entities might not imply equality of electric interaction 'strengths' of $e^\pm$, $\mu^\pm$, and $\pi^\pm$. For this reason, it appears to me that, at the time of this writing, such various measurements are not measurements of only '$m_{\pi^\pm}$', but include dependences on interaction 'strengths'.

The m_{π^-} by $\gamma \rightarrow e^+e^-$ pair energy measurement (Kenneth M. Crowe and Robert H. Phillips, ref. 15) used an internal target in a cyclotron, with relatively short π^- beam transport distance, compared to other m_{π^-} experiments referenced herein. In view of, from L 17.7, likely $\tau_{\pi_L^\pm} < \tau_{\pi_H^\pm}$, it

Table 3.

Some Comparisons with Reports of Observations of Various Particle Properti

	① x	② ℒ I_x	③ ℒ𝒜𝒟 (Def 21[a])	④ See Also	⑤ ℒ 'b'	⑥ ℜ L	⑦ ℒ𝒜ℰ 'Rest Information' I''_x ean	⑧ ℒ𝒜𝒞𝒱 $I''_x E_1$ (MeV) ean	⑨ ℜ𝒜 $m_x c^2$ (MeV)ea
1	γ	1	W1, E/E_1	L 17.2	NO	NO	$m_x \overset{\blacktriangle}{=} 0$	$\overset{\blacktriangle}{=} 0$	<7×10⁻
2	ν_e	2	N2,2,+, E/E_1	L 17.3	+1	+1	$I''_x \approx 2(E/E_1)^{1/2}$	5.058570742×	< 60 e
3	$\bar{\nu}_e$	2	N2,2,−, E/E_1		−1	−1	$=2I'^{1/2}_x$	$10^{-17}E_x^{1/2}$ ($I''_x E_1$ and E_x in eV	
4	e^+	3	E2,3,−,+,+	T 14−C14.16	−1	−1	$I''_x \approx 3.6^{1/3}F^{-1}I_U^{1/3}$ with	(INPUT) 0.511003	
5	e^-	3	E2,3,+,+,−	L 17.4	+1	+1	$F=(32\pi^2(\ln(2I''_x))^2)^{1/3}$		
6	α^{-1}		as F	Tbl 1	$F^3\exp(F^3/32\pi^2)^{1/2}$ $\approx 7.2\hbar c/m_e^2 G$			$F \sim 137.0387173$ (number)	$\alpha^{-1} \sim$ 137.038 137.036
7	$\nu^{f}_{\mu(L,H)}$	6	N2,6,+, E/E_1	L 17.6	$+b_\mu$	+1	$I''_x \approx 2(5E/E_1)^{1/2}$ $=2\sqrt{5}\ I'^{1/2}_x$ or	(1.131130805 or 1.599660525) $\times 10^{-16}E_x^{1/2}$	<0.65
8	$\bar{\nu}^{f}_{\mu(L,H)}$	6	N2,6,−, E/E_1		$-b_\mu$	−1	$\approx 2(10E/E_1)^{1/2}$ $=2\sqrt{10}\ I'^{1/2}_x$	($I''_x E_1$ and E_x in eV)	
9	μ_L^+	7	C2,7, −, +, +	L 17.5	$-b_\mu$	−1	$I''_x \approx 11.76^{1/3}\ I_U^{1/3}$	103.9047318	——
10	μ_L^-	7	C2,7, +, +, −		$+b_\mu$	+1	$(\sim 2.27406273 I_U^{1/3})$		
11	μ_H^+	7	C2,7, −, −, +		$-b_\mu$	−1	$I''_x \approx 12.25^{1/3}\ I_U^{1/3}$	105.3282612	105.659
12	μ_H^-	7	C2,7, +, −, −		$+b_\mu$	+1	$(\sim 2.305218146 I_U^{1/3})$		
13	π^0	12	C3, 12, 0, 0, 0	L 17.9	NO	NO	$I''_x \approx (15\sqrt{3}\ )^{1/3}\ I_U^{1/3}$ $(\sim 2.961765221 I_U^{1/3})$	135.3267071	134.964
14	π_L^+	13	C3, 13, 0, +, +	L 17.7	NO	NO	$I''_x \approx \left(\frac{557.7\sqrt{3}}{34}\right)^{1/3} I_U^{1/3}$	139.4206181	compos value 139.56
15	π_L^-	13	C3, 13, 0, +, −		NO	NO	$(\sim 3.051364705 I_U^{1/3})$		
16	π_H^+	13	C3, 13, 0, −, +		NO	NO	$I''_x \approx \left(\frac{1859\sqrt{3}}{112}\right)^{1/3} I_U^{1/3}$	139.9716934	
17	π_H^-	13	C3, 13, 0, −, −		NO	NO	$(\sim 3.063425560 I_U^{1/3})$		
18	$I''_{\pi_L^\pm} - I''_{\pi^0}$		$\approx ((557.7\sqrt{3}/34)^{1/3} - (15\sqrt{3}\ )^{1/3}) I_U^{1/3}$ $(\sim 0.089599484 I_U^{1/3})$					4.093911016	compos value 4.60
19	$I''_{\pi_H^\pm} - I''_{\pi^0}$		$\approx ((1859\sqrt{3}/112)^{1/3} - (15\sqrt{3}\ )^{1/3}) I_U^{1/3}$ $(\sim 0.101660339 I_U^{1/3})$					4.644986361	
20	$I''_{\pi_H^\pm} - I''_{\pi_L^\pm}$		$\approx ((1859\sqrt{3}/112)^{1/3} - (557.7\sqrt{3}/34)^{1/3}) I_U^{1/3}$ $(\sim 0.012060855 I_U^{1/3})$					0.551075345	see te
21	$\tau_{\mu_L^\pm} \ll \tau_{\mu_H^\pm}$								see te
22	$\tau_{\pi_L^\pm} < \tau_{\pi_H^\pm}$								see te

⑩ 𝒜 ractional Difference[b] (⑧−⑨)/⑧	⑪ ℒ𝒜ℰ Least I_y'' for L 17.10 Interaction with x, for x Satisfying Conditions for L 17.10[b,c]	⑫ ℒ𝒜𝒞𝒱 Least such $I_y'' E_1$ (MeV)	⑬ ℒ L 17.10 Interaction of x and $p^{\pm}$ or $\overset{(-)}{n}$	⑭ ℛ 'Strong Interaction' of x and $p^{\pm}$ or $\overset{(-)}{n}$	(A) I_x/I_x' or (B) I_x/I_x'': ⑮ ℒ𝒜ℰ	(A) I_x/I_x' or (B) I_x/I_x'': ⑯ ℒ𝒜𝒞𝒱
consistent	(not rest[e])	—	NO	NO	(A) E_1/E	(A) (E in eV) $6.397\times 10^{-34}/E$
consistent	(not rest[e]) (not rest[e])	— —	NO NO	NO NO	(A) $2E_1/E$	(A) (E in eV) $1.279\times 10^{-33}/E$
(INPUT)	$I_y'' \geqslant \sim \frac{48}{3.6^{2/3}} F^2 I_U^{1/3}$ ($\sim 383759.795\, I_U^{1/3}$)	$\sim 1.7534 \times 10^7$	NO NO	NO NO	(B) $7.5^{1/3} F/I_U^{1/3}$	(B) 3.7557×10^{-39}
0.6035×10^{-6}	(from ref. 6) (1963) (mostly from spectroscopy and QED theory)					
19.537×10^{-6}	(ref. 4) (1973) (including Josephson junction $2e/h$ using BCS theory)					
consistent	(not rest[e]) (not rest[e])	— —	NO NO	NO NO	(A) $6E_1/E$	(A) (E in eV) $3.838\times 10^{-33}/E$
——	$I_y'' \geqslant \sim \frac{144}{11.76^{2/3}} I_U^{1/3}$ ($\sim 27.84566608\, I_U^{1/3}$)	~ 1272	NO NO	— —	(B) $(175/6)^{1/3}/I_U^{1/3}$	(B) 4.3098×10^{-41}
0.0031446	$I_y'' \geqslant \sim \frac{144}{12.25^{2/3}} I_U^{1/3}$ ($\sim 27.09807452\, I_U^{1/3}$)	~ 1238	NO NO	NO NO	(B) $28^{1/3}/I_U^{1/3}$	(B) 4.25156×10^{-41}
0.0026765	$I_y'' \geqslant \sim \frac{120}{(15\sqrt{3})^{2/3}} I_U^{1/3}$ ($\sim 13.67980757\, I_U^{1/3}$)	~ 625	YES	YES	(B) $\left(\frac{192\sqrt{3}}{5}\right)^{1/3} \Big/ I_U^{1/3}$	(B) 5.6727×10^{-41}
0.0010628	$I_y'' \geqslant \sim 132\left(\frac{34}{557.7\sqrt{3}}\right)^{2/3} I_U^{1/3}$ ($\sim 14.17704424\, I_U^{1/3}$)	~ 648	YES YES	(YES?) YES	(B) $\left(\frac{442}{3.3\sqrt{3}}\right)^{1/3} \Big/ I_U^{1/3}$	(B) 5.965×10^{-41}
0.0028784	$I_y'' \geqslant \sim 132\left(\frac{112}{1859\sqrt{3}}\right)^{2/3} I_U^{1/3}$ ($\sim 14.06563257\, I_U^{1/3}$)	~ 643	YES YES	(YES?) YES	(B) $\left(\frac{1456}{11\sqrt{3}}\right)^{1/3} \Big/ I_U^{1/3}$	(B) 5.9415×10^{-41}
0.124670268						
0.008759199						
onsistent	reinterpretation of report of Haddock + (1959) (ref. 16)					
onsistent						
onsistent?)						

(Script abbreviations: 𝒜pproximate, 𝒞alculated, can except as noted, ℰxpression for, ℒikely, ℛeported; 𝒱alue for)

continued

Footnotes to Table 3

Input Values for Approximate Calculation (SI units):

$G \sim 6.672 \times 10^{-11}$ m^3/(k sec^2)

$c \sim 299792458$ m/sec

$\hbar \sim 1.0545887 \times 10^{-34}$ j sec

$m_e \sim 9.109534 \times 10^{-31}$ k

1 eV$\sim 1.6021892 \times 10^{-19}$ j (for energy unit conversion)
(from Cohen and Taylor (1973), ref. 4)

Intermediate Values (see also Tables 1 and 2):

$F \sim 137.0387173$ (number)

$I_U^{1/3} E_1 \sim 45.69123378$ MeV (1 MeV$=10^6$ eV ('eV'))

$E_1 \sim 6.397284490 \times 10^{-34}$ eV

$I_U^{1/3} \sim 7.142285739 \times 10^{40}$

(Comparison values from N. Barash-Schmidt+ (1976), ref. 8, except as noted)

Notes: [a](Excerpted from Def 21: see Def 21 for details)

W I_x, I'_x	('Wave')
N n, I_x, b, I'_x	('Nonbound')
E n, I_x, b, d, f	('Extended')
C n, I_x, b, d, f	('Substantially Circular and Localized')

[b]Intermediate numerical values are stated as used in approximate calculations presented herein. The last digit of the value stated for F, the last one to two digits of all other 10-digit numbers, and to various amounts fractional differences appearing in column 10, are arithmetically suspect and not significant. See also Table 2 for 'error'.

[c]For 'Energy', multiply I''_x or I''_y entries in this column by E_1; for 'Mass', by E_1/c^2.

[d]If computational method from adseparability be approximately applicable ($v = v_m$).

[e]Does not satisfy conditions used in L 17.10.

[f]'(L, H)' subscript because not yet (12 June 1977)clear to me. (See text.)

appears noteworthy that their reported value for $m_{\pi^-} \sim (272.7 \pm 0.3) m_e \sim (139.3506 \pm 0.153)$ MeV/c^2 is ~ 0.07 MeV/c^2 less than the value approximately calculated for $m_{\pi_L^\pm}$, a fractional difference of ~ 0.00050 of the calculated value. This relatively close correspondence might suggest that, with $\tau_{\pi_L^\pm} < \tau_{\pi_H^\pm}$, pion beam transport distance might have been an important but unrecognized variable, significantly affecting other measurements of the average mass of 'surviving' charged pions.

There was one measurement, reported by John B. Czirr (ref. 17), which was referenced by UCLBL, but which appears to have not been included in calculating their recommended value (1976) for $m_{\pi^\pm}$ (appearing in Table 3, column 9), which was based on flight times of neutrons from $\pi^- + p^+ \rightarrow \gamma$ or $\pi^0 + n$, and thus not as dependent on assumptions regarding $\pi^\pm$ charge interaction strength, but with relatively long π^- beam transport distance. Czirr warned that his interpretation included various assumptions about systematic errors, discussed more fully hereinafter, and stated that certain features of his reported neutron time of flight distribution were not fully accounted for. However, Czirr's report appears, at the time of this writing, to be one of only two reports (with ref. 15) referenced by UCLBL for $m_{\pi^\pm}$ not strongly subject to problems of interpretation relating to $\pi^\pm$ charge interaction strength.

Czirr reported $m_{\pi^-} \approx (139.69 \pm 0.41)$ MeV/c^2, and, by neutron time of flight measurements in the same system, $m_{\pi^-} - m_{\pi^0} \approx (4.6056 \pm 0.0055)$ MeV/c^2, in which 'error' appears understated, as noted more fully hereinafter.

Using Czirr's formulas, substituting from UCLBL 1976 tables $m_p \approx 938.2796$ MeV/c^2 and $m_n \approx 939.5731$ MeV/c^2, and from CT (ref. 4) $c \approx 299792458$ m/sec, yields $m_{\pi^-} \approx 139.7043$ MeV/c^2 and, from such m_{π^-}, $m_{\pi^0} \approx 135.10353$ MeV/c^2, and $\Delta m = m_{\pi^\pm} - m_{\pi^0} \approx 4.6007$ MeV/c^2; or, substituting from UCLBL 1976 $m_{\pi^-} \approx 139.5688$ MeV/c^2 yields $m_{\pi^0} \approx 134.96517$ MeV/c^2 and $\Delta m \approx 4.60363$ MeV/c^2. (Such recalculations were under the same assumptions as stated by Czirr: see discussion hereinafter of Czirr's stated 'error'.)

From Table 3 lines 16 and 17, column 8, and the recalculated value (m'_{π^-}) from Czirr's data and formulas for m_{π^-}, the approximate calculated value for $m_{\pi_H^\pm}$ and m'_{π^-} differ by ~ 0.2674 MeV/c^2, or ~ 0.0019 of $m_{\pi_H^\pm}$ as approximately calculated, about 0.652σ per Czirr's report. In view of the questions regarding charge interaction strength, and the consistency to within 0.2% and 0.652σ with such report of such measurement not depending strongly on $\pi^\pm$ charge strength, from such report, and from reference 15 as discussed hereinbefore, the approximate calculated values for $m_{\pi_L^\pm}$ and $m_{\pi_H^\pm}$ appear, at the time of this writing, to be substantially allowed by these various reported prior observations.

If $\pi^{\pm}_{L,H}$ exist, various reported values for $m_{\pi^{\pm}}$ would be for unspecified mixture(s) of $\pi^{\pm}_{L}$ and $\pi^{\pm}_{H}$. From Link 17.11, and Table 3 columns 13 and 14, likely both $\pi^{\pm}_{L}$ and $\pi^{\pm}_{H}$ could participate in reactions of the form $\pi^{\pm}_{L,H}+p^{\mp}\rightarrow\pi^{0}+n$. From Link 17.9, likely m_{π^0} is approximately describable as one rest mass. With approximate values for m_{π^0} or $m_{\pi^{\pm}_{L,H}}$, and m_p and m_n, by measuring π^0 and/or n velocity from such reactions, one might obtain approximate values for $\Delta m_L = m_{\pi^{\pm}_{L}} - m_{\pi^0}$ and $\Delta m_H = m_{\pi^{\pm}_{H}} - m_{\pi^0}$.

I have looked through some prior published reports (refs. 10–30), and, at the time of this writing, I have found only one report, by Roy P. Haddock, Alexander Abashian, Kenneth M. Crowe, and John B. Czirr (ref 16, 'Haddock+'), which appears to report observation distinguishing between Δm_L and Δm_H.

Haddock+ report time of flight of neutrons from $x^{-}+p^{+}\rightarrow\pi^{0}+n$, where x^{-} was reported as π^{-}. With reference to their figure 2c (p. 479, ref. 16), two peaks may be seen in number N of neutrons per $\sim$4 nsec (1 nsec$=10^{-9}$ sec) time of flight 'bin' plotted against time of flight over 12.44 feet (3.791712 m) distance. The earlier peak appears in bins between 408 and 444 nsec, and, with a vernier caliper, was read as containing 252 counts, and the later peak, in bins between 460 and 468 nsec, as containing 48 counts. 'Background' may be approximately read as the average of counts (256) in all other (19) bins (between 380 and 500 nsec flight time), as $B\sim13.47368\pm0.0625$ counts per 4 nsec bin. (If one were to estimate 'background' for the later peak as the average of 8 adjacent bins, 4 on each side, the value so obtained would be $\sim$13.375 per bin, not substantially different at the statistical accuracies involved.)

From this, the earlier peak appears to contain $\sim 252-9B\sim130.73$ counts more than background in 9 bins, and the later peak appears to contain $48-2B\sim21$ counts more than background expected in two bins, an 'excess' $>4(2B)^{1/2}$.

If one were to treat such detected neutrons as Poisson distributed, one could ask what the probability P would be of seeing $\geqslant 48$ counts from average background of $\bar{n}\approx 2\times 13.47368$ counts, approximately expressed as the sum

$$P\approx\sum_{j=48}^{100}\frac{e^{-\bar{n}}\bar{n}^{j}}{j!},$$

where 100 was conveniently chosen so that size of each subsequent term in such series would be $<10^{-10}$ of P. Using a Hewlett Packard Model 67 programmable (pocket) calculator (for specificity as to read only memory programs, serial number 1610A00754), I found $P\approx1.596856\times10^{-4}$. One can also ask what, if $\bar{n}$ were 48, would be the probability of seeing 27 or fewer counts (rounding $2B$ up to 27 for convenience in calculating), by setting $\bar{n}=48$ and summing over $j=0$ to 27; such sum appeared to be $\approx 7.0395\times 10^{-4}$. From these two probabilities, it would seem that such later peak is likely statistically significant, at a 'confidence level' so defined >0.999.

Letting $c=1$ for the following steps 1–8, in domain of adseparability, for $x^{\pm}+p^{\mp}\rightarrow\pi^0+n$ substantially at rest, neglecting xp binding energy,

1. From A-Theorem 6 and A-Def 22, in domain of adseparability, one can choose a 'frame of reference' such that momenta $p_{\pi^0}\approx p_n\approx p$. ('center of momentum' 'system', ps are momentum magnitudes)
2. From A-Theorem 6 and A-Def 22, in domain of adseparability, energies E and rest masses m,

$$E_{\pi^0}+E_n\approx m_x+m_p.$$

3. From A-Cor 2.1, A-Theorem 6, and A-Def 22, in domain of adseparability, energy, mass, and velocity in units with $c=1$,

$$E_n\approx\frac{m_n}{\sqrt{1-v_n^2}}.$$

4. From A-Cor 2.1 and 2.2, A-Def 22, and 1,

$$E_{\pi^0}^2-m_{\pi^0}^2\approx p^2\approx E_n^2-m_n^2$$

$$E_{\pi^0}^2\approx E_n^2-m_n^2+m_{\pi^0}^2.$$

5. From 2 and 4, eliminating E_{π^0} and solving for E_n,

$$E_n^2-m_n^2+m_{\pi^0}^2\approx(m_x+m_p-E_n)^2$$

$$E_n^2-m_n^2+m_{\pi^0}^2\approx(m_x+m_p)^2-2(m_x+m_p)E_n+E_n^2$$

$$2(m_x+m_p)E_n\approx(m_x+m_p)^2+m_n^2-m_{\pi^0}^2$$

$$E_n\approx\frac{(m_x+m_p)^2+m_n^2-m_{\pi^0}^2}{2(m_x+m_p)}.$$

6. From 3, solving for v_n,

$$\sqrt{1-v_n^2}\approx\frac{m_n}{E_n}$$

$$1-v_n^2\approx\left(\frac{m_n}{E_n}\right)^2$$

$$v_n^2\approx 1-\left(\frac{m_n}{E_n}\right)^2$$

$$v_n\approx\left(1-\left(\frac{m_n}{E_n}\right)^2\right)^{1/2}$$

7. From 5 and 6,

$$v_n\approx\left(1-\left(\frac{2m_n(m_x+m_p)}{(m_x+m_p)^2+m_n^2-m_{\pi^0}^2}\right)^2\right)^{1/2}$$

8. Time interval Δt for n with velocity v_n to travel distance d is approximately describable by d/v_n.
9. From 8, and writing c explicitly (letting $v'=cv$), difference δt in times of arrivals of neutrons from $x_1+p\rightarrow\pi^0+n$ and $x_2+p\rightarrow\pi^0+n$ is approximately describable by

$$\delta t\approx d\left|(v_n'(x_1))^{-1}-(v_n'(x_2))^{-1}\right|/c.$$

10. From 7 and 9,

$$\delta t\approx\frac{d}{c}\left|\left(1-\left(\frac{2m_n(m_{x_1}+m_p)}{(m_{x_1}+m_p)^2+m_n^2-m_{\pi^0}^2}\right)^2\right)^{-1/2}\right.$$

$$\left.-\left(1-\left(\frac{2m_n(m_{x_2}+m_p)}{(m_{x_2}+m_p)^2+m_n^2-m_{\pi^0}^2}\right)^2\right)^{-1/2}\right|.$$

Substituting from UCLBL 1976 $m_p \approx 938.2796$ MeV$/c^2$ and $m_n \approx 939.5731$ MeV$/c^2$, and letting $m_{\pi^0} \approx 135.3267071$, $m_1 = m_{\pi_L^\pm} \approx 139.4206181$, and $m_2 = m_{\pi_H^\pm} \approx 139.9716934$, MeV$/c^2$, as approximately calculated as presented hereinbefore, the π_L^- neutron arrival peak would follow the π_H^- neutron arrival peak by δt of about 10.4995 nsec per meter of neutron flight distance. From Haddock+'s Figure 2c, the smoothed curve for the earlier peak appears centered on about 424 nsec, and from Table 1 run 2, noting the typographical error therein (434.34 was printed instead of 424.34), was reported as having occurred at $\sim(424.34 \pm 0.79)$ nsec. From its approximately equal division between the 460 to 464 and 464 to 468 nsec bins, the second peak appears to have occurred at ~464 nsec. At 12.44 ft (3.791712 m), the expected time interval between π_H^- and π_L^- neutron peaks would be ~39.811 nsec, as compared with an observed time interval of $\sim(39.66 \pm$ (at 1σ) about 1) nsec. Thus, the earlier peak appears to correspond to π_H^-, the previously recognized 'π^-', and, with the mass difference noted hereinbefore, the later peak appears to correspond (at >0.999 'confidence level') to existence of π_L^-, present, in Haddock+'s experiment, in an amount of ~13.8% of the total amount of $\pi_{L,H}^-$.

From Link 17.7, and Table 3, a $\pi_{L,H}^\pm$ mixture with ~14% $\pi_L^\pm$ would have average rest mass ~139.8945428 MeV$/c^2$, larger than the (composite) value as stated (not recalculated) from Czirr's report by ~0.20454 MeV$/c^2$, ~0.499σ stated by Czirr, or ~0.0014621 of such approximate calculated average for such mixture.

Czirr has suggested[17] that much of the width of the (first) peak might be due to reaction $\pi^- + p^+ \to \pi^0 + n$ occurring before stopping of the $\pi^- p^+$ system's motion (from π^-'s momentum) was substantially complete. With such interpretation of peak width, the relative narrowness of the second peak might suggest that $\pi_L^- + p^+ \to \pi^0 + n$ reaction occurs on the average later in the stopping process than $\pi_H^- + p^+ \to \pi^0 + n$. (With detection of a π^0 decay γ starting flight timing, this would not have substantially affected peak position.) This might be interpretable, in terms of d coding conservation and complementation, as suggesting that $p^\pm$ has $d = +1$ only, like $e^\pm$, so that $\pi_H^\pm p^\mp$ reaction could proceed by 'cancellation' of complementary d codes (-1 in $\pi_H^\pm$ and $+1$ in $p^\pm$), but that $\pi_L^\pm p^\mp$ reaction would involve some other (slower) transcription process with both d codes $= +1$.

As noted in Link 17.8 step 3, $\delta 3,2,13 > 0$. One might look for collisional excitation from C3,13,0,d,f to C2,13,?,d,f ('?' because not (yet?), at the time of this writing, formally clear to me). From Link

17.1 and Definition 18 and 21, $Y(13)=(2\times 1)+(6\times 2)+(5\times 3)=29$,

$$m_{\mathrm{C}2,13,?,d,f} \approx \left(\frac{29}{1+\frac{d}{169}} \right)^{1/3} I_U^{1/3} E_1 / c^2,$$

and, with $I_U^{1/3}E_1$ as in Table 3, $m_{\mathrm{C}2,13,?,+,f} \approx 140.1021539$ MeV$/c^2$ and $m_{\mathrm{C}2,13,?,-,f} \approx 140.6559233$ MeV$/c^2$. If such collisional excitation occurs, four peaks in neutron velocity distribution might be expected, related to C3, 13, 0, $+$, f, C3, 13, 0, $-$, f, C2, 13, ?, $+$, f, and C2, 13, ?, $-$, f, spaced successively, in order of increasing mass, starting with that associated with $\pi_L^{\pm}$, (C3, 13, 0, $+$, f), by $\sim$0.5510753, $\sim$0.1304605, and $\sim$0.5537694 MeV$/c^2$; the two higher mass peaks might have lower intensity, and the highest mass peak might be roughly as broad (from incomplete stopping?) as the peak listed second. Such higher mass peaks, if observed, might provide a relatively sensitive test of $Y(N)$ for $E_j = jE_1$, $j = 1, 2, 3$.

There is not any later peak in Haddock+'s Figure 2d (ref. 16), for their neutron flight distance of 17.5 ft (5.334 m), but such (narrow) peak would, from the time interval in Figure 2c multiplied by 17.5/12.44, be expected to have occurred slightly off the right edge of the plot in Figure 2d. Therefore, that such Figure 2d does not show a second peak appears not significant with respect to existence of π_L^-.

Other prior literature reviewed also appears consistent with existence of $\mu_L^{\pm}$ and/or $\pi_L^{\pm}$:

(1) Panofsky+[18] (1951): $\Delta m = m_{\pi^-} - m_{\pi^0} \sim (10.6 \pm 2.0) m_e$, by γ distribution width from π^0 decay: resolution as stated appears insufficient for use of their report as a test of existence of $\pi_L^{\pm}$.

(2) Chinowsky+[19] (1954): $\Delta m \sim (8.8 \pm 0.6) m_e$, by γ angular correlation from π^0 decay in flight: composite of $m_{\pi_H^-} - m_{\pi^0}$ and $m_{\pi_L^-} - m_{\pi^0}$; insufficient resolution, but they noted that their 'experimental resolution' was poorer than they had calculated (if on basis of one charged pion).

(3) Crowe+[15] (1954): $m_{\pi^-} \sim (272.7 \pm 0.3) m_e$, by $\pi^- + p^+ \rightarrow n + \gamma$ γ spectrum: insufficient resolution (see ref. 15, Figure 23 p. 482); data interpretation assumed one value for m_{π^-} (see ref. 15, eq. 24, p. 482); but note discussion hereinbefore of beam transport distance in relation to, from Link 17.7, likely $\tau_{\pi_L^{\pm}} < \tau_{\pi_H^{\pm}}$.

(4) Barkas+[14] (1956): $m_{\pi^+} \sim (273.3 \pm 0.2) m_e$, $m_{\pi^-} \sim (272.8 \pm 0.3) m_e$, $m_{\mu^+} \sim (206.9 \pm 0.2) m_e$, by ratios of track lengths in emulsions:

insufficient resolution (e.g., see ref. 14, Figure 8, p. 792); interpretation assumed one value each for m_{π^+} and m_{π^-}.

(5) Cini+[20] (1958): theoretical discussion about hypothesis of two different π^0s with different masses, indicating that such hypothesis would be inconsistent with scattering data. Such theoretical argument appears to not apply to the possibility of more than one π^+ or π^-.

(6) Cassels+[21] (1959): by γ angular correlation, on assumption of only one value of $\Delta m = m_{\pi^-} - m_{\pi^0}$, $\Delta m \sim (8.90 \pm 0.14) m_e$. Report also specifically addressed question of more than one value for Δm, in the context of a search for a second π^0, and such data may be considered instead in the context of a search for two distinct values for m_{π^-}. From ref. 21, Figure 4 (p. 94), such experiment would not have been sensitive to presence of 14% π_L^-s with mass difference noted hereinbefore.

(7) Gettner+[22] (1959): $\Delta m \sim (9.6 \pm 0.5) m_e$, by $\pi^- + p^+ \rightarrow \pi^0 + n$ with measurement of neutron velocity by time of flight: insufficient statistical significance (see ref. 22, Figure 3, p. 472).

(8) Hillman+[23] (1959): $m_{\pi^-} - m_{\pi^0} \sim (9.01 \pm 0.08) m_e$, by neutron time of flight: such report does not state numbers of events observed, and Hillman+'s data plot (Figure 2c, p. 892) does not have any numbers or units on the 'counts per unit flight time' coordinate. From visual inspection of such plot, such data so reported does not appear to have sufficient statistical significance or resolution to be useful in searching for 14% π_L^-; however, in view of the breadth of the first peak (from about 150 to about 160 nsec flight time) and their flight path length ($\sim$1.38 m), the broad distribution from about 160 to about 170 nsec would seem consistent with a later peak corresponding to existence of π_L^-, with $\delta t \sim 14.8$ nsec. Such report, without statement of numbers of events, appears insufficient to support reliance on the error stated therein of less than 0.89%.

(9) Samios[24] (1960): $m_{\pi_-} - m_{\pi^0} \sim (4.69 \pm 0.07)$ MeV$/c^2$, by π^0 momentum, by measurements on double e^+e^- pairs: insufficient resolution and statistics (see ref. 24, Figure 2, p. 159).

(10) Selove+[25] (1960): search for a second π^0, by neutron velocity; using reinterpretation, with one π^0, as search for π_L^-: insufficient resolution to see 14% π_L^- (see ref. 25, Figure 5, p. 597).

(11) Czirr[17] (1963): $m_{\pi^-} - m_{\pi^0} \sim (4.6056 \pm 0.0055)$ MeV$/c^2$, by neutron time of flight, over distances d of 17, 47, and 67 ft (1 ft$=$0.3048 m). In each case, with reference to Figures 6–9 (ref. 17, p. 346),

Czirr's reported time windows would not have included a later peak due to π_L^- neutrons, with $\delta t(\text{nsec}) \approx 10.4995\ d$ (meters).

Another use for Czirr's data might have been comparison with γ angular correlation data. γ angular correlation measurements in effect would yield a composite value from $m_{\pi_L^-}$ and $m_{\pi_H^-}$: if such composite value were the same, to sufficient accuracy, as Δm from earlier peaks visible in Czirr's report, such correspondence might allow existence of later peaks to be substantially excluded. Czirr reported $m_x - m_{\pi^0} \sim (4.6056 \pm 0.0055)\ \text{MeV}/c^2$, a stated error of about 1% of $m_{\pi_H^-} - m_{\pi_L^-}$, an error ostensibly small enough for such use. However, Czirr's report stated (paragraph 1, p. 346) that it was assumed that all asymmetries in velocity distributions of detected neutrons arose from the shape of the intrinsic resolution of equipment used, except for a relatively small scattering effect. Such assumption appears to have excluded from his report of calculations of error all effects of stochastic variation in arrival of neutrons, such as shifts in the apparent mean flight time associated with stochastic fluctuation in the number of neutrons arriving earlier or later in the distribution. In paragraph 2 of page 346, Czirr's report stated that the low velocity side of the neutron distribution at 47 ft was recorded over only one-third of the time span from the peak to the limit of the 'spread'; which one-third appears, from Czirr's Figure 8, to have been about 35 nsec, indicating that half the width of the full distribution was $\sim$105 nsec, and the full width $\sim$210 nsec. Czirr reported (Table 1, p. 348) that 125 'significant' neutrons were so detected. 'Switching' arrival of one neutron from one side to the other of such velocity distribution may be roughly estimated as corresponding to a difference in arrival time of about 100 nsec, and, thus, a difference in the mean arrival time for the set of 125 neutrons of about 0.8 nsec. Czirr stated (p. 349) as the mean velocity of such neutrons between 17 and 47 ft $v_n = 0.029384$ ft/nsec (8956243.2 m/sec), corresponding to time of flight of $\sim$1020.96379 nsec. Using Czirr's reported expression (p. 348) for $\Delta m = m_x - m_{\pi^0}$,

$$\Delta m \approx m_{\pi^-} - \left((m_{\pi^-} + m_p)^2 + m_n^2 - \frac{2m_n(m_{\pi^-} + m_p)}{\left(1 - v_n^2/c^2\right)^{1/2}} \right)^{1/2},$$

from such velocity Czirr reported (p. 349) $\Delta m = (4.6030 \pm 0.0102)$ MeV/c^2. If one recalculates v_n for flight time decreased by 0.8 nsec, $v_n' \sim 8963266.575$ m/sec, and, substituting in such expression, Δm appears larger by about 0.0052554 MeV$/c^2$ from 'switching sides' of one neutron; the expected fluctuation on each side, about $\pm(125/2)^{1/2}\sim 8$, would lead to fluctuation in Δm several times larger than all other errors combined, and several times larger than Czirr's stated error of ± 0.0102 MeV$/c^2$ for such portion of his reported measurement.

Czirr stated, for flight from 17 to 67 ft, $v_n = 0.029400$ (sic) ft/nsec ($\sim 8.96112\times 10^6$ m/sec), from which he reported $\Delta m = (4.6043757\pm 0.0065)$ MeV$/c^2$. Extrapolating from 47 to 67 ft, 'switching sides' of one neutron may be roughly estimated as corresponding to a difference in arrival time of about 142.6 nsec; noting that Czirr's Table 1 indicated that 33 'significant' neutrons were detected at 67 ft, this would correspond to a difference in mean arrival time of about 4.32 nsec. Recalculating from Czirr's data for flight from 17 to 67 ft, but with one neutron 'switched', Δm appears different by ~ 0.0171033 MeV$/c^2$, or about 2.63 times Czirr's stated error of ± 0.0065 MeV$/c^2$ for that portion of his reported measurement. Expected fluctuation in each 'side' would be roughly $\pm(33/2)^{1/2}\sim 4$; such total fluctuation in that part of Czirr's reported measurement may thus be roughly estimated as of the order of 10 times Czirr's stated error for such portion. 'Background' counts would only increase such statistical error.

Combining these, omission of stochastic contributions to asymmetries in observed neutron velocity distributions appears to have led to an unrealistically low estimate of error. Curves superposed on Figures 6–9 (ref. 17, p. 346) appear to have incorporated such assumption of symmetry, and to be unrealistically narrow.

For these reasons, Czirr's report[17] appears to be not useable in considering existence of π_L^-.

(12) Vasilevsky+[26] (1964): search for a second π^0, by γ angular correlation; reinterpreting such search, with one π^0, as a search for π_L^-: although such report stated an upper limit of 0.5% on any 'second component' with $m_{\pi^-} - m_{\pi^0}$ smaller by ~ 2 MeV$/c^2$, it was not apparent what statistical confidence level might have been involved, nor how sensitive such test would have been to $\sim 14\%$ π_L^-, with $m_{\pi_L^-} - m_{\pi^0}$ only about 0.55 MeV$/c^2$ less than $m_{\pi_H^-} - m_{\pi^0}$. Thus, such report appears to not exclude existence of π_L^-.

(13) Vasilevsky+[27] (1966): $m_{\pi^-} - m_{\pi^0}$, as being (4.6034 ± 0.0052) MeV$/c^2$, by γ angle correlation, using same data as in ref. 26; as measurement of a composite value for $m_{\pi_L^-}$ and $m_{\pi_H^-}$, on assump-

tion of only one π^-. With regard to such report (returning to the derivation sequence):

11. From adseparability, writing θ_L as least included angle between the two γs, from $\pi^0 \rightarrow 2\gamma$ while moving with velocity v_{π^0}, θ_L is approximately given by $\theta_L \approx 2\cos^{-1}(v_{\pi^0}/c)$.
12. Substituting step 11 in step 7, and solving for θ_L,

$$\theta_L \approx 2\sin^{-1}\left(\frac{2m_{\pi^0}(m_x+m_p)}{m_{\pi^0}^2-m_n^2+(m_x+m_p)^2}\right).$$

13. Solving 12 for m_{π^0},

$$m_{\pi^0} \approx \frac{m_x+m_p-\left(\left((m_x+m_p)\cos(\theta_L/2)\right)^2+\left(m_n\sin(\theta_L/2)\right)^2\right)^{1/2}}{\sin(\theta_L/2)},$$

where x might be π_L^-, π_H^-, or approximately some statistical combination thereof.

14. Substituting m_{π^0}, m_{π^-}, m_p, and m_n, from UCLBL 1976 (ref. 8), in 12, $\theta_L \approx 2.731265895$ radians $\approx 156.49°$, and $180° - \theta_L \approx 23.51°$.
15. Suppose one were to disregard all errors related to calibration; geometry; electronics; timing; background; the effect that, for a given angular window, a γ pair whose angle bisector does not lie along v_{π^0} (and which thereby has opening angle $\theta > \theta_L$) could have been produced by decay of a π^0 with a wider range of velocity than if the angle bisector lies along v_{π^0}; all other systematic errors; and all errors propagated from input parameters: in such case, one might estimate the 'error' by assuming that detection of N events in an angular window $\Delta\theta$ would provide angular 'error' $\sim\Delta\theta/N^{1/2}$, combining such errors from Vasilevsky+'s reported 3 cm and 8 cm diameter targets into a composite (smaller) angular 'error' $\delta\theta$, and calculating the corresponding 'error' of Δm.
16. From ref. 27, Figure 1 (p. 281) and paragraph 2 of p. 281, and ref. 26 Figure 2 (p. 884), Vasilevsky+'s staggered two-row arrays of 1500 counters each appear to have provided $\sim$750 resolved elements on a 17-sided polygon $\sim$2 m across, or a resolution width x approximately describable by $x \sim 2\pi/750 \sim 0.00837758$ m. Noting $\theta > \theta_L \approx 156.49°$, the angular windows $\Delta\theta_3$ and $\Delta\theta_8$, for $\sim$3 and $\sim$8 cm diameter (D) targets, may be estimated crudely, neglecting all angles out of the plane of such polygons of counter arrays, by combining the angular

spread obtained by projecting from one counter through the target onto the array of counters viewing the other γ of a pair, with the range spanned by two counters (one for each γ), as $\Delta\theta \sim (2x+2D)/r$, $r \sim 1$ m (see ref. 27, Figure 1). From this, $\Delta\theta_3 \sim 4.397746°$, and $\Delta\theta_8 \sim 10.12732°$.

17. From Vasilevsky+'s stated numbers of counts with each target, with assumptions as in step 15, $\delta\theta \sim ((\Delta\theta_3/52000^{1/2})^{-2} + (\Delta\theta_8/16000^{1/2})^{-2})^{-1/2} \approx 0.018749139°$.
18. Substituting $\theta_L \pm \delta\theta$ in step 13, using values as in step 14, for $\Delta m = m_x - m_{\pi^0}$, $\Delta m(\theta_L + \delta\theta) \sim 4.5990646$ MeV$/c^2$, and $\Delta m(\theta_L - \delta\theta) \sim 4.6095390$ MeV$/c^2$, or an 'error' on $\Delta m(\theta_L) \sim 4.6043$ (from step 14) of $+\sim 0.005239$, $-\sim 0.005235$, MeV$/c^2$, which, rounded to four decimal places, would be ± 0.0052 MeV$/c^2$, equal to the 'error' stated by Vasilevsky+. (Using Vasilevsky+'s 4.6034 as an input instead of UCLBL 1976's 4.6043 would not substantially alter this 'error' estimate.)

This correspondence suggests that Vasilevsky+'s stated error did not include effects of various 'error's from other sources listed hereinbefore, various of which might involve exceeding their stated error of about $\pm 0.11\%$. Therefore, such report appears insufficient to support reliance on the error stated therein.

From discussion herein of Czirr's[17] and Vasilevsky+'s[27] reports and 'error's, comparison of Czirr's Δm for one peak with Vasilevsky+'s Δm for composite $\pi^-_{L,H}$ appears not sufficiently accurate to exclude existence of π^-_L. Also, $m_{\pi^-} - m_{\pi^0}$, and m_{π^0}, appear substantially less well known than previously reported in references 8 (UCLBL 1976), 17, and 27.

(14) Hyman+[28] (1967): masses and momenta by ranges in helium bubble chamber: insufficient resolution and statistics, assumed one μ^+, one μ^-, one π^+, and one π^- (e.g., see ref. 28, Figure 1, p. 378).

(15) Shafer[10] (1967): m_{π^-} by crystal diffraction of mesic atom X-rays: assumed only one peak present in each spectrum (paragraph 2, p. 1458, ref. 10); statistics and resolution appear insufficient to evaluate a 14% π^-_L component (e.g., see and compare Figures 5 and 6, p. 1457, ref. 10), for which offset might have been expected as, in calcium, ~ 0.2 turns, and, in titanium, ~ 0.16 turns, of 'sine screw'.

(16) Booth+[29] (1967): momentum of μ^+ from π^+ decay, by magnetic spectrometer: insufficient resolution and statistics (e.g., see Figure 2, p. 40, ref. 29); although single μ^+ and π^+ appear to have been

assumed, data appears to substantially allow a bimodal distribution (same Figure 2).

(17) Booth+[30] (1970): $m_{\pi^+} \sim (139.548 \pm 0.025)$ MeV$/c^2$, from ref. 29 μ^+ momentum: see subcomment (16) hereinbefore.

(18) Backenstoss[11] (1970): m_{π^-} by π^- mesic atom spectroscopy, using germanium diode detectors: depiction in Figure 1 (ref. 11, p. 480) of apparatus, although scale is not given and other beam transport is not sufficiently described, suggests substantial flight distance after incoming pion beam loses velocity in carbon 'brake', with consequently longer flight time in their 'rest frames' for moving particles. For $\tau_{\pi_L^-}$ substantially less than $\tau_{\pi_H^-}$, this might reduce the fraction of π_L^-s present to substantially less than 14%, making such spectra (e.g., Figure 2, p. 483, ref. 11) so observed relatively insensitive to existence of π_L^-. Resolution with respect to presence of a relatively small component of π_L^- was not described sufficiently for such report to substantially exclude (or support) existence of π_L^-.

(19) Backenstoss+[12] (1971): method and apparatus similar to those in ref. 11, but using radioisotope radiation sources for calibration: see subcomment (18) hereinbefore. Report did not include quantitative description of spectral line shapes, nor details of beam transport.

(20) Backenstoss+[13] (1973): reevaluation and correction to theory used in ref. 12: see subcomments (18) and (19) hereinbefore.

Searches for $e^\pm$ from $\pi^\pm$ decay likely presumed some $e^\pm$ momentum spectrum, for event identification and selection of data from background; absent specific searches for the decay modes mentioned hereinbefore, such experiments might seem to not bear on question of existence of $\mu_L^\pm$ and/or $\pi_L^\pm$.

For experimental searches for $\pi_L^\pm$, relatively short flight times (in particles' frames of reference) appear desirable; with variable flight time (e.g., by moving carbon 'brake' to change distance traveled at relatively low velocity), dependence on flight time of relative numbers detected might be used to measure relative lifetimes of $\pi_L^\pm$ and $\pi_H^\pm$.

From literature reviewed, existence of π_L^- appears supported at >0.999 'confidence level'. Further neutron time of flight measurements appear warranted.

From Link 17.7, $m_{\pi_H^\pm} - m_{\pi_L^\pm}$ appears to depend on $I_{\pi_{L,H}^\pm}$ through the $((I_x - 2)1.5\sqrt{3}\,)^{1/3}$ and $(1/(1 + d/I_x^2))^{1/3}$ factors. The second of these factors is used in approximate description of $e^{\pm}$, $\mu^{\pm}_{L,H}$, and $\pi^{\pm}_{L,H}$, in approximately describing effects on Theorem 13 interaction among photons associated with such entities. Thus, $m_{\pi^{\pm}_{H}} - m_{\pi^{\pm}_{L}}$ appears to offer a test of treatment of 'charge' hereinbefore, and of $I_{\pi^{\pm}_{L,H}}$, substantially independent of $Y(N)$.

From Link 17.7, fractional mass 'splitting' z approximately describable by such factor with $d = +, -$ is approximately describable by

$$z(I_x) \approx \left(\frac{1}{1-\frac{1}{I_x^2}}\right)^{1/3} - \left(\frac{1}{1+\frac{1}{I_x^2}}\right)^{1/3}.$$

(Note that z appears substantially independent of the 'unsplit' 'mass', and of distinction between approximate description as a two or three 'photon' entity.)

Substituting,

$$z(12) \approx 0.0046297455, \quad z(13) \approx 0.0039448447, \quad z(14) \approx 0.0034014066,$$

$$\frac{z(12)-z(13)}{z(12)} \approx 0.1479348703,$$

and

$$\frac{z(13)-z(14)}{z(13)} \approx 0.1377590606.$$

These approximate values may be compared with Haddock+'s report[16] Figure 2c: A 13.7% change in $m_{\pi^{-}_{H}} - m_{\pi^{-}_{L}}$ would change δt roughly proportionately, by ~5.48 nsec. Haddock+'s Figure 2c time bin size was reported as ~4 nsec. Thus, likely if $I_{\pi^{\pm}_{L,H}}$ were larger or smaller by 1 bit, the later peak would have been an entire bin earlier or later. The later peak appears to be substantially entirely within only two bins. Thus, Haddock +'s report appears to substantially exclude $I_{\pi^{\pm}_{L,H}} \neq 13$, and to substantially allow $I_{\pi^{\pm}_{L,H}} = 13$. From Links 17.1–17.7, likely value for $I_{\pi^{\pm}_{L,H}}$ was obtained by a chain of reasoning using I_x for all types of 'particles' but π^0 appearing in Table 3 Column 1. In this way, this correspondence between $m_{\pi^{\pm}_{H}} - m_{\pi^{\pm}_{L}}$ and Haddock+'s report appears to test I_x for all such particles, and to partially test treatment hereinbefore of 'charge' sign coding.

From UCLBL 1976 report (ref. 8), m_{π^0} appears to have been obtained from $m_{\pi^{\pm}}$ and $\Delta m = m_{\pi^{\pm}} - m_{\pi^0}$, with $m_{\pi^{\pm}}$ and Δm having been measured, not m_{π^0}. $I_{\pi^{\pm}_{L,H}}$ and I_{π^0} were treated approximately as differing only by presence of 1 bit coding 'charge' sign in $I_{\pi^{\pm}_{L,H}}$. For this reason, comparison of approximately calculated ('Δm_c') and reported ('Δm_r') values for Δm appears to offer a partial test of the treatment of 'charge' sign coding, and 'd' coding, substantially independent of effects approximately described by $Y(N)$. Various reports cited in UCLBL 1976 of measurements of Δm do not appear to depend strongly on assumed $\pi^{\pm}$ charge interaction strength.

For convenience in using comparison of Δm_c with Δm_r as a partial test of treatment hereinbefore of 'charge' sign coding and 'd' coding, in the following Table 4 appear each combination of $d = +, -, 0$ for C3, 12, 0, d, 0 (π^0) and C3, 13, 0, d, f ($\pi^{\pm}_{L,H}$), approximate values for Δm_c for each such pair, and comparison to Δm_r obtained from compilation UCLBL 1976, expressed both in MeV$/c^2$ and as a fraction of Δm_r. Each cell in the matrix in Table 4 lists, from top to bottom, Δm_c, $\Delta m_c - \Delta m_r$, and $(\Delta m_c - \Delta m_r)/\Delta m_r$. For your convenience, $(m_c - m_r)/m_c$ for each approximately calculated mass is also tabulated below the corresponding m_c. (all 'masses' in MeV$/c^2$)

Note that $\pi^{\pm}_{L,H}$ and π^0 masses are *not* themselves treated as separate measurements as reported, and that, therefore, both $\pi^{\pm}$ and π^0 mass reports would be considered in terms of questions about assumptions as to $\pi^{\pm}$ charge interaction strength. Δm_r is the reported value comparison with which is addressed in Table 4. 'X' appears next to C3, 12, 0, +, 0, C3, 12, 0, −, 0, and C3, 13, 0, 0, ±, because for each of these d and f coding appear inconsistent (see Def 21, $d = 0$ iff $f = 0$). 'Y' appears in the only two matrix cells for which d and f coding appear consistent. (Fractional differences listed in cells of Table 4 have the reported value in the denominator, for convenience in comparing entries in different cells of the matrix, as distinct from use of the calculated value in the denominator for entries in Table 3 Column 10.)

From Table 4, it appears that approximate descriptions obtained in Links 17.9 (π^0, C3, 12, 0, 0, 0) and 17.7 ($\pi^{\mp}_H$, C3, 13, 0, −, f) provide Δm_c substantially closer to Δm_r than the next closest (and likely not properly coded) (C3, 12, 0, +, 0 and C3, 13, 0, 0, ±) by a factor of ~ 1.866.

UCLBL 1976 cites eight reports of Δm, which were combined in a weighted average to provide a recommended value therein of $\Delta m_r \approx (4.6043 \pm 0.0037)$ MeV$/c^2$; the more precise weighted average value of 4.604337436 obtained from such data was used for Δm_r in calculation of comparisons tabulated in Table 4.

Are any of the values of Δm_c listed in Table 4 allowed by reports of observation prior to and near this writing? The smallest $|\Delta m_c - \Delta m_r| \approx 0.0406489$ MeV$/c^2$, which would be ~ 11 times UCLBL 1976's stated

Table 4. (approximate values; please see text)

'$\pi^{\pm}$' / 'π^0'	C3,13,0,+,± 139.4206179 −.001062842	X C3,13,0,0,± 139.6950687 0.000903888	C3,13,0,−,± 139.9716934 0.002878392
X C3,12,0,+,0 135.0148933 0.000373243	4.4057246 −0.1986128 −0.04313603	4.6801754 0.0758380 0.016470983	4.9568001 0.3524627 0.076550138
C3,12,0,0,0 135.3267071 0.002676538	Y 4.0939108 −0.5104266 −0.11085778	4.3683616 −0.2359758 −0.05125677	Y 4.6449863 0.040648864 0.008828385
X C3,12,0,−,0 135.6414215 0.004990522	3.7791964 −0.8251410 −0.17920951	4.0536472 −0.5506902 −0.11960249	4.3302719 −0.2740655 −0.05952334

Δm_r used as 4.604337436 (please see text)

'error'. With reference to p. 62 of UCLBL 1976 (ref. 8), it may be seen that only two of the eight reports of Δm referenced there were cited as having stated error $< \pm 0.03$ MeV$/c^2$. If the other six reports are combined by average $\approx(\Sigma_i \text{ value}_i \times (\text{error}_i)^{-2})/(\Sigma_j (\text{error}_j)^{-2})$ and error $\approx(\Sigma_i (\text{error}_i)^{-2})^{-1/2}$, the resulting value would be $\approx$4.60157 MeV$/c^2$, and such composite (smaller) error would be $\approx \pm 0.01978$ MeV$/c^2$; smallest listed $|\Delta m_c - \Delta m_r|$ would be $\sim$2 times such composite 'error', or substantially smaller if a 7–14% $\pi_L^{\pm}$ with $\pi_H^{\pm}$ mixture, suggested by Haddock+'s[16] Figure 2c as noted hereinbefore, were considered. Thus, it appears that Δm_c as obtained from Links 17.7 and 17.9 would be substantially allowed by those six reports of Δm.

The two reports cited in UCLBL 1976 (ref. 8) stating smaller 'error' for Δm are Czirr's[17] (discussed hereinbefore with respect to $m_{\pi^{\pm}_{L,H}}$), stating as 'error' ± 0.0055 MeV$/c^2$, and Vasilevsky+'s[27], stating as 'error' ± 0.0052 MeV$/c^2$: UCLBL 1976's stated 'error' appears to be as small as it does mostly because of these two reports' stated 'errors'.

As noted hereinbefore, both such reports appear insufficient to support 'errors' stated therein, and their stated 'errors' are not used herein in evaluating approximate results presented hereinbefore. (My first reason for looking for misestimation of 'error' in such reports was the discrepancy between them and my estimate of Δm. This provided a rather pointed test of this approach: in testing an idea, in human terms, there is quite a difference between having agreement with all the available reports, in which case one might have pointlessly 'fitted' the 'data', and having agreement with only some of the available reports and finding subsequently that all discrepant reports appeared significantly discrepant only because they overstated their accuracies.)

In view of the foregoing, Δm_c (calculated) obtained as discussed herein from approximate descriptions of π^0 as C3,12,0,0,0 (L 17.9) and $\pi^{\pm}_{L,H}$ as C3,13,0,$\pm$, f (L 17.7) appears substantially allowed by such knowledge prior to the time of this writing.

The composite error ($\sim\pm 0.01978$ MeV$/c^2$) in Δm_r (reported) from the remaining six reports, other than those of Czirr and of Vasilevsky+, appears sufficiently small to substantially exclude, as deviating by more than $\sim 11.9\sigma$, 'calculation' of Δm, without inclusion of treatment of the $I_{\pi^{\pm}_{L,H}}$ 'charge' sign bit, as $m_{C3,13,0,0,f} - m_{C3,12,0,0,0}$. In view of this, treatment of 'charge' sign coding as affecting mass appears substantially allowed by this comparison of Δm_c with Δm_r.

Note that treatment presented herein of pions is not (yet?), at the time of this writing, offered here as having been explicitly formally tested for adequacy (as in A-Def 2).

I found Czirr's more detailed report of his apparatus, data, and analysis appreciably more helpful for retrospective analysis than the much shorter report of Vasilevsky+. Because of the possibility of changes in apparent significance of observation(s), and/or error, it appears to me, at the time of this writing, that inclusion of sufficient detail for subsequent analysis is a significant consideration in publication of research results, and that the format of letter journals near this writing may, at the time of this writing, often be not satisfactory for such purpose, although possibly useful for publication of a preliminary summary of a more detailed report in preparation and referenced therein.

An experiment such as that reported by Czirr, but with better instrumentation and statistics, capable of precision and absolute accuracy $\lesssim$ 0.25% of $m_{\pi^-_{L,H}}$ at $\geqslant 6\sigma$ 'confidence level', might be interesting, both as a check of the status of the approximate treatment herein of $m_{\pi^{\pm}_{L,H}}$, and as a test, with comparison to mesic atom measurements, of $\pi^- p^+$ charge interaction strength. I am not, at the time of this writing, aware of a formal reason to expect that conservation of charge coding would necessarily be equivalent to conservation of charge interaction strength. In view of the conceptual and formal structure presented hereinbefore, such an experiment might offer a relatively direct way of addressing the question of what is conserved in charge conservation, and the concept of charge.

As noted hereinbefore, comparison of reported (but apparently not recognized (Haddock+, ref. 16)) value for $m_{\pi_H^-} - m_{\pi_L^-}$ with $m_{C3,13,0,-,-} - m_{C3,13,0,+,-}$, and of reported values for $m_{\pi_{L,H}^\pm} - m_{\pi^0}$ with $m_{C3,13,0,\pm,f} - m_{C3,12,0,0,0}$, appear to test treatment appearing hereinbefore of 'charge' sign coding substantially independently of treatment presented hereinbefore of spectrum of photon states in U. Comparison of $\delta 3,2,I_x > 0$ with reports of values for $m_{\pi_{L,H}^\pm}$, m_{π^0}, and apparent absence of any pion-like entity with 'mass' less than m_{π^0}, appears to provide a test for photon state spectrum (including various multiplicities) in U, substantially independent of treatment presented hereinbefore of 'charge' sign. From Definition 22, $\delta 3,2,I_x$ depends monotonically on I_x for $10 \leqslant I_x \leqslant 14$. From Link 17.8, absence of relatively slowly decaying pion-like entity with mass $< m_{\pi^0}$ appears likely approximately describable by $\delta 3,2,I_x$ changing sign between $I_x = 11$ and $I_x = 12$.

From Definition 22,

$$\delta 3,2,11 \approx -5.515694 \times 10^{-3},$$

$$\delta 3,2,12 \approx +2.467004 \times 10^{-4},$$

$$\delta 3,2,13 \approx +4.864564 \times 10^{-3}.$$

Involved in these δs are $Y(11) = 23$, $Y(12) = 26$, and $Y(13) = 29$, in comparison (before cancellation of like terms in E_1) with multiples of E_1 not involving $Y(I_x)$. From L 17.1, for $11 \leqslant I_x \leqslant 13$, $Y(I_x)$ involves single occupancy of all different 'photon' states in U with 'energy' E_1 (two states) and 'energy' E_2 (six states), and single occupancy of three (in $Y(11)$), four (in $Y(12)$), or five (in $Y(13)$) 'photon' states in U with 'energy' $E_3 = 3E_1$. Thus, so testing L 17.8 would seem to test $Y(N)$ with respect to both 'energies' $E_j = j\ E_1$ for $j = 1, 2, 3$, and multiplicities of distinct photon states for $j = 1, 2$.

In view of this, the qualitative feature of apparent absence of relatively slowly decaying pion-like entity with mass $< m_{\pi^0}$ would appear to substantially allow derivation presented hereinbefore of $Y(N)$, for $N \leqslant 13$, as being accurate to within ~0.552%.

From A-Definition 22 and Definition 21,

$$m_{C3,11,0,0,0} \approx \left((11-2)1.5\sqrt{3}\,\right)^{1/3} I_U^{1/3} E_1 / c^2;$$

using $I_U^{1/3} E_1 \approx 45.69123378$ MeV obtained as presented hereinbefore, $m_{C3,11,0,0,0} c^2 \approx 130.6564991$ MeV, about 4.6702080 MeV less than $m_{C3,12,0,0,0} c^2$ approximately calculated hereinbefore in relation to π^0.

Various reports of searches for a 'second' π^0 have appeared, as noted more fully hereinbefore, including those of Gettner+[22], Haddock+[16],

Hillman + [23], and Selove + [25]. From all such reports, and from compilations such as UCLBL 1976 (ref. 8), it appears that there is not among those noted hereinbefore any substantial report of any pion-like entity with mass less than m_{π^0}, and that presence of a substantial proportion of relatively slowly decaying entities somewhat like π^0, but with mass less by ~4.67 MeV/c^2, might have been noticed if they existed. The confidence level, from reports relating to this qualitative feature, and the accuracy of this test of $Y(N)$ based on such qualitative feature if it exists, appear in some sense substantially independent of each other. Thus, various reports noted hereinbefore appear to substantially support the qualitative feature that there not be any relatively slowly decaying pion-like entity with 'mass' $< m_{\pi^0}$, and such qualitative feature appears to substantially allow derivation presented hereinbefore of $Y(N)$, $N \leqslant 13$, as being accurate to within ~0.552%.

Such accuracy, of within ~0.552%, would seem sufficient to substantially exclude spectra in prior theory of a flat 3-space rectangular cavity with 'hard' boundary conditions (square root of sum of squares of relatively small integers), and of a flat 3-space spherical cavity with 'hard' boundary conditions.

In view of this, comparison of approximate calculations presented hereinbefore with reports as noted of observations relating to pions appears to provide tests of treatment presented hereinbefore of effects of 'd' coding iff 'charge' sign coding, and of $Y(N)$, which was obtained in the axiomatic form, substantially separately. This was one reason for discussing these comparisons before those relating to $\mu^{\pm}_{L,H}$ and $e^{\pm}$, which comparisons involve treatment presented hereinbefore of 'charge' sign and $Y(N)$ together.

3.5.8.2 Table 3, Items 9–12, and 21 (Muons). In Table 3, for smaller I_x, treatment of the 'charge' sign bit involves larger fractional differences in m_x: for example, we can approximately calculate m_x for $I_x = 7$, and compare $d = +, 0, -$ values for $m_{C2,7,b,d,f}$ with UCLBL 1976 recommended value for m_μ (from μ mesic atom spectra), as appears in Table 3, items 11 and 12, column 9.

In Table 5, $\Delta = m_{C2,7,b,-,f} - (m_{\mu^\pm}$ as reported, ref. 8), and approximate multipliers therefor, are included for convenience in comparing entries for $m_{C2,7,b,d,f} - (m_{\mu^\pm}$ as reported).

Note that comments hereinbefore on assuming that $\pi^-_{L,H}p^+$ charge interaction strength is the same as for e^-e^+ or e^-p^+ apply also to $\mu^-_{L,H}p^+$, and that I am not, at the time of this writing, aware of a formal reason for regarding such charge interaction strengths as equal. For this reason, μ

Table 5: $m_{C2,7,b,d,f}$ and Reported (ref. 8) Value for $m_{\mu^{\pm}}$, in MeV$/c^2$

Approximate values, $I_U^{1/3}E_1 \approx 45.69123378$ MeV$/c^2$

$m_{C2,7,b,d,f}$	$m_{C2,7,b,d,f} - m_{C2,7,b,-,f}$	$m_{C2,7,b,d,f} - m_{\mu^{\pm}}$ reported
$d = -$		
$\sim$105.3282612	0	$-\sim$0.3312 $(=\Delta)$
$d = 0$		
$\sim$104.6068121	$-\sim$0.721	$-\sim$1.0527 $(\sim 3.178\Delta)$
$d = +$		
$\sim$103.9047318	$-\sim$1.4235	$-\sim$1.7547 $(\sim 5.3\Delta)$

mesic atom spectra as used in obtaining UCLBL 1976 recommended value for m_{μ^-} appear to measure a combination of $m_{\mu^-_{(H)}}$ and $\mu^-_{(H)}p^+$ interaction strength.

As noted with respect to $m_{\pi^{\pm}_{L,H}}$, measurement of $m_{\mu^{\pm}_H}$ in a way not strongly dependent on $\mu^{\pm}_H$ charge (and/or magnetic) interaction strength to $<0.3\%$ at 6σ might be interesting. $e^{\pm}$ velocity spectrum from decay might be considered for such measurement. Comparison of a value so obtained for $m_{\mu^{\pm}_H}$ with μ mesic atom spectra might address experimentally questions about $\mu^{\overset{(+)}{\mp}}_H p^{\overset{(\mp)}{\pm}}$ charge interaction strength and what is conserved in charge conservation.

Note also that treatment presented herein of $m_{\mu^{\pm}_{L,H}}$ is not (yet?), at the time of this writing, offered here as having been explicitly formally tested for adequacy (as in A-Def 2).

With these cautionary notes, comparison presented in Tables 3 and 5 between $m_{C2,7,b,d,f}$ and $m_{\mu^{\pm}}$ as reported seems to suggest that C2, 7, b, 0, f, and C2, 7, b, +, f are substantially less satisfactory as approximate descriptions of $\mu^{\pm}_H$ than C2, 7, b, −, f. Because of the question of $\mu^{\pm}_H$ charge interaction strength, the reports noted hereinbefore for $m_{\mu^{\pm}_H}$ do not appear, at the time of this writing, formally sufficient to exclude the approximate calculated $m_{C2,7,b,-,f}$ value presented hereinbefore in relation to $m_{\mu^{\pm}_H}$. Looking for $\mu^{\pm}_L$ might be worthwhile; $\mu^{\pm}_L$ might appear as (a) 'resonance' in $e^{\pm}4\gamma$ production. That $\mu^{\pm}_L$ seems to have not been recognized in such reports seems consistent with $\tau_{\mu^{\pm}_L} \ll \tau_{\mu^{\pm}_H}$, and thus with decay preserving 'd' coding and in $e^{\pm}$ $d = +1$ only. (I have a partial recollection, from within a few years before the time of this writing, of a news item reporting that some researchers thought they had found a muon related $e+4\gamma$ decay scheme but had rejected it as having a total mass energy somewhat less than that of the previously recognized muon(s) (likely $\mu^{\pm}_H$); I have not succeeded in finding this item again.)

Likely $e^{\pm}$ has $d=+1$ only (Cor 14.5, 14.8, 14.15, 14.16, L 17.4, Def 21). With this relatively closer agreement with observation for $m_{\mu_H^{\pm}}$ as $m_{C2,7,b,-,f}$ and $m_{\pi_H^{\pm}}$ as $m_{C3,13,0,-,f}$, reasoning presented hereinbefore in support of approximately calculating $m_{\mu_H^{\pm}}$ and $m_{\pi_H^{\pm}}$ in such way (L 17.5 and 17.7) might be 'turned around', and 'd' coding regarded as offering insight into 'why' $\pi_H^{\pm}$ decays to $\mu_H^{\pm}$ much more often than 'directly' to $e^{\pm}$.

Thus, these approximately calculated masses for $\pi_H^{\pm}$, π^0, and $\mu_H^{\pm}$, such picture of such $\pi_H^{\pm}$ decay 'branching ratio', and $\tau_{\mu_L^{\pm}} \ll \tau_{\mu_H^{\pm}}$, all appear substantially allowed by such knowledge prior to the time of this writing, and might be regarded also as further tests apparently substantially satisfied by treatment presented hereinbefore of 'charge' sign (and, in the case of $m_{\mu_H^{\pm}}$, approximately $Y(N)$, although it is not (yet?), at the time of this writing, clear to me whether one might need to consider other bits in $I_{\mu_{L,H}^{\pm}}$ in describing $m_{\mu_{L,H}^{\pm}}$).

3.5.8.3 Table 3, Items 7 and 8 (Muon Neutrino and Antineutrino). As may be seen from Table 3, with for such tests only energies $<10^{12}$eV available, comparison of approximately calculated values for $m_{\nu_{\mu(L,H)}}$ and $m_{\bar{\nu}_{\mu(L,H)}}$ with UCLBL 1976 reported values for such items does not appear to provide a particularly sensitive test of the picture presented herein. Approximate description herein of $\nu_{\mu_{(L,H)}}$ and $\bar{\nu}_{\mu_{(L,H)}}$ would seem to be in some respects more sensitively tested by use thereof in deriving approximate description of, and approximate masses for, $\pi_{L,H}^{\pm}$ and π^0. Description hereinbefore of $\nu_{\mu_{(L,H)}}$ and $\bar{\nu}_{\mu_{(L,H)}}$ appears, to the extent tested as presented hereinbefore, substantially allowed by knowledge prior to the time of this writing.

3.5.8.4 Table 3, Items 4–6 (Positron, Electron, and α^{-1}). From the foregoing, it would seem that, in considering the significance of the closeness of agreement of calculated $e^{\pm}$ properties with report(s) of observation, we might regard the treatment herein of 'charge' sign and $Y(N)$ as having been tested separately from $e^{\pm}$.

I chose m_e as an input to these calculations, to avoid relying on prior reports which I had read of measured values for Hubble's constant H, which reports appeared to state value(s) for H substantially less precisely than did CT (ref. 4) for m_e, G, c, or $\hbar$ (see discussion following Cor 14.15).

By combining expressions for F and m_e, I eliminated H, obtaining the relation among G, c, $\hbar$, m_e, and F appearing in item 6 of Table 3. Although m_e does not appear in Table 3 as a quantity calculated for explicit comparison, the approximate expression for m_e was used, with the value for F calculated from such expression as in item 6, to obtain the value for $I_U^{1/3}E_1$ of $\approx$45.69123378 MeV used in calculating approximate values for

$m_{\mu^{\pm}_{L,H}}$, m_{π^0}, $m_{\pi^{\pm}_{L,H}}$, and various mass differences, and corresponding approximate values for I_x/I'_x or I_x/I''_x; thus, such approximate expression for m_e is tested by comparison of F with α^{-1} and of each such approximate calculated mass with report(s) of observation.

The 'sharpest' test of approximate description herein of $e^{\pm}$ may be from the relation obtained therefrom among G, c, $\hbar$, m_e and F, where F^{-1} appears as the 'strength' of the 'electrostatic' interaction between a pair of electrons and/or positrons, corresponding to $\alpha=(e^2/\hbar c)\{\mu_0 c^2/4\pi\}$ (SI units), comparing F with α^{-1}. As noted in item 6 of Table 3, the value so approximately calculated for F differs fractionally from values for α^{-1} recommended by Cohen and DuMond (ref. 6), apparently based mostly on atomic spectra interpreted using 'quantum electrodynamics' theory, by $\sim -0.6035\times10^{-6}$, not statistically significant (see Table 1); and by Cohen and Taylor (ref. 4), apparently involving reliance on a value for $2e/h$ obtained from Josephson junction measurements interpreted using BCS superconductivity theory (see Table 1 and discussion following Cor 14.15), by $\sim +19.537\times10^{-6}$. For various reasons, noted in more detail following Corollary 14.15 and in relation to Table 1, it seems to me that the difference between such approximate calculated value for F and Cohen and Taylor's recommended value for α^{-1} does not, at the time of this writing, appear to warrant rejecting the treatment presented herein of $e^{\pm}$. (Please see also Section 6 question 38.)

As noted more fully in discussion following Corollary 14.15, relations between F and such two values recommended for α^{-1} might suggest that BCS superconductivity theory, in terms of which such Josephson effect measurements would seem to have been interpreted, might be not exact.

In the expression for F, in items 4 and 5 of Table 3, as

$$F\approx\left(32\pi^2(\ln(2I''_x))^2\right)^{1/3},$$

the I''_x term may be traced through $(m_x c^2/E_1)$ to $E_1=\hbar c/R_U$, or R_B as it appears in Corollary 14.3, where R_B appears as a limit of integration over a 3-space region. In this way, R_U appears in calculating properties of $e^{\pm}$ not only through the spectrum of photon states in U, as in calculating approximate values for 'masses' of relatively localized forms in relation to $m_{\mu^{\pm}_{L,H}}$, m_{π^0}, and $m_{\pi^{\pm}_{L,H}}$, but also through 3-space calculation relating to extended interaction in Corollary 14.3 (and 14.5), which underlies calculation, in Corollary 14.8, of F. Thus, approximate agreement between F and report(s) of measured values for α^{-1} appears related to, and to allow, approximate partial description of an observer's universe as having a 3-space 'linear size' R_U.

Use of G to obtain R_U, and thus the relation among G, c, $\hbar$, m_e, and α^{-1} (as F) appearing in Table 3 item 6, also appears to make comparison between F and α^{-1} a partial test of Theorem 10 through Theorem 13, and the preliminary test of Corollary 14.15 using H appears to be a partial test of Corollaries 10.2 and 10.3. Thus, to the extent tested as presented hereinbefore, all of these appear substantially allowed by knowledge prior to the time of this writing.

3.5.8.5 Table 3, Items 2 and 3 (Electron Neutrino and Antineutrino), and 'Weak Interaction'. Although upper limits reported by UCLBL 1976 (ref. 8) for m_{ν_e} and $m_{\bar{\nu}_e}$ appear lower than for $m_{\nu_{\mu(L,H)}}$ and $m_{\bar{\nu}_{\mu(L,H)}}$, they do not appear to provide a particularly sensitive test of description herein of ν_e and $\bar{\nu}_e$. Description herein of ν_e and $\bar{\nu}_e$ would seem to be in some respects more sensitively tested by use thereof in deriving approximate descriptions of, and values relating to, $\nu_{\mu(L,H)}$, $\bar{\nu}_{\mu(L,H)}$, $\mu^{\pm}_{L,H}$, π^0, and $\pi^{\pm}_{L,H}$, comparison of which with reports of observations has been discussed hereinbefore.

Description hereinbefore of ν_e and $\bar{\nu}_e$ appears, to the extent tested as presented hereinbefore, to be substantially allowed by knowledge prior to the time of this writing.

With reference to Corollaries 14.15, 15.7, 15.11, and 15.12, Links 17.3 and 17.6, and Definition 20, likely for $\nu_{\mu(L,H)}$ and $\bar{\nu}_{\mu(L,H)}$, with electron as least massive bound entity in U,

$$|a_f| \sim |a_e| \frac{\sqrt{2}\, E^{1/4} E_-^{1/4}}{m_e^{1/2} c} \left(\frac{0.9\, Y(3)}{Y(6)} \right)^{1/6},$$

which, with $E_- = Y(6)E_1$, L 17.1 and 17.4, E_1 as obtained hereinbefore, and $m_e c^2 \approx 511003.4$ eV, would be approximately, for E in electron volts, $|a_f| \sim |a_e|\ 1.492287314 \times 10^{-11}\ E^{1/4}$. This appears as an approximate ratio of 'interaction strength' for $e^{\pm}$ and $\overset{(-)}{\nu}_{\mu(L,H)}$. As noted hereinbefore, such 'interaction strength' would be respectively in each particle's 'rest frame', and neutrino momentum 'graininess' might affect 'cross section', rate, and 'phase space' calculations.

For example, in $\pi^{\pm}_H \rightarrow \mu^{\pm}_H \overset{(-)}{\nu}_{\mu}$, in domain of adseparability, writing E for energy and P for momentum magnitude, E_ν may be approximately obtained from A-Corollary 2.1 and A-Definition 22, by solving

1. $(m_{\mu_H} c^2)^2 = E_\mu^2 - P_\mu^2 c^2$;
2. (noting $E_\nu \approx E_+ \gg E_-$) $P_\nu \approx E_\nu / c \approx P_\mu$;

and

3. $m_{\pi^{\pm}_H} c^2 = E_\mu + E_\nu$

for E_ν:

substituting from 2 $P_\mu \approx E_\nu / c$ in 1,

$$m_{\mu_H}^2 c^4 \approx E_\mu^2 - E_\nu^2;$$

solving 3 for E_ν, and substituting,

$$m_{\mu_H}^2 c^4 \approx \left(m_{\pi_H^\pm} c^2 - E_\nu\right)^2 - E_\nu^2;$$

solving for E_ν,

$$m_{\mu_H}^2 c^4 \approx m_{\pi_H^\pm}^2 c^4 - 2 m_{\pi_H^\pm} c^2 E_\nu + E_\nu^2 - E_\nu^2,$$

$$E_\nu \approx \frac{m_{\pi_H^\pm}^2 c^4 - m_{\mu_H}^2 c^4}{2 m_{\pi_H^\pm} c^2}.$$

Substituting from UCLBL 1976 (ref. 8) $m_{\pi_H^\pm} c^2 \approx 139.5688$ MeV and $m_{\mu_H} c^2 \approx 105.65948$ MeV (subject to cautionary notes hereinbefore as to accuracies of such numbers), and calculating, $E_\nu \approx 29.79005416$ MeV. Substituting this value for E_ν and solving for $|a_f / a_b|$, ('b' for 'bound'), $|a_f / a_b| \sim 1.102478858 \times 10^{-9}$, or, inverting, $|a_b / a_f| \sim 9.070468724 \times 10^8$. In some sense, this might be regarded as approximately describing what appears to have been described in prior theory(s) in terms of the ratio of 'strengths' of such theory(s) 'electromagnetic' and 'weak' interactions.

Noting that adequate (as in A-Def 2) treatment might also take into account any effects of momentum 'graininess' and 'phase space', this ratio for $|a_f / a_b|$ might be roughly compared with ratio of decay rates for $\pi_H^\pm \to \mu_H^\pm \overset{(-)}{\nu}_{\mu(H)}$ and $\pi^0 \to 2\gamma$, which appear to have been treated in prior theory(s) as involving respectively 'weak' and 'electromagnetic' 'processes'. From UCLBL 1976 (ref. 8), 'mean life' of $\pi_H^\pm$ is $\approx 2.603 \times 10^{-8}$ sec, and of π^0 is $\sim 8.28 \times 10^{-17}$ sec, from which such ratio of $\pi_H^\pm$ decay rate to π^0 decay rate appears to be $\sim 3.1809 \times 10^{-9}$, about 2.885 times such $|a_f / a_b|$. This crude comparison is included as seeming to suggest rough compatibility of this ratio for $|a_f / a_b|$, to the extent so tested, with such reports of observations. I have not yet, at the time of this writing, understood formally and in detail such decay processes. This report is of work in progress, and not yet, at the time of this writing, complete. (see also Section 6 question 35).

From A-Corollaries 4.8, 5.1, 6.1, and 7.3, and Link 17.3, likely $\overset{(-)}{\nu}_e$ propagating in one direction (not a 'standing wave') would be approximately describable in terms of two B_0 ψs, with only one 'helicity' for ν_e and only the opposite 'helicity' for $\bar{\nu}_e$, a result traceable to likeliness of only

1 bit of information being accessible within $\overset{(-)}{\nu}_e$. This appears substantially allowed by those prior reports of which I am at the time of this writing aware of measurements, and substantially consistent with such prior 'two component' theories of $\overset{(-)}{\nu}_e$.

Corollary 15.11 appears to approximately partially describe one class of process involving production of neutrinos or antineutrinos, and such approach was used in obtaining approximate descriptions of ν_e, $\bar{\nu}_e$, $\nu_{\mu(L,H)}$, and $\bar{\nu}_{\mu(L,H)}$, and of internal information, 'structure', and 'mass' of μ_L^+, μ_L^-, μ_H^+, μ_H^-, π^0, π_L^+, π_L^-, π_H^+, and π_H^-.

In view of comparisons presented hereinbefore of various of these approximate descriptions with reports of observations,

Correspondence Link 17.12. *The physical picture, and formal structure based thereon, presented hereinbefore, appear, at the time of this writing, to the extent tested as presented hereinbefore, substantially able to subsume approximate description of all neutrinos and antineutrinos reported prior to the time of this writing as observed, and of phenomena apparently regarded in prior theory(s) in terms of such theory(s) 'weak interaction(s)'.*

●

3.5.8.6 Table 3 Item 1 (Photon). Although the upper limit reported (UCLBL 1976) for m_γ appears lower than for m_{ν_e} or $m_{\bar{\nu}_e}$, it does not appear to provide a particularly sensitive test of description herein of γ. Description herein of γ was used, in the axiomatic form, in deriving approximate description of all other items in Table 3 column 1, and in many other ways hereinbefore, and, to the extent tested as presented hereinbefore, appears substantially allowed by knowledge prior to the time of this writing.

3.5.8.7 Table 3, Columns 5 and 6, and 'Lepton Number'. From comparison of columns 5 and 6 of Table 3, 'b' and 'L', it would appear that coding one bit of information as 'b' or 'b_μ' corresponds, for the entities listed in Table 3, to the conventional designation 'lepton', although, as noted following Link 17.5 and discussed further in Section 6 question 21, I do not yet, at the time of this writing, have a clear formal understanding of the relation (if any) between 'b_μ' and 'b' of $e^\pm$.

3.5.8.8 Table 3, Columns 11–14, and 'Strong Interaction'. L(ink) 17.10 interaction appears to provide description of subsuming I_x of x by Theorem 13 interaction between x and y. One remarkable feature of L 17.10

interaction is that whether such interaction might occur appears to depend on whether I_y'' (or m_y) is large enough. From approximate description in Link 17.10 of least I_y'' as a function of I_x, I_x'', and x's 'structure', and likely approximate descriptions appearing hereinbefore for xs which might be observable substantially at rest by O_U, I obtained likely approximate descriptions for least I_y'' for L 17.10 interaction between x and y (L 17.11), which appear in Table 3 column 11. Expressing $I_y'' E_1$ in terms of $I_U^{1/3} E_1 \approx$ 45.69123378 MeV, obtained as noted hereinbefore, I found corresponding approximate numerical values for such least $I_y'' E_1$ $(m_y c^2)$, which appear in Table 3 column 12.

Although, for smaller values of $I_y'' E_1$ appearing in Table 3 column 12, neglect of y spatial extent in Link 17.10 might involve inaccuracies of several percent, such inaccuracies appear substantially smaller than differences between such values for least $I_y'' E_1$, $(m_y c^2)$, and values from UCLBL 1976 (ref. 8) for $m_p c^2$ and $m_n c^2$ for proton ('p') and neutron ('n'). Thus, whether ('YES') or not ('NO') L 17.10 interaction would be likely between $x = e^+$, e^-, μ_L^+, μ_L^-, μ_H^+, μ_H^-, π^0, π_L^+, π_L^-, π_H^+, or π_H^-, and ($p^{\pm}$ or $\overset{(-)}{n}$), (Table 3 column 13), appears substantially independent of such approximation; and also appears substantially independent of the relatively smaller 'mass' difference between proton and neutron.

From comparison between columns 13 and 14 of Table 3, whether or not L 17.10 interaction would be likely between x and (p or n) appears to correspond to prior reports of which I am at the time of this writing aware of whether ('YES') or not ('NO') (Table 3 col. 14) x and (p or n) have been observed to interact in phenomena apparently regarded in prior theory(s) of which I am at the time of this writing aware in terms of such theory(s)'s 'strong interaction(s)'. As may be seen from Links 17.10 and 17.11, and various numerical values involved, L 17.10 interaction between pions and (protons and/or neutrons ('nucleons')) might appear therein as only being likely over a range of distances roughly comparable to a pion's 'size'.

With reference to discussion hereinbefore, following Link 17.9, and A-Definitions 12 and 22, coding of internal information as 'charge' sign appears to involve an entity with nonzero 'rest mass'. In view of this, and of Link 17.7, likely L 17.10 interaction with $\pi_{L,H}^{\pm}$ could transfer 1 bit of internal information coded as 'charge' sign.

As noted following Link 17.10, L 17.10 interaction appears to involve loss of separate 'identity' of the particle ('x') whose internal information was so subsumed.

From A-Definition 1 and Postulates 1, 2, and 3, everything consistent with a representation of information must be included as the 'physical system' representing such information. This appears to correspond to various notion(s) in (various) prior theory(s) of 'virtual states', with such prior theory(s)'s notion(s) of 'unobservability' of 'virtual states' subsumed in this physical picture by requirement of communicability ('observability', A-Def 5) of information (physical systems representing information).

In this context, Links 17.10 and 17.11 and Table 3 columns 11–14 appear to approximately partially describe pion-nucleon interactions, and, with such basis for a 'virtual pion' concept, the physical picture and formal structure presented hereinbefore thus appear substantially able to subsume the basis underlying treatment in various prior theory(s) of such theory(s)'s 'strong interaction(s)', for example, between nucleons.

In view of the approximate results and comparisons tabulated in Table 3, and these various considerations,

Correspondence Link 17.13. *The physical picture, and formal structure based thereon, presented hereinbefore, appear, at the time of this writing, to the extent tested as presented hereinbefore, substantially able to subsume description of phenomena apparently regarded in prior theory(s) in terms of such theory(s)'s 'strong interaction(s)'.*

•

3.5.8.9 Table 3, Columns 15 and 16. As noted in discussion preceding Chain 17, various expressions and values for I_x/I_x' or I_x/I_x'' (Table 3, columns 15 and 16) appear to suggest that, in comparison only with prior theories of which I am at the time of this writing aware, likely approximate calculations presented herein of 'particle' 'masses' might lie outside the conceptual domain of applicability of all such theory(s) prior to the material presented herein, and, at the time of this writing, might to that extent lie peculiarly within the domain of the physical picture presented herein and formalism based thereon.

3.5.9 Apparent Correspondence

In view of all material presented hereinbefore,

Correspondence Link 17.14. *The physical picture, and formal structure based thereon, presented hereinbefore, appear, at the time of this writing, to the extent tested as presented hereinbefore, substantially allowed by knowledge prior to the time of this writing.*

•

('substantially' because likeliness arguments appeared in some comparisons)

3.6 BREAK OF DEVELOPMENT

Where does one break such development as this to make part of the material available to others who might be interested? Various prior theories appear to have postulated four 'forces' or 'interactions': gravity, electromagnetic ('both' 'electric' and 'magnetic') interaction, 'weak interaction', and 'strong interaction'. Is having addressed all four a useful criterion? Or is it that summer has come? (and now, as I edit this, fall (1976) has come) (and, as I check this, winter (early 1977) has come) (and, as I 'proof galleys', spring (1977) has come)

Section Four

Comments

I have collected into this section various thoughts which occurred to me in relation to the physical picture and formal structure presented herein, and some explanatory material, some of which did not seem to belong with only one part of the formal treatment appearing hereinbefore.

4.1 RELATIVISTIC AND NONRELATIVISTIC THEORIES

Physics is based on observations of and operations on an environment. From this, it might seem for physics necessary to deal with all, and sufficient to deal with only, communication possible to an observer. For understanding (in terms of information) the consequences of various physical theories, it has sometimes been helpful to hypothesize an idealized observer, of finite extent, whose attributes depend on the theory considered.

One fundamental philosophical principle underlying theories of physics which allow more than one observer is the following:

Principle 3. *A finite observer can have its external state varied by a process without affecting its internal state if and only if all laws and constants of physics as seen by it are not changed by the process.*

('its' and 'it' used generically)

A motivation for this philosophical principle is the concept 'There, but for…., go I'. One might seek insight into the structures of various physical theories by examining the kinds of external changes recorded in the internal history of a finite observer, according to that theory.

For example, in nonrelativistic classical mechanics (NRCM), an observer watching a test mass in a finite box would be able to keep an infinitely accurate record of its acceleration as a function of time. Thus, two such observers who started together would always know their velocities and positions, infinitely precisely, referenced to (as a 'frame of reference') their common starting time and place. Therefore, NRCM does not require that physical laws and their constants be independent of position or velocity of the observer: it can always tell from its own history what has happened.

An observer in special relativistic classical mechanics (SRCM) sees the same form for the laws of physics, and the same value for the speed of light, independent of any velocity which it can attain (in that theory, $v < c$). This is philosophically proper if and only if the observer does not have a way to directly measure its velocity (e.g., by anisotropy of c, or graininess), *and is devoid of any internal history which would allow it to know its velocity.*

In this respect, special relativity appears improper: it requires independence of velocity, but admits continuous infinitely precise measurements, which would allow a complete acceleration history to be kept internally, within a closed box of finite size. In this regard, Principle 3 appears illustrated by consistency between Corollary 8.16 and A-Theorem 6 as to velocity at which separability of internal and external information, and special relativity, break down, obtained by consideration of observer's possibility of current observation (Cor 8.16) and internal history (A-Theorem 6). More generally, if a theory requires no dependence on a variable, it appears reasonable to ask that a way be presented by which that external aspect of an observer might be changed without its noticing.

Similarly, one consequence of Corollary 8.16 or A-Theorem 6 is that, if you shake a particle in a small enough box gently enough, it doesn't rattle.

Or, if a theory requires isotropy, it should allow an observer to be rotated without its noticing.

I have also used the principle that measurements of length are only made locally. A remote observer relies on messages from local observers for length measurements. In keeping with Principle 3, such local observations by a local observer are accepted as valid if the local observer can not have in internal history record of changes which would invalidate them (Principle 4).

NRQM has as its classical limit NRCM. Correspondence can be expressed by replacing NRCM Poisson brackets with NRQM operator commutators, e.g., $\dot{\mathbf{p}} \propto [\mathbf{H}, \mathbf{p}]$. The classical limit can be obtained by letting Planck's constant h go to zero. As h approaches zero, discrete quantum states get closer together, and small disturbances of the system become increasingly effective for inducing transitions among states of the system. Thus, as the classical limit is approached, the quantum mechanical system displays increasingly the ability to respond to and integrate arbitrarily slow acceleration, and thus, in the classical limit, displays the history retention characteristic of NRCM.

However, in NRQM, the probability that a small disturbance will cause a transition directly from one state to another is proportional to the *square* of the disturbance (square of 'matrix element' for coupling between the two states).

In NRQM, suppose that the disturbance were a small acceleration of a closed finite box. In such box, all quantum states are discrete. The acceleration required to reach a velocity v in time t is v/t. Thus, neglecting discreteness of such box's states in a finite universe, the probability of transition per unit time from any state directly to any other is proportional to v^2/t^2, and the total probability of transition is proportional to v^2/t. In such approximation, in the limit of long time to reach v, no transitions would be induced directly between the two states. Thus, neglecting effects of finiteness of universe, the pair of states would not retain a history by which the velocity could be related to an initial condition. This absence of velocity history seems an essential aspect of a relativistic theory.

In NRQM, neglecting finiteness of enclosure, this absence of velocity history would be retained for sums over a finite number of discrete intermediate states by which the system might go from an initial to a final state. However, in NRQM, mass particle states in a closed box are more closely spaced at higher energies. Extending sums over virtual intermediate states to infinite energy would include virtual states almost all of which were (fractionally) more closely spaced than any arbitrarily small spacing (e.g., in energy or momentum). For any given time t to reach velocity v, one could find infinitely many virtual intermediate states spaced so closely as to respond to a disturbance of magnitude v/t. These closely spaced virtual states would appear to provide a 'channel' by which the initial state could go to a final state with total transition probability proportional to v, for any t. Thus, for any finite enclosure, extending sums over intermediate states to infinite energy makes the system appear to retain its acceleration history, and thus precludes forming a relativistic theory.

Summing over infinitely many intermediate states with finite maximum energy also leads to a nonrelativistic theory: clustering of intermediate states (e.g., near a limiting energy) could provide a velocity history preserving 'channel'; uniform distribution of infinitely many states over a finite energy range would also require that density of states be infinite, making velocity history appear to be preserved.

Suppose one were to arrange for all the states in the system to be so constructed as to allow an infinite sum over such states to have the absence of acceleration history characteristic of a relativistic theory. These states must be nowhere dense. Thus, infinite sums over contributions adequately describable as nonzero from such states can be expected to diverge.

Or, one might choose an upper limit to the sums over intermediate states, e.g., by choosing a cutoff energy (mass). This would make the theory have the absence of acceleration history required for a relativistic theory. However, in the absence of specified computational rules and/or tables giving a unique value for each such cutoff, such a procedure does not satisfy the philosophical requirements for physical theories.

Or, one might choose states related in such a way as to form a relativistic theory, and then replace divergences with values derived from observation. This might also yield a relativistic theory, but one whose axiom set might include tables of numbers. Such a procedure seems to obscure the physics involved.

In constructing the nonaxiomatic form (included in Section 2) of this theory, I arranged the order in which steps relating to time development and acceleration were performed to display explicitly the null effect of gentle acceleration on internal information. In Lemma 4.1, Theorem 4, and its corollaries, I constructed a nonrelativistic photon particle representation (NRPPR), using an information argument for a system nearly at rest. In Theorem 7, I showed that NRPPR applies in a system moving with $v \ll c$. These became the foundation for Lemma 8.1.

Specifically neglecting effects of finite size of universe, as noted, in Lemma 8.1, I used NRPPR and Theorem 7 to show that gentle acceleration to $v \ll c$ would induce transitions at most as $(v/c)^2$. Then, by applying the nonrelativistic approximation at this step, I obtained the result that the internal information would not contain a history of acceleration. With this, one could then proceed in many ways to a relativistic theory, and the way presented in such nonaxiomatic form may not be the most elegant.

For example, one might consider the null effect on internal state of very slowly accelerating a system containing an experiment embodying every law and constant of physics. For example, gently accelerating a Michelson interferometer would not shift observed fringes; therefore, the speed of light is constant, independent of v. Similarly, all the laws of physics must have the same form and constants in the slowly moving system. This shifts the frame of reference of NRQM to the moving system, allowing the process to be repeated, leading to a still faster moving system. This then justifies neglecting terms of order $(v/c)^2$ and higher, and leads to Einstein's special theory of relativity in an 'infinite enclosure'.

One could then continue with the derivation of PODIC, and might find some simplification of several of Theorem 8's lemmas and corollaries.

By proceeding through Theorems 4, 7, and 8, with their lemmas and corollaries, I avoided using infinite sums over intermediate states in describing gentle acceleration of a closed box. This helped keep the development physical, leading to a relativistic theory. In such development, gentle acceleration is the process by which the velocity of a system can be changed without changing its internal information, thus displaying explicitly the transformation process whose existence is necessary for a philosophically acceptable foundation for special relativity and its requirement of the covariance of physical laws with constant speed of light.

Finiteness of the amount of information which can be represented by finite energy in a finite box is also in keeping with the inability to retain a complete history of its acceleration in an infinite enclosure. Thus, this finiteness of contained information is related to formulation of laws of physics in a way which does not depend on acceleration history and its time integrals, velocity and position.

It is interesting to note that, by requiring absence of acceleration history for at least one process for producing velocity change, special relativistic classical mechanics is in some respects conceptually closer to quantum mechanics than to nonrelativistic classical mechanics. 'Graininess', or discreteness, is a way of not responding to small disturbances. Also, absence of acceleration history offers a way to change position without changing internal information, a necessary feature for a theory which demands relativity of position, and independence of laws and constants with respect to position. In this sense, relativity of position is *not* a property of nonrelativistic classical mechanics.

By deriving special relativity from information separability, one obtains a conceptual basis for defining a limitation on its domain of applicability. In Corollaries 8.5–8.16, it appears that, for a test system moving fast enough (or, from A-Theorem 6, accelerated to such velocity) in a finite enclosure, internal and external information are not separable. Thus, special relativity does not describe physics in such systems. From the numerical examples, it seems clear why special relativity appears to work as well as it does.

Note also discussion following A-Theorem 1: that local velocity of local propagative communication be $\leqslant c$ does not appear hereinbefore as a formal statement about transporting information other than by transmission through a succession of adjacent local regions. Note also, from Corollaries 10.2–10.4, that not local communication appears to have features involving R_U, a 'global' property of U. w states in U are states in U 'as a whole'. In view of these and other considerations, communication (including, like Theorem 9 and Corollary 9.1, of 'objects') over distance d might not be restricted to occur only at or after time interval $\Delta t \geqslant d/c$, and (the concept of) 'causality' might be defined (only?) in U's rest frame.

In the axiomatic form appearing in Section 3, 'special relativity' and a limitation on its domain of applicability are obtained formally together, rather than by the two-step procedure, appearing in the nonaxiomatic form, of first neglecting, and subsequently including, effects of finite size of universe. In the axiomatic form, adseparability, including special relativity, was treated in terms of effects on internal information of accelerating a finite system in a finite universe U. Such treatment depended on having a formal statement of spectrum of states in U, which I first obtained by the axiomatic approach appearing in Section 3. That such substantially different formal treatments led to results differing only by appearance of '$\approx$' in that of Corollary 8.16, with respect to domain of applicability of separability of internal and external information and concepts of special relativity, suggests that such formal treatments substantially conform to the same physical picture, and, thus, suggests that each such treatment substantially conforms to such physical picture. Because degree to which a formal structure conforms to a physical picture does not appear to be explicitly directly tested by comparison of results derived in such formal structure with observation, this substantial agreement between such different formal treatments appears significant, as a test of conformity between formal structure and physical picture, in supporting use of formal structure presented herein in linking substantial agreement of calculated results with observation to support use of the physical picture presented herein in providing 'insight'.

4.2 TREATMENT OF GRAVITY

This theory treats gravity in a way which differs philosophically from both Newton's and Einstein's theories. Newton included approximately in his theory of mechanics effects of masses on each other, by giving a rule for calculating 'force' between two masses, in which appeared a constant relating force, mass squared, and distance squared. No reason was given for this force. The constant was obtained by measurement.

Einstein formulated a theory of gravity, in which 'space' is distorted about a mass. Any small enough body moving through this distorted space follows a path which appears to an observer substantially with it as a locally approximately straight line. Thus, a reason is given for behavior of the affected body. However, no reason is given for distortion of space about mass, and (a) constant relating distortion and mass was found by correspondence with (approximate measured value as for) Newton's rule.

The picture presented herein for gravity treats motion of an affected body in a way which philosophically resembles Einstein's approach, in that a small enough freely moving affected body can be regarded as moving in the same way that would appear as a straight line in an undistorted region. However, in this picture, a reason is given for the amount and nature of distortion of 'space' about mass, as a consequence of subsuming information by nearness. Thus, the fundamental symmetry of gravity, with respect to being affected and being a 'source', seems more nearly apparent. With this picture, amount of distortion about mass is related to the mass in a definite way, derivable from size of, and mass in, the containing 'box' ('universe' U). This derived interaction can be recalculated in units of a local observer near a massive body, leading in effect to a 'reduced' value for G. This, a 'throat' (with a 'horizon') does not appear for the gravitational interaction. This is also significant in relation to information conservation: gravity does not produce any region in U which cannot exchange information bidirectionally with the rest of U.

4.3 INERTIA

The picture presented herein gives a 'reason' for inertia, in part as related to why time is a useful parameter for describing various (local propagative) '*processes*'. Were it not for information (local propagative) communication *rates*, to what extent would it be useful to partly characterize various (local propagative) 'processes' by 'inertia'(s)?

4.4 REALITY

The three postulates and derivation which I used seem in essence primarily a statement of what is meant by *observation*. This suggests that any observer, by the nature of observation, would see the 'physics' 'generated' by those postulates, and, therefore, would see the same physics, however expressed, as any other observer. Thus, it might seem that 'subjective reality' of physics would be the same for all observers, which seems to be the meaning of the concept of *objective reality*.

This is consistent with Principle 3, that all laws and constants of physics be invariant under all changes which do not affect an observer's internal information, that is, that every observer does physics on information accessible to that observer, as illustrated by, for example, the consistency noted hereinbefore between Corollary 8.16 and A-Theorem 6.

Thus, this physical picture might provide a basis for a concept of objective reality, and thus might provide a tool for consideration of such concepts as truth, falsity, truth-telling, lying, sanity, insanity, 'semantics' and 'semiotics', first communication with another observer ('stranger'), and 'subjective reality' 'experienced' by artificial 'intelligent' entities. (See also following Comments 4.5 and 4.6.)

4.5 INTROSPECTION

As noted hereinbefore, it would seem that, to an observer, information appears conserved, communicable, and finitely accessible. However, introspection (contemplation) would seem to not involve perception of communication as such. Thus, one might consider as postulate candidates, for adequate description of introspection, only Postulates 1 and 3: information is conserved and finitely accessible.

One might, for stating such postulates, denote observation (or its equivalent ('oie')) by use of the word 'appears' oie, and introspection oie by use of the word 'seems' oie, as:

for observation:

information appears conserved, communicable, and finitely accessible;

for introspection:

information seems conserved and finitely accessible.

The four dimensionality of spacetime in which an observer can adequately describe observed events appears to have depended, in derivation presented herein, on Postulate 2. As in A-Corollary 1.1, existence of an observer involves 1 'existence' bit, so, as in A-Theorem 1, adequate description of phenomena perceived in introspection (by an 'introspector' qua introspector) requires (exactly) two independent variables, which, following the general line of reasoning in A-Theorem 1 and its corollaries, might be regarded as everywhere ('globally') flat, as orthogonal, as homogeneous, as including one variable adequately describable as unidirectional and not interchangeable with the other, and as having such other variable adequately describable as isotropic.

For convenience, one might call the 2-space spanned by such two variables 'introspection space' ('inner space', 'introspace', 'inspace', 'i-space', etc.).

In view of introspection space involving only two variables and spacetime ('observation space', 'obspace', 'external space(time)', 'exspace', 'espace', 'expace', 'o-space' ('o' for 'observation'), 'c-space' ('c' for 'communication'), etc.) involving four variables, I have not yet, at the time of this writing, understood formally in detail whether the apparently unidirectional inspace variable corresponds to (the usual) exspace 'time'; nor, there being only two variables in inspace, whether the apparently unidirectional inspace variable must 'always' be the same one of the two.

As examples, one might consider the apparently unidirectional inspace variable as an 'introspection time' ('*ti*', 'intime', 'i-time', etc.), and the other inspace variable as an index ('*xi*') for describing information processing. Some kind(s) of 'looping' might involve 'resetting' *ti* according to condition(s) on *xi*, subject to the requirement that there 'always' be one inspace variable (without specifying which) that appears unidirectional. In some description, such index *xi* might correspond to 'position' along the 'tape' of a 'Turing machine'.

In view of these various considerations, an observer's description of another entity's introspection might be formally different from the introspecting entity's description of such introspection.

This might offer some insight into the 'subjective' distinction between 'mind' and 'body', in a way consistent with to all other observers ('objective') absence of any such distinction. (see also Section 6, questions 5, 43, 44, and 47)

4.6 FREE WILL

This might be stated more nearly explicitly: were an observer in possession of a full description of a closed system including the observer, could such observer, from the laws of physics, predict in detail all of the future in that system? Implicit in this is the assumption that there would be some way to process such data. From the preceding analysis, in any closed system, transcription of such complete information would use up all the ability of the system to represent information, (all energy (including mass) present in the system). Thus, the required experiment cannot be performed in any closed system containing the observer and separately reported to the observer. However, an observer would seem free to treat explicitly (apply the laws of physics to) a complete measurement on any smaller subsystem: adding a separate observer enlarges a subsystem, and predictions about the subsystem might then be contained in the larger combined system. Another being might be regarded as part of such a subsystem. Thus, it would seem that any being is allowed by the preceding analysis to treat every other being as fully determined, while regarding self as either free or determined (as might be 'desired'). In view of these considerations, it might seem that every observer's 'physics', as limited to only (all) statements necessary (and sufficient) to describe all (possible) observation (including introspection?), would not include the statement that such observer is fully determined, and the 'concept' of 'predetermination' might seem so formally excludable from such 'physics'.

4.7 PLANCK'S CONSTANT AND PHOTONS

(I constructed the informal discussion presented in this subsection years before I reached the axiomatic form presented in Section 3, in which these concepts appear as an expression of finite accessibility of information. I have included it here for explanation: this reasoning was among the steps I took in the course of looking for a picture based on information, and is offered in the spirit of sharing an attempt at understanding, from which you might reason differently than I did.)

To me, a puzzling aspect of our universe has been that stars, much bigger than a wavelength of light, and atoms, much smaller than a wavelength of light, both emit light which is detected in 'lumps' whose energy is proportional to the frequency of the light.

Note that, up to and including Corollary 10.1, the nonaxiomatic form of this theory, as presented herein, did not use properties of propagating electromagnetic energy, only of electromagnetic energy in allowed states in a box.

From Theorem 10, size and information are related. A small enough lens far enough from a star or atom would focus incident electromagnetic energy from either substantially the same. This lets us relate conceptually and 'informally' the frequency of light from an extended source to a minimum energy which can be exchanged with such light, as follows:

In Theorems 4 and 4′, and as applied in Theorem 14, I showed that Planck's constant, the speed of light, and the linear size R_B of containing box B can be related to minimum energy E_1 to represent 1 bit of information in B, in any way using electromagnetic energy and/or mass particles, by (exact in axiomatic form) $E_1 = \hbar c / R_B$. In Theorem 10, I showed that localization to a smaller linear size r in B represents more information, as R_B / r. From Maxwell's equations, an electromagnetic wave with frequency f can be focused to $c/2\pi f$ linear extent. From Corollary 1.1, information carried in the wave is conserved during focusing. Therefore, information in the focused wave is represented in a region of linear size $c/2\pi f$. Thus, Corollary 1.1 and Theorem 10 require that the smallest amount of energy which can be exchanged with a propagating electromagnetic wave of frequency f be that amount which could represent 1 bit of information in the smallest region to which such a wave can be focused, which is

$$\frac{R_B E_1}{\left(\frac{c}{2\pi f}\right)} = \left(\frac{2\pi R_B E_1}{c}\right) f.$$

Thus, energy is exchanged in 'lumps' proportional to the frequency of the wave. Thus, photons. (A formal and conceptual treatment of photons appears in Section 3, in the axiomatic form.)

If we treat information as basic, Planck's constant may be eliminated from the nonaxiomatic form of the theory by this substitution. If B is regarded as always in a much larger 'enclosure', from PODIC, such a form would be substantially unaffected by changes q in scale (linear size R_B) of B.

Expressing R_B in terms of local Hubble's constant H, as $R_B = c/H$, Planck's constant is given by $\hbar = E_1/H$, $(h = 2\pi\hbar)$, where E_1 is the (minimum, and in axiomatic form definitional of 'energy') energy per bit of information, thus expressing Planck's constant in terms of only relatively locally observable properties of our universe.

This might offer some insight into the meaning of Planck's constant, and of photons. Planck's constant appears as a way of relating an information formulation and an energy formulation.

Expressing the intuitive picture of the photon differently, taking information as basic, we can regard an electromagnetic wave with a frequency as defining a length. Amount of information represented by localization is inversely proportional to length, that is, defining length L represents information, as $2\pi R_B/L$. Emitting an electromagnetic wave with wavelength L requires emitting at least that information (energy) represented by defining length L. Similarly, detection of information from such an electromagnetic wave represents reception of the information that the wavelength is L. Thus, information (energy) can be introduced into, or extracted from, an electromagnetic wave with wavelength L in not less than a certain amount, independent of emitter or detector.

In the axiomatic form, such 'quantization' appears as a conceptual and formal consequence of finite accessibility of information, leading to expression, in A-Theorem 4, in terms of wavelength and bits of information, of least information exchange 'lump' size; reexpressed, in A-Corollary 4.3, in terms of frequency and bits of information.

In the axiomatic form, information is taken as basic, and Planck's constant has only the role of part of a calibration constant for measuring equipment reading in energy units instead of bits of information.

4.8 REPRESENTING MEASUREMENTS

From relation between 'energy' and information, and finite accessibility of information (and 'energy') in any closed system, we can restrict admissible representations for physical measurements. For example, in classical mechanics, position is represented by three real variables. Each such variable appears without restriction on amount of information represented, and thus purports to represent an infinite amount of information. From relation between energy and information, it appears that the amount of information represented by mass of a particle in any finite system is (adequately describable as) finite. Thus, it appears proper to reject, as nonphysical, position representations using unrestricted real numbers. It appears appropriate to use a finite set and amounts of information commensurable on such finite set, and/or a suitable concept of 'adequate description' (A-Def 2). A more accurate representation might be to use numbers whose accuracy, for example, numbers of digits, were related to the mass of the particle whose position was being described.

One might similarly consider prior use of unrestricted complex numbers for projections onto 'quantum mechanical' states. Each is effectively a pair of real numbers, again without restriction. Also, in quantum mechanics, unrestricted real numbers are used to label relative phase between states. Predicted values for measurements often have the form of probabilities, represented by unrestricted real numbers. Without a suitable concept of 'adequate description', such description would also appear inappropriate.

A somewhat different problem arises in defining continuous functions, such as probability densities. Corresponding to such density distributions has, prior to the time of this writing, been the concept of operators definable at each point, labeled by continuous variables. Not any physical detection process can correspond to such a hypothetical system: for all space to be filled with such detectors, each must have a transcription volume not bigger than its assigned location, so a finite amount of energy would be unable to record a single bit into such a detector; also, unambiguously recording even a single bit in one such detector to the exclusion of all others would represent an infinite amount of information.

Similarly, use of integrals unrestricted to finite information in defining inner products, projections, etc., seems misleading, in that functions corresponding to observations and appearing in such integrals become less precisely defined as the mesh of the integral is made smaller. Thus, the notion of completeness of decomposition and reconstruction in Hilbert space seems misleading. Hilbert space may be too 'big' for relativistic quantum mechanics.

Formal sufficiency of 'adequate' description has as one consequence that formally sufficient descriptions can be expressed with finite amounts of information. Although integrals, roots, fractions, logarithms, and powers appear in various approximate descriptions herein, such expressions are taken to represent only an amount of information sufficient for such description to be adequate. For example, in describing a system with total 'energy' (including 'mass') E, an amount $I = E/E_1$ of information would suffice for adequate description (see also Comments 4.6 and 4.9). As noted hereinbefore following A-Corollary 4.1, where '=' signs appear herein, they are to be taken as denoting 'equal' for mathematical manipulation and 'adequately describable as equal' for comparison with observation and possibility of satisfaction of expression in (a) universe with finite size and finitely accessible information.

Several (approximate partial) description(s) appearing hereinbefore have form(s) which might be regarded as in part 'consequence(s)' of describing representation of finite amount(s) of information:

1. In Corollary 14.3, step 16, (p. 128), 'shell thickness' y and least radius r_1 are both taken as $\lambda/2\pi$. In the axiomatic form, from A-Theorems 1, 3, and 4, $\lambda/2\pi$ appears as the linear size of the smallest 3-space region into which 1 bit could be transcribed from w with least amplitude >0 and wavelength λ. From Corollary 14.3, step 15, the interaction under consideration would be larger for smaller y and/or r_1. In Corollary 14.3, m_0 would be smaller for stronger interaction. So, for least m_0, least y and r_1 were sought. In view of these considerations, in the axiomatic form, the taking in step 16 of y and r_1 as $\lambda/2\pi$ might seem formal (and/or 'adequate' as in A-Def 2).
2. In Corollary 14.8, step 5, (p. 134) (two 'photon' relatively localized substantially circular configuration as used also in Link 17.5 ('C2' in Def 21)), and Link 17.7, step 26, (p. 249) (three 'photon' relatively localized substantially circular configuration as used also in Link 17.9 ('C3' in Def 21)), Theorem 13 interaction of photons is treated as Theorem 13 interaction of 'lumps'. In both cases, what is sought is (a) description of such interaction(s) in 3-space region(s) having linear size(s) so small that only 1 bit of information could be transcribed into such region(s) from each w respectively considered. If such regions were divided into smaller regions, there would not be enough available information to provide description of 'densities' in such smaller regions.

In view of these considerations, in the axiomatic form, from A-Theorems 1, 3, and 4, in such instances, treatment of 'photons' as 'lumps' might seem formal (and/or 'adequate' as in A-Def 2).

These considerations also appear related to significance of comparison between approximate calculated value for F and reported value(s) for α^{-1}.

3. In Link 17.10, step 15, (p. 260), integrals appear in approximate description of (a) Theorem 13 interaction between two already formed entities: in this case, amount(s) of information available appear sufficient for transcription of 1 bit of information into 3-space region(s) with linear size substantially smaller than the linear size(s) of the 3-space region(s) which the relatively less localized entity is adequately describable as contained in, and the accuracy sought in such approximate description is only a few percent. For these reasons, such approximate description appears to be not excluded as an approximation by the physical picture and formal structure presented hereinbefore.

In view of these various considerations as to form of (approximate) description, various comparisons between approximate calculated results and reports of observations, for example, appearing in Table 3, items 2–20, appear as partial tests also of these concepts as to form(s) of (approximate) description(s) consistent with finite amount(s) of available information. (See also question 79, Section 6.)

Various 'problems' in measurement theory for quantum mechanics would seem traceable to attempting to conceptually separate communication, as if into transmission and reception, and treating only reception as if it were a separable process. As noted hereinbefore, there is not any experimental procedure in which only either transmission or reception can occur without the other. If one regards 'experiment' as an observer O_U communicating with itself through its 'universe' U, for each such O_U there would seem to not be such problems of 'measurement' in the conceptual and formal structure presented herein.

One remarkable property of information is that it is quantifiable by a number without 'dimensions' (e.g., length, time, mass). For this reason, it would seem that mathematical operations such as powers (and roots), exponentiation, and logarithms can have as both input and output amounts of information. Such amounts of information can be converted to other units, by, for example, dividing into λ_1 $(=2\pi R_U)$ to find 'length', dividing into λ_1/c to find 'time interval', multiplying by E_1 to find 'energy', or multiplying by E_1/c^2 to find 'mass'. Such mathematical operations appear in $I_U^{1/3}$ (e.g., as used in 'particle' 'masses', in $I_U^{1/3}E_1/c^2$)

and in F (in '$\ln(2I''_{e^\pm})$' term), and might also appear in $K^0_{\text{Long, Short}}$ 'mass splitting' (e.g., if approximately partially describable by $I''_{K^0_L} - I''_{K^0_S} \sim (I''_{K^0_{L,S}}/2)^{2/3}$). (please see hereinafter also Comment 4.9, and, in relation to $K^0_{L,S}$, question 52, Section 6)

4.9 FINITE ACCESSIBILITY OF INFORMATION

That information be finitely accessible (Postulate 3) might substantially change much of prior mathematics, including, for example, various concepts relating to induction, infinities, infinitesimals, calculus, Gödel's theorem, infinite dimensional vector and/or function spaces including Hilbert and Banach spaces, the 'axiom of choice', compactness, completeness, convergence, continuity, limits; statements including 'for each', 'for every', 'for all'; meaning(s) of solutions of equations; etc.

One might think of finite accessibility of information in terms of a movable 'window', with optionally part of the total amount of information retainable or 'exchangeable': at the time of this writing, it is not clear to me whether different observers' universe(s) contain the same information. (See also Section 6 question 47.)

From Postulates 1, 2, and 3, and A-Theorems 3 and 4, in observer O_U's universe U, information is communicable (and representable) only in amounts adequately describable as finite positive integer numbers of bits. Each bit is representable by, or can specify, one out of two alternatives, a binary choice. Therefore, all choices are adequately describable as integer numbers of binary choices. This might provide a context for Aristotle's logic (two valued ('either-or') logic), and for the concept, related to 'verifiability', that a result which could not have been otherwise does not contribute to verification ($\log_2 1 = 0$), that is, is meaningless. Also, the bit might thus appear as a unique natural unit of amount of information, which appears especially interesting in view of information being treated as a 'pure' number which can be operated on in various (mathematical) ways (e.g., powers, roots, exponentiation, logarithms, etc.) and still appear as information. (See also Comment 4.8 hereinbefore, and Section 6 question 66.)

4.10 TUNNELING AND VIRTUAL STATES

One might view 'quantum mechanical tunneling' in terms of size of neighborhoods in phase space associated with distinguishable elements which can represent information. Information appears to be always exactly conserved: there is not any 'virtual' information. Limitations on virtual states of systems may be calculated by specifically requiring that such excursions not remove the system from the neighborhood associated with an information code element. That is, the system 'can do' everything that can't be observed.

Observations are done on a system's information. (From this also follows behavior approximately described in prior theory(s) by such theory(s)'s 'uncertainty relations'.) Calculations are done by including all system configurations compatible with that information. Were it only necessary to include some of those configurations, designating which ones to include would represent information; and the others might be thought of as available to represent information. This would not conserve information. Thus, *all* configurations compatible with the information must be included. That is, if it's not forbidden, it's mandatory. In this way, conservation of information appears to offer some insight into why 'virtual states' appear in ('quantum mechanical') calculations.

4.11 HUBBLE'S CONSTANT

From Corollary 10.3, the radius of our information 'box' appears as $R_U \approx c/H$. Note that this might not be the same as saying that every universe has the same contents. For example, it appears that an observer anywhere would see a surrounding information 'box', with 'wall' $\approx c/H$ distant, as in the approximate calculation of the electron mass (especially Cor 14.3), in which each electron appears to be at the center of its universe.

One might imagine our universe as a spherical 3-surface in 4-space, with a 'size', with 3-'wall' equidistant from everywhere.

Or, one might imagine traveling in a 'fog', always surrounded by the same size of accessible information (communication) region, never reaching an 'edge' of an observer's own universe.

(See also Comments 4.8 and 4.9 on finite accessibility of information, and Section 6 question 47.)

4.12 PHONONIC AND PHOTONIC SYSTEMS

Although, as developed herein, this theory has involved properties of photons, many features of this theory might appear in systems based on phonons. This might have practical applications. For example, electron coupling in or for superconductivity might be approached in this way. Phononic information processing 'structures' might be 'synthesized' within, or on the surface(s) of, substantially homogeneous solid, for example, amorphous or crystalline, media, or other systems capable of supporting 'propagation' and/or 'modes', which might include networks, arrays of (e.g., shift and/or logic) registers, etc.

The condition that the 'wall' be equidistant from everywhere might be (approximately or adequately as in A-Def 2) met for surface waves on a sphere. More intricate topologies (surface or solid connectivities) might also be interesting.

One might speculate on phononic 'universes' in extended media, for example, neutron stars.

Similarly, photons in dielectric media might be used, for example, for information processing. For example, photons might be dielectrically 'guided' in a thin film on the surface of a sphere, (as) for an 'equidistant from everywhere' boundary condition, or on or in topologically simpler or more intricate systems. Photonic 'structures' might be 'synthesized' in such systems.

'Losses' encountered in media might be compensated by suitable amplification processes, such as stimulated emission of radiation, traveling wave amplification, etc., for photons and/or phonons.

In amplifying media, such as pumped media with suitable population inversions, extensive photonic systems might be constructed. For the two-dimensional case, traveling wave amplification might be used to sustain phononic and/or photonic systems.

Combinations of phononic and photonic systems might be interesting and/or useful, both in themselves, and for application in input and/or output of systems of either type. For example, phononic systems might be used to affect photons, for information transfer.

As a technically potentially interesting example, by constructing informational structures in suitable extended media, one might achieve information processing whose intricacy was not related to hardware intricacy in the same way as with present electronic information processors, which might greatly simplify hardware and/or extend the range of feasible processing. The 'computer' might then comprise one or more cyclic, progressive, recursive, and/or other information structures, for example, represented by photons and/or phonons in a relatively homogeneous medium.

One might speculate on processes of this sort also in extended regions of dilute gas, for example, pumped by an ultraviolet star.

4.13 POSTULATES AND CONSTANTS

It is not clear to me, at the time of this writing, whether the postulate set which I presented in Section 3 is minimal. In some ways, various of the postulates seem conceptually related: what would it mean to conserve something which was infinitely accessible (Postulates 1 and 3)? How could an observer learn of any information which was not communicable (is Postulate 2 necessary)? Might it be sufficient to postulate only that information is conserved (Postulate 1); or so postulate and state the amount (combining Postulates 1 and 3 and a constant)? Or, might one postulate the existence of an observer (or introspector), and proceed by considering (treating?) such observer's (introspector's) universe ('internal universe')? (Would an entity filling U be capable only of introspection?) Might I_U be shown formally to 'always' appear the same, and be 'eliminated'?

Although I stated them separately for ease of reference, Postulates 1, 2, and 3 might be stated together as 'Information is conserved, communicable, and finitely accessible.', a somewhat less verbose form. (or, 'Information: (1) conserved, (2) communicable, (3) finitely accessible.'? or without parenthetical numbering? or as adjectives preceding 'information'? (a longish comment on brevity))

Some of such considerations one might call philosophy. For calculations, one uses also a suitable set of constants, such as I_U, R_U, c; or, for convenience in using instruments calibrated in energy (mass) units, I_U, R_U, c, E_1; or I_U, R_U, c, $\hbar$; or G, c, $\hbar$, m_e; or G, $\hbar$, m_e, α; or various others. With suitable sets of postulates and constants, one has also technology and engineering.

As noted in the introduction to Section 3, part of the purpose of axiomatizing the formal structure based on the physical picture presented herein was to display more starkly the nature of such physical picture. Although the postulate set which I chose involves three postulates with a total of ten words (in English as stated), I suspect that such postulate set was sufficiently small to be of some help for such purpose, and I did not wish to have to treat formally various questions noted hereinbefore in this subsection before making available to others this report of work in progress.

You might wish to choose other postulate sets and test their sufficiency.

In some sense, it is especially satisfying to relate aspects of physics to information. Information would seem to be an especially simple and elegant concept, stateable without reference to dimensionality or units: *result or specification of choice*. Experiment would seem (adequately) describable as transmission and reception of information, and physical 'laws' as statements about the information channel, and/or about coding.

4.14 'WHY DO THINGS FALL?'

This is mostly to talk about how doing this has felt.

Although much of the material presented in Sections 2 and 3 has been icily formal, that was one of the few tools I had thought of to try to keep from fooling myself into thinking that I had already understood when I hadn't yet, and even so, I wonder...

One would like of a theory one to one correspondence with 'reality'. Late in this development presented, finding that the two ostensibly significantly discrepant $m_{\pi_H^\pm} - m_{\pi^0}$ reports only seemed significant because of insufficiently supported statements of accuracy brought some relief and excitement, and finding a bump on Haddock+'s Figure 2c where a 'π_L^- peak' would be brought a feeling of wonder.

But, what questions have I been addressing behind all the formality? Before going on to more relatively formal discussion, I would like to share with you the smile which has sometimes come (I'm smiling as I write this (and type this)) with realizing that I have been addressing questions which puzzled me when I was smaller (when I realized that I wanted to understand action at a distance, gravity and magnetism, and that school would be an interruption in my learning process, I was standing in the hallway near the pantry in the house where we lived on Shepherd Street between 13th and 14th streets Northwest in Washington, D. C.: there was a table in that hallway which I remember as then about eye height, which only comes a little above mid thigh on me more recently. The answer 'gravity' for why things fall seemed to me about then not an answer but another name for ignorance; later, general relativity seemed to not contain any reason for space to be curved, and thus not an answer (in this picture, space curvature appears from subsumption of external information by nearness to 'mass')).

Some are questions which might bring a smile to you, too:

'Why do things fall?' ('Theorem 13' interaction?)

'How high is up?' (R_U?)

'What is time?', 'Why is time different?' (A-Theorem 1, A-Cor 1.1, 1.2?)

'What is mass?' (A-Def 12, 21, 22?)

'Why does space have three dimensions?' (A-Theorem 1, A-Cor 1.1, 1.2?)

'Why does a moving object keep moving?', 'What is inertia?' (Theorem 9, Cor 9.1, Theorem 16?)

'What is energy?', 'Why can energy and matter be interconverted?' (A-Def 21 and 22, and nonaxiomatic treatment in Lemma 4.1 and Theorems 4 and 4′?)

'Why does special relativity "work"?' (absence of acceleration history and equivalent status of 'naive' observers (us?)?)

'What is light? a photon? (or "are photons?")' (A-Theorems 1, 3, 4?)

'Why are there electrons?' (Theorem 14–Cor 14.16?)

There have been times when doing this has been exciting, times when it has felt strange (why me? can I really understand this?), times tedious (renumbering internal references after inserting new material, typing this 'manuscript').

I looked for one factor of two in my living room and down by the river, and finally found it in my bedroom and quickly put it where it seemed to belong (Cor 8.16 steps 7 and 8).

I chose a nomenclature, 'A-separable' where herein I have used 'adseparable', and used it for about half a year before I realized that, if it were to be spoken aloud, it might produce unwanted confusion with the word 'aseperable', of opposite meaning, and chose 'adseparable'; maybe much aloneness...

But, recently, thinking about 'How high is up?' has brought a smile, and 'Why do things fall?', wonder that anything should ever be understandable.

Section Five

Some Predictions and Experiments Which Might Be Interesting

5.1 PREDICTIONS

By 'prediction', I mean in this Section 5 *only* statement of (derived) result(s) while I am not, at the time of this writing, aware of what I regard as unambiguous, and unambiguously comparable, prior result(s) of prior direct experimental test(s) of such (derived) result(s).

In those cases displayed hereinbefore for which reports of observations were cited herein, this theory, as presented herein, appears to agree closely with, or be substantially allowed by, such knowledge prior to the time of this writing. Although, at the time of this writing, I have not (yet?) developed many formal details, this theory appears to offer several results for which I am not yet, at the time of this writing, aware of what seem to me to be sufficient observations (please see also questions 50, 51 (especially 51.7), and 52, Section 6):

1. From Theorem 8 and its corollaries (PODIC) or A-Theorem 6 (adseparability), in domain of applicability of adseparability, 'quantum mechanical' spreading of a box would be not observable by an observer contained therein.

2. From Corollary 13.3, there is not any region of our universe so affected by gravity as to preclude exchange of information with the rest of our universe. (there is not any gravitational 'black hole')

3. From Theorem 13 and Corollary 14.16:

- 3.1 Electrons and positrons are affected equally by gravity (both fall in the same direction);
- 3.2 Electrons and positrons are equally effective as 'sources' for gravity.

4. From Corollaries 10.2–10.4, and Theorems 13 and 14 and their corollaries, Hubble's constant is $\approx 9.7191 \times 10^{-19}/\text{sec}$.

5. From Theorem 13, and 4, our universe has average mass (energy) density $\approx 3.3799 \times 10^{-27}$ kg/m^3.

6. From Corollaries 14.10 and 14.16, several (l) electrons or positrons may be in the same state (both momentum and 'spin'), with total mass (energy) in that state roughly proportional to, and of the order of, $l^2 m_e$. Thus, Fermi–Dirac statistics would not correctly describe electrons or positrons packed so closely that their Fermi level was $\gtrsim m_e c^2$.

7. From A-Theorems 3 and 4 and A-Definition 22, photon energy levels in our universe are adequately describable by $E_j = jE_1$, where j is a finite integer > 0.

8. From Link 17.5, in $\mu_H^{\pm} \to e^{\pm} \overset{(-)}{\nu} \overset{(-)}{\nu}$, both neutrinos are electron neutrinos and/or antineutrinos (see also Section 6 question 21 p. 329, and item 11 of 'Predictions').

9. From Links 17.2–17.9, m_γ, m_{ν_e}, $m_{\bar{\nu}_e}$, $m_{e^\pm}$, $\alpha^{-1} \approx F$, $m_{\nu_{\mu(L,H)}}$, $m_{\bar{\nu}_{\mu(L,H)}}$, $m_{\mu^\pm_{L,H}}$, m_{π^0}, $m_{\pi^\pm_{L,H}}$, $m_{\pi^\pm_{L,H}} - m_{\pi^0}$, and $m_{\pi^\pm_H} - m_{\pi^\pm_L}$ are approximately as listed in Table 3 column 8, and there are not any pion-like particles with mean life as long as τ_{π^0} approximately describable by C3, I_x, 0, d, f with $3 < I_x < 12$.

10. From Links 10, 11, and 13, and Table 3 columns 11–14, likely all particles with 'rest mass' greater than proton's rest mass interact by 'strong interaction' with proton and neutron.

As possibilities rather than relatively specific results:

11. From Links 17.4 and 17.5, as noted in Section 6 question 21, decay of an ensemble of many $\mu_H^{\pm}$s might be approximately partially describable by $\mu_H^+ \to e^+ (2 - x_1) \nu_e x_1 \bar{\nu}_e$ and/or $\mu_H^- \to e^- (2 - x_2) \bar{\nu}_e x_2 \nu_e$, where $x_1 \sim 1/2 \sim x_2$ (see also Section 6 question 21 as to whether $x_1 \neq 1/2 \neq x_2$).

12. From Link 17.5 and Table 3, there might be a 'light muon' $\mu_L^{\pm}$ $(C2,7,b,+,f)$ with mean lifetime much less than that of $\mu_H^{\pm}$, and mass $m_{\mu_L^{\pm}} \approx 103.9047318$ MeV$/c^2$, decaying substantially entirely by $\mu_L^{\pm} \to e^{\pm}4\gamma$.

13. From Link 17.7 and Table 3, there might (as appears indicated at >0.999 'confidence level') be a 'light charged pion' $\pi_L^{\pm}$ $(C3,13,0,+,f)$ with mass $m_{\pi_L^{\pm}} \approx 139.4206181$ MeV$/c^2$, decaying mostly by $\pi_L^{\pm} \to \mu_L^{\pm} \overset{(-)}{\nu}_{\mu(L)}$ followed relatively soon thereafter by $\mu_L^{\pm} \to e^{\pm}4\gamma$, or by $\pi_L^{\pm} \to e^{\pm} \overset{(-)}{\nu}_e 8\gamma$, etc.

14. From Corollaries 14.4, 14.8, 14.9, 14.12–14.16, and Links 17.4, 17.5, 17.7, and 17.14, 'electric' interaction strength ('charge' magnitude in electrostatic and/or electromagnetic interaction) of various 'particles' might not be the same, and 'charge' interaction strength in e^+e^-, $e^{\pm}p^{\mp}$, $\mu_{L,H}^{\pm}p^{\mp}$, and $\pi_{L,H}^{\pm}p^{\mp}$ might not be the same.

EITHER **15**. From Conjectures 1 and 1.1 (Appendix 1), processes involving making or 'absorbing' neutrinos or antineutrinos are seen by local observers as affected by gravity,

OR **16**. From A-Theorems 3 and 4, Corollaries 14.15, 14.16, 15.11, and 15.12, Links 1, 3, and 6, A-Definition 22, and Conjectures 2–2.2 (Appendix 1), for particles (e.g., neutrinos or antineutrinos) with energy E, interaction strength approximately describable by 'charge' strength $|a_f|$, and representing internally amount I of information, in a 'gravitational field' approximately describable as 'linear' and having 'acceleration' g, if

$$\left|\frac{a_f}{a_b}\right| \leqslant \sim \frac{\sqrt{2}\, E_1^{1/4} E^{3/8}}{(g\hbar)^{1/8} m_b^{1/2} c^{7/8}} \left(\frac{3.6}{Y(I)}\right)^{1/6},$$

then vertical propagation of such particle might be observable (statistically) only within vertical distance $\pm d$ of the 'source',

$$d \sim \left(\frac{\hbar c^3}{gE}\right)^{1/2},$$

and such particles may be trapped in shells about gravitational sources.

With reference to the immediately preceding list, results 2, 4, and 6 might be amenable to astrophysical comparison. For example, result 6 might be testable in tests of neutron star models.

Result 2, for example, Corollaries 13.1, 13.2, and 13.3, represents a qualitative change from a previous picture of the behavior of large, dense, masses. In such previous picture, accretion of mass led to unstoppable collapse; thus, in such picture, only a relatively narrow range of size, or other properties, of massive object might provide large gravitational red shift(s) with respect to a remote observer, so such object(s) seemed relatively unlikely.

In Corollary 13.1, it appears that an observer near a large mass, and subject to its $q(r)$, would in effect see G as smaller. Thus, accretion of mass would lead to progressively larger $q(r)$ and red shift, without quickly leading to unstoppable gravitational collapse. Thus, one natural fate of large masses might lead to objects with relatively large $q(r)$ red shifts, and such objects might be relatively common. For example, they might be formed instead of globular clusters, or in some other way related to formation of galaxies. Dark objects of this type might be a fate of previously luminous matter, and might represent a substantial part of the mass of our universe. From Corollary 13.2, collision of two such objects might release a large part of their mass energy, and thus be a relatively highly energetic process. (See also Section 6 question 51, especially 51.3.)

In this theory, such objects appear to be a likely result of mass accretion, and, thus, might be an interesting model for quasars. For example, quasars might be large, massive, relatively common objects, much or most of whose red shifts were due to $q(r)$ effects[31]; or might involve collisions of such objects.

Highly energetic processes in nuclei of nuclear galaxies might involve presence of one or more such massive objects, for example, onto which the fall of other matter might allow efficient conversion of mass to energy and/or 'reversion' of iron to hydrogen. (See also Section 6 question 51.)

Results 1, 3.1, 6, 8–14, and 16 might be amenable to laboratory experiment. Result 7 might be relevant for models of 'elementary particles' other than photons, ν_e, $\bar{\nu}_e$, e^+, e^-, $\nu_{\mu_{(L,H)}}$, $\bar{\nu}_{\mu_{(L,H)}}$, μ_L^+, μ_L^-, μ_H^+, μ_H^-, π^0, π_L^+, π_L^-, π_H^+, and π_H^-; and for studies of our universe, including interpretation of measurements of 'background' radiation previously described in terms of a 'black body' temperature defined with the assumption of a 'hard (interacting) wall cavity' spectrum as in nonrelativistic quantum mechanics, of about 3°K (see also questions 40 and 50 Section 6).

Result 10 might be relevant to various prior theory(s)'s models for such theory(s)'s 'weak interaction(s)' as 'mediated' by a particle hypothesized in such theory(s) as being more massive than a proton but interacting only by such theory(s)'s 'weak interaction', as likely substantially inconsistent with such hypothesis.

5.2 EXPERIMENTS

Table 6 is intended as a partial list of experiments, or, in some cases, possibly merely reexaminations of data from experiments, which might be interesting in the context of the physical picture and formal structure presented herein. (some of these might also be interesting for other reasons)

Table 6. Some Experiments Which Might Be Interesting

	Item	Question	To about	By..., NOT...	As test of...
1	α^{-1}	as F from $F^3 \exp(F^3/32\pi^2)^{1/2} \approx 7.2\hbar c/m_e^2 G$	$\sim$2–3 ppm/σ	by electron 'atom' (e.g., e^+e^-) spectroscopy, NOT using e/h from Josephson effect in any way (e.g., including definitions of units), NOT mesic atom spectroscopy	treatment herein of $e^{\pm}$ and 'charge'; might also be indirectly a test of BCS superconductivity theory underlying calculations, from Josephson effect measurements, of e/h, and of various recommended values for 'physical constants' (e.g., α, m_e, $\hbar$, etc.)
2	G	(for use in item 1 and other approximate calculations presented herein)	$\sim$100–200 ppm/σ		
3	$\mu_L^{\pm}$	$\mu_L^{\pm} \to e^{\pm} 4\gamma$ $\tau_{\mu_L^{\pm}} \ll \tau_{\mu_H^{\pm}}$ $m_{\mu_L^{\pm}} \approx 11.76^{1/3} I_U^{1/3} E_1/c^2$ $(m_{\mu_H^{\pm}} - m_{\mu_L^{\pm}}) \sim 1.4235\ \text{MeV}/c^2$	$\sim$0.2–0.3%/σ	by various 'particle physics' experimental techniques, including as a step in item 7 of this Table 6, NOT mesic atom spectroscopy	$\mu_L^{\pm}$ as approximately describable as C2, 7, b, +, f; 'b' code; 'why' $\pi_H^{\pm} \to \mu_H^{\pm} \overset{(-)}{\nu}_\mu$ more often than 'directly' to $e^{\pm} \overset{(-)}{\nu}_e$; with $m_{\mu_H^{\pm}}$, $I_{\mu_{L,H}^{\pm}}$ by $(1 \pm 1/I_x^2)^{-1/3}$ factor; treatment of 'charge' sign and 'rest mass'
4	$\mu_H^{\pm}$	$m_{\mu_H^{\pm}} \approx 12.25^{1/3} I_U^{1/3} E_1/c^2$	$\sim$600 ppm/σ	by $e^{\pm}$ 'momentum' spectrum, NOT mesic atom spectroscopy	with comparison to mesic atom spectroscopy, as test for 'meaning' of 'charge conservation' as conservation of coding or of 'interaction strength' (see also item 3 of this Table 6)

5	$\mu_H^\pm$	$\mu_H^+ \rightarrow e^+(2-x_1)\nu_e x_1 \bar{\nu}_e$ $\mu_H^- \rightarrow e^-(2-x_2)\bar{\nu}_e x_2 \nu_e$ $x_1 \sim 1/2 \sim x_2$	x_1 and x_2 to ~ 5–$10\%/\sigma$	by ν_e and $\bar{\nu}_e$ reaction with counting of resulting e^+s and e^-s, and absence of $\nu_{\mu(H)}$ and/or $\bar{\nu}_{\mu(H)}$ by absence or relatively small amounts of μ_H^+s and/or μ_H^-s	$I_{\mu_{L,H}^\pm}$, $I_{\overset{(-)}{\nu}_e}$, $I_{e^\pm}$; 'b' code; 'why' $\pi_H^\pm \rightarrow \mu_H^\pm \overset{(-)}{\nu}_{\mu(H)}$ more often than 'directly' to $e^\pm \overset{(-)}{\nu}_e$; 'lepton number'; treatment of 'charge' sign and 'rest mass'
6	π^0	τ_{π^0}	$\sim 1\%/\sigma$	NOT by 'Primakoff effect'	
7	$\pi_L^\pm$	$\pi_L^\pm \rightarrow \overset{(-)}{\nu}_{\mu(L)} \mu_L^\pm \searrow e^\pm 4\gamma$ $\pi_L^\pm \rightarrow e^\pm \overset{(-)}{\nu}_e 8\gamma$ (mass) $m_{\pi_L^\pm} \approx \left(\frac{5577\sqrt{3}}{340}\right)^{1/3} I_U^{1/3} E_1/c^2$ $(m_{\pi_H^\pm} - m_{\pi_L^\pm}) \sim 0.551$ MeV$/c^2$ (life-time) $\tau_{\pi_L^\pm}$	~ 0.05–$0.08\%/\sigma$	by various 'particle physics' experimental techniques, NOT mesic atom spectroscopy (see also items 8 and 10 of this Table 6)	$I_\gamma, I_{\overset{(-)}{\nu}_e}, I_{e^\pm}, I_{\overset{(-)}{\nu}_{\mu(L)}}, I_{\mu_{L,H}^\pm}, I_{\pi_L^\pm}$; treatment of 'charge' sign and 'rest mass'; 'why' $\pi_H^\pm \rightarrow \mu_H^\pm \overset{(-)}{\nu}_{\mu(H)}$ more often than 'directly' to $e^\pm \overset{(-)}{\nu}_e$; with $m_{\pi_H^\pm}$, $I_{\pi_{L,H}^\pm}$ by $(1 \pm 1/I_x^2)^{-1/3}$ factor, coding
8	$\pi_L^\pm$ (decay modes, codes)	$\pi_L^\pm \rightarrow e^\pm \overset{(-)}{\nu}_e 8\gamma$ $\pi_L^+ \rightarrow e^+ 3\nu_e 2\bar{\nu}_e$ $\pi_L^- \rightarrow e^- 2\nu_e 3\bar{\nu}_e$		by various 'particle physics' experimental techniques; ν_e and $\bar{\nu}_e$ reactions with counting of resulting e^+s and e^-s	$I_{\nu_e}, I_{\bar{\nu}_e}, I_{e^\pm}, I_{\pi_{L,H}^\pm}$; 'why' $\pi_H^\pm \rightarrow \mu_H^\pm \overset{(-)}{\nu}_{\mu(H)}$ more often than 'directly' to $e^\pm \overset{(-)}{\nu}_e$; coding
9	$\pi_H^\pm$	$m_{\pi_H^\pm} \approx \left(\frac{1859\sqrt{3}}{112}\right)^{1/3} I_U^{1/3} E_1/c^2$	~ 400 ppm$/\sigma$	by $\mu_H^\pm$ 'momentum' spectrum, OR by n recoil velocity in $\pi_H^- + p^+ \rightarrow n + \gamma$, NOT mesic atom spectroscopy, NOT assuming $\mu_H^\pm$ has same 'interaction strength' ('electric' or 'magnetic') as $e^\pm$ (see also item 10 of this Table 6)	with comparison to mesic atom spectroscopy, as test for 'meaning' of 'charge conservation' as conservation of coding or of 'interaction strength' (see also items 4 and 7 of this Table 6)

(*continued*)

Table 6. *(continued)*

Item		Question	To about	By..., NOT...	As test of...
10		$m_{\pi_L^\pm} - m_{\pi^0} \approx \left(\left(\frac{5577\sqrt{3}}{340} \right)^{1/3} - (15\sqrt{3})^{1/3} \right) I_U^{1/3} \frac{E_1}{c^2}$	$\sim$0.01–0.02%/σ	by neutron velocity distribution, in $x + p^+ \to \pi^0 + n$, $x = \pi_L^-$ or π_H^-; might use mesic atom spectroscopy ('splittings') (see Table 3 items 18–20 for approximate numerical values)	$\sim 8.76\sigma$ to $\sim 4.38\sigma$ test of approximate calculated value for $m_{\pi_{L,H}^\pm} - m_{\pi^0}$ (see also Table 3 items 18–20); $m_{\pi_H^-} - m_{\pi_L^-}$ to $\sim$3%/σ might provide a $\sim 4.6\sigma$ test of $I_{\pi_{L,H}^-}$ by $(1 \pm 1/I_x^2)^{-1/3}$ factor (please see text)
		$m_{\pi_H^\pm} - m_{\pi^0} \approx \left(\left(\frac{1859\sqrt{3}}{112} \right)^{1/3} - (15\sqrt{3})^{1/3} \right) I_U^{1/3} \frac{E_1}{c^2}$	$\sim$0.01–0.02%/σ		
		$m_{\pi_H^\pm} - m_{\pi_L^\pm} \approx \left(\left(\frac{1859\sqrt{3}}{112} \right)^{1/3} - \left(\frac{5577\sqrt{3}}{340} \right)^{1/3} \right) I_U^{1/3} \frac{E_1}{c^2}$	$\sim$1.5–3%/σ		
		$\tau_{\pi_L^\pm} / \tau_{\pi_H^\pm}$ (please see text)	$\sim$10%/σ		
11	H	$H \approx \frac{G m_e^3 \alpha^{-3} c}{3.6\hbar^2}$ (α^{-1} as F?)	$<5\%/\sigma$	astronomical observation and modeling (at the time of this writing, I have not (yet?) thought through in detail how this might be done)	Cor 10.2 and 10.3, Theorem 13, Theorem 14–Cor 14.15
12	ρ_U	$\rho_U \approx \frac{3H^2}{4\pi G} \approx \frac{G m_e^6 \alpha^{-6} c^2}{17.28\pi\hbar^4}$ (α^{-1} as F?)	$<5\%/\sigma$	(see this column, item 11)	Cor 10.2 and 10.3, Theorem 13, Theorem 14–Cor 14.15
13	$Y(N)$	'background' radiation spectrum and opacity of U? (see also questions 40 and 50, Section 6)	?	various radiation detectors (microwave, infrared, etc.): if 'NRQM blackbody spectrum', then if $Y(N)$ as described hereinbefore, U must be substantially absorptive and/or emissive for such radiation in distance $< R_U$ (see also questions 40 and 50, Section 6)	$Y(N)$; might be search for part of ρ_U; might (with comparison to assumed NRQM 'blackbody' temperature) offer some indication of source (and source 'temperature') for such radiation (see also questions 40 and 50, Section 6)

14	$m_{K^0_{L,S}}\left(\begin{matrix}\text{'Long'}\\ \text{'Short'}\end{matrix}\right\}\text{ half-life}\Big)$ $m_{K^0_L}-m_{K^0_S}$ is $I''_{K^0_L}-I''_{K^0_S}\approx\left(\frac{I''_{K^0_{L,S}}}{2}\right)^{2/3}$? $\left(I''_x=\frac{m_x c^2}{E_1}\right)$	$\sim 0.1\%/\sigma$ $\sim 0.1\%/\sigma$	various 'particle physics' experimental techniques, NOT assuming 'charge' 'interaction strength' for other than $e^{\pm}$.	for consideration of $K^0_{L,S}$ 'model(s)', E_1, R_U, I_U, etc. (see Section 6 question 52)
15	$\overset{(-)}{\nu}_e$ $\overset{(-)}{\nu}_{\mu(L,H)}$ in gravity 'g': is propagation observably substantially isotropic (1); or, is observation, and is 'flux', substantially confined (2) in vertical range with half-height $\lvert d\rvert\sim\left(\frac{\hbar c^3}{gE}\right)^{1/2}$? (please see Appendix 1)	to within factor of $\sim(2\pi)^{\pm 1}$	isotropy (1), or anisotropy (2), of $\overset{(-)}{\bar{\nu}}_e$ flux about nuclear reactor on or near earth, or of particle accelerator $\overset{(-)}{\bar{\nu}}_e$ or $\overset{(-)}{\bar{\nu}}_{\mu(L,H)}$ beam spread and/or directional sensitivity to angular deflection in vertical or horizontal directions, comparison of $\overset{(-)}{\bar{\nu}}_e$ and $\overset{(-)}{\bar{\nu}}_{\mu(L,H)}$ beam shapes; absence (1) or presence (2) of daily vertical shift in such beam(s); presence (1) or absence (2) of solar neutrinos (if any) at or near earth (please see Appendix 1)	(1) Conj 1–1.1, and (2) Conj 2–2.2; consequences if case (2) appear to suggest that experimental resolution of this question may be warranted (please see Appendix 1)

(See also question 51 (especially 51.7), Section 6 hereinafter.)

Section Six

Questions

In an undertaking of this type, I find that wherever I might choose to break the development to make part of the material available to others who might be interested, many questions remain. I have listed some (not offered here as being all subgrouped by content and/or ordered sequentially by time of conception) in this Section 6:

1. Formal derivation of Dirac's equation and/or propagator theory. With Corollary 4.4 and Theorem 14 and its corollaries, A-Corollary 2.1, A-Theorems 4 and 5 and their corollaries, and A-Definition 22, one can recognize many terms in Dirac's equation, or its form after Foldy and Wouthuysen's transformation. From Corollary 4.4, the 'jitter frequency' is just $(E_+ + E_-)/h$. The propagation vector k has magnitude $(E_+ - E_-)/\hbar$. With A-Definition 22, these various forms may be found in A-Corollary 2.1, which appears to substantially subsume such description. From Theorem 14–Corollary 14.16, A-Corollaries 2.1 and 2.2, and A-Theorems 4 and 5 and their corollaries, Dirac's four component 'spinors' appear as an alternative way of expressing the four internal states labeled by 2 bits of the internal information of the least massive bound B_0 (electron, positron), apparently corresponding to particle versus antiparticle ('charge' sign) and 'statistically' to 'spin $\frac{1}{2}$', as noted in examples 3 and 4 following A-Corollary 5.1.

An information context might seem to offer considerable insight into why Dirac's equation 'works'. By placing it in this context, one might make (approximately) unnecessary the ad hoc insertion of the mass and charge of the electron, because they appear (approximately) derived elsewhere in such context.

2. Particle lifetimes: might one consider details of transcription among modes populated by ensembles of particles, and/or progressive changes in phase within an entity? For example, due to different Theorem 13 $q(r)$ effects (on 'different' parts thereof)?

3. In view of apparent correspondence of various features of fast B_0s with properties of neutrinos, several questions seem interesting:

3.1 What is the physical significance of the various bits of internal information in muon neutrinos? How is such information coded? Might it correspond to placing the down shifted photon in (a) different state(s) than electron neutrinos in U?, or to using more than one photon where the electron neutrino might use only one?

3.2 With Conjectures 2–2.2, what happens to neutrinos near a planet's center if the planet rises so far in its sun's gravitational field that they could not have their (with Conjecture 2) necessary red shift even at the planet's center? Do they react more rapidly with each other or with matter when they are far into a forbidden region, and thus release their energy near the planet's center? (e.g., $\nu + \bar{\nu} \rightarrow 2\gamma$) If so, each year several percent of the energy released by radioactive decay in a relatively large fraction of the planet might be transported as neutrinos and released as heat near its center, providing energy to drive convective processes near the center with substantially higher velocity and smaller scale than might previously have been considered energetically feasible. Might this be related to planetary magnetism? Or, (less likely?) might neutrinos tunnel from the planet's center into shells about the sun, with their (with Conjecture 2) necessary red shift(s)? If so, some of the energy which was previously thought to be carried off approximately isotropically by neutrinos might be removed to such solar shells.

3.3 With Conjectures 2–2.2, might shifting of solar neutrino shells in the sun be one way in which positions of planets in the solar system might influence solar processes, e.g. sun spots, rates of energy release?

3.4 With Conjectures 2–2.2, might neutrinos (crudely?) focus opposite a source in a gravitationally trapped shell? Or refocus near their source?

4. Other particles; stable versus unstable particles, lifetimes. Particle multiplicities seem in some respects reminiscent of binary coding multiplicities, as internal information is added. In relation to defining space time position by particle and antiparticle mutual annihilation, might Barker codes for internal information be interesting, for correlation between coded internal information and its 'time reflection' to be relatively sharply defined?

5. How are observer's properties and capabilities related to its mass energy and size? It might be interesting to explore behavior of systems with velocity $v \approx c$, noting that finite mass M in (a) universe imposes an upper velocity limit for each observer, such that M (or $M/2$?) might be used as the observer's relativistic energy at v. If observation of a specific property required an observer with mass m, might failure of that property be unobservable for $v > v_0$ such that relativistic $m(v_0)$ were approximately M (or $M/2$?)? ('$M/2$': something left relative to which v might be defined?) Also for history keeping: e.g., for amount I_S of information accessible in adseparable system S containing observer O having access to longest wavelength λ of irad w with amplitude >0 adequately describable as in S; noting 1 bit for existence of O; is longest historical time period $\mathfrak{T}_{L,S}$ in S related to (e.g., $\overset{\text{A}}{=}$ or $\overset{\text{A}}{<}$?) $2^{I_S-1}\lambda/c$? might there be consequences of such history limitation? e.g., on conservation of complementary internal codes in entities for times longer than that? and/or, with respect to concept of simultaneity? and/or 'transportability' (in space and/or time) of 'self contained' time 'standard'? in all U (e.g., related to $2^{I_U-1}\lambda_1/c$)? And, is longest historical time period of least observable time intervals in S (a 'full history', $\mathfrak{T}_{F,S}$) related to $2^{I_S-1}\lambda/(I_S-1)c$? in all of U (e.g., related to $2^{I_U-1}\lambda_1/(I_U-1)c$)? or to I_S-2, noting also $\log_2(\text{card}\{k\})=1$? (noting Postulates 1, 2, and 3) might this offer a basis for a concept of existence longer than (recordable?) history? and (thus?) a concept of 'eternal'?

6. Indirect addressing of objects. Suppose one were to store information about an object's position at a place different from the object's location. One might then direct an observer to go to the region of the object, observe where it was, and go to it; or, one might direct an observer to go to the region of the object's address, (observe where the address was and go to it) read the object's address, and go to the object's position without further observation. In the second case, the observer would arrive at the object by indirect addressing. More steps of indirect addressing might be included, giving addresses of addresses. In what sense might two (widely separated) objects be near each other if their (indirect) addresses were near each other? Or if an object were near another object's (indirect) address? Or, similarly, might objects near each other in some sense be far apart if their indirect addresses were far apart, or one were far away?

7. Would breakdown of special relativity as in Corollary 8.16 or A-Theorem 6 be observable? If CPT is equivalent to Lorentz covariance, might some particle experiment(s) show CPT breakdown? ('CPT' as 'charge', 'parity', and 'time' 'reversals')

8. Might momentum information of objects interact in a way analogous to gravitational interaction of objects' position information? From relativistic covariance, one might suspect so.

9. Might statistics of interaction with a 'sea' containing finitely many faster than light particles or time-reversed observers lead to short time fluctuation in Planck's constant? Is this meaningless? (see also question 48)

10. Might h depend on v for $v \approx c$? Might this appear as part of 'break-down' of adseparability and SR? and/or (otherwise?) be meaningless?

11. Might astronomical sources seen in opposite directions be matched with distances ('red shifts') and evolutions to appear as the same sources seen in opposite directions on a 3-surface of a 4-ball? For example, with total red shift for one circumferential length, back to the observer? Is there a dip in apparent numbers of 'objects' near 1/2 circumference distance? Or might red shift be total for 1/2 circumference distance, with corresponding apparent distribution of sources (described as a function of 'distance')?

12. What is magnitude of charge of particles other than electrons and positrons? Is it the same as for electrons?

13. Might formal treatment of each bit of $I_{\mu^{\pm}_{L,H}}$ (likely 7 bits) provide a somewhat different (approximate?) calculation for $m_{\mu^{\pm}_{L,H}}$?

14. What is the physical significance (if any?) of each bit of internal information in $\overset{(-)}{\nu}_{\mu_{(L,H)}}$, $\mu^{\pm}_{L,H}$, π^0, and $\pi^{\pm}_{L,H}$?

15. Might one think of momentum conservation in scattering ('collisions') in terms of heterodyning of associated up and down-shifted *ws*?

16. Why do the three postulates tested 'work'?

17. Might one approximately describe any entities (e.g., proton, neutron) as coupled (perpendicular?) diametrical modes in U? other 'hadrons'?

18. Properties of modes in a 'soft' dielectric varying index of refraction ball? (1,2,3,... 'dimensions'?)

19. Does $\mu^{\pm}_{H}$ decay (mostly?) with electron (positron) emitted along an axis perpendicular to the 'plane of orbit' of its associated two photons, e.g., because of twofold symmetry of complementary representation of information?

20. Might in $e^{\pm}$ $d=+1$ only fix a 'sense of gyration' of $e^{\pm}$'s associated two *ws*? Might this offer some insight into asymmetry of direction of emission of $e^{\pm}$ with respect to $\mu^{\pm}_{H}$ or neutron 'spin' 'axis', as related to a 'sense of gyration' of the 'source' system which would correspond, with a particular 'direction of view', to that coded in the emitted entity?

21. Does the b coding in $e^{\pm}$ and $\mu^{\pm}_{H}$ have a $1/I_x^2$ dependence for some of its effects? If so, then one might suspect that, noting $I_{e^{\pm}}$ likely $=3$ and $I_{\mu^{\pm}_{H}}$

likely $=7$, a $1/49$ contribution might appear as a branching ratio between alternative modes with $1/9$ contributions, with, in $\mu_H^{\pm} \rightarrow e^{\pm} \overset{(-)}{\nu}_e \overset{(-)}{\nu}_e$, one of the neutrinos fixed as neutrino or antineutrino by coproduction with positron or electron, and the other statistically distributed in a way corresponding to representing a $1/49$ contribution by balancing $1/9$ contributions. If so, in decay of ensembles of μ_H^{+}s or μ_H^{-}s with $\mu_H^{\pm} \rightarrow e^{\pm} \overset{(-)}{\nu}_e \overset{(-)}{\nu}_e$, might the numbers $n(\nu_e)$ and $n(\bar{\nu}_e)$ be roughly, in $\mu_H^{+} \rightarrow e^{+} \overset{(-)}{\nu}_e \overset{(-)}{\nu}_e$, $n(\nu_e) \lesssim 3n(\bar{\nu}_e)$, and, in $\mu_H^{-} \rightarrow e^{-} \overset{(-)}{\nu}_e \overset{(-)}{\nu}_e$, $n(\bar{\nu}_e) \lesssim 3n(\nu_e)$? Or, noting $I_{\overset{(-)}{\nu}_e}=2$, balancing $1/4$ contributions?

22. Might operation of a nuclear reactor or occurence of a nuclear explosion at or near a (with Conj. 2–2.2) conjugate focus of a neutrino detector affect measurements with such detector? e.g., of presence or absence of solar neutrino flux at or near the earth?

23. What happens to (if any) consistency requirement on '*d*' coding in different entities? Can entities with different numbers of associated 'photons' supply or complement '*d*' coding to each other?

24. Might one look for 'resonances' with (multi) photon and/or other particle decays conserving amount and/or representation of internal information I_x in spectra approximately describable by C2,I_x,b,d,f and/or C3,I_x,b,d,f?, or E types? (see Def 21)

25. Calculation of extended interaction as seen by observer at location other than B_0? or in other 'geometries', for example, 3-surface of 4-ball? (as compared to calculation of extended interaction in Cor 14.3?) Does such alternative calculation have any use or distinct physical meaning?

26. Might 'graviton' be adequately describable as (2?) coupled 'photons' (*w*s) propagating (at velocity *c*) in the same direction? and not able to interact as photons? like behavior of Corollary 14.3 extended interaction mode? Might 'gravitation(al) radiation' be produceable?, by 'shaking off' part of 'both' two 'photons' together of part of a system's externally equivalent B_0 (as in externally equivalent B_0 ψ (A-Cor 5.1, A-Theorem 6))? Or, is 'time lag' zero for 'gravity' seen at distance >0 from 'source'?

27. Are there any observable consequences of ranges allowed by 'is adequately describable as', for example, of a B_0's n_+ (see discussion near A-Cor 4.2 (p. 211) and question 49)?

28. Might there be observable effects, e.g., traceable through stellar evolution calculations, of the two conjectures in Appendix 1? with Conjectures 2–2.2, effects near gravitational minimum in star due to sweeping in and release of neutrino carried energy? by planetary or binary star gravitational potential changes? and/or due to loss of mass, e.g., in nova or supernova?

29. What would happen to a 'particle' at its maximum velocity in *U*?

(from A-Cor 2.1 and 2.2, and A-Theorem 6, internal and external information not separable?)

30. Might one consider a minimum 'age' of an observer O_U's universe U in terms of equivalence of Corollary 8.16 and A-Theorem 6 with respect to maximum velocity at which adseparability and special relativity apply, as the longest time required for gentle accelaration of any system in U to such velocity for such system? (subject to total (simultaneously?) accessible information ('energy') limitation?)

31. Might some statistical astrophysical questions as to probabilities of formation and numbers of such entities as ring galaxies be resolved by treatment herein of 'cosmological red shift' as not involving recessional velocity, with much longer time for formation of such entities and related 'degree' of correlation of mass distribution in U than suggested in previous theory(s) involving an expanding universe with such red shift attributed to recessional velocity?

32. Is the earth gravitationally shielded (Conjecture 2–2.2) from (very?) high energy particles, for example, electrons?

33. Might one describe some 'particles' in terms of (perpendicular?) combinations of relatively localized two and/or three photon entities as in Definition 21?

34. What do G, h, α, m_e, and/or other 'particle' 'masses' look like in a strong gravitational 'field', and/or with substantial Theorem 13 $q(r)$ effect(s)?

35. Might masses and lifetimes of mesons be understandable in terms of a succession of stepped 'wells', like the $\delta 3, 2, I_x$ 'well' that pions seem to exist in? Partly as a 'barrier penetration' process?

36. Is it always possible to reconstruct any entity that has existed? an observer? an introspector? would this involve 'giving up' information that such observer's or introspector's existence had been discontinuous? (see also questions 43 and 44)

37. Is 'b' a separate internal bit? comparing $\overset{(-)}{\nu}_e$ and $\overset{(-)}{\nu}_{\mu_{(L,H)}}$?

38. Is 'electrostatic' and/or 'electromagnetic' interaction between e^+ and e^-(not?) the same as that between p^+ and e^-? are hydrogen measurements of α directly related to e^+e^- measurements of α? e.g., CT (ref. 4 p. 695) approximate value for α^{-1} obtained from positronium 'hyperfine structure' of $\sim$137.0387(17)(13 ppm) (see discussion following expression 22.12 of CT, on p. 695 of ref. 4, and discussion herein preceding Cor 14.16) Might 'equal mass' condition in Corollary 14.9 (as in steps 14–16) be satisfied by (a) component(s) of particle(s) more massive than $e^\pm$? Might relation between 'relative phases' of two interacting 'particles' with different ('rest') masses represent 1 bit of information? and (noting Postulate 1) reasoning as in Corollary 14.14 steps 10–12, 14, and 15 be applicable in case of unequal masses? (see also questions 78 and 80)

39. Is total number I_U of bits of information accessible in 'our' universe U odd or even? might such 'oddness' or 'evenness' of I_U or of I_S accessible in irad system S in U be related to 'parity conservation' or 'parity nonconservation'?

40. What is spectral density for radiation from objects in equilibrium with photons in states $E_j = jE_1$ as in A-Theorems 3 and 4 in U? what would their temperature appear to be if interpreted on assumption that NRQM blackbody spectra described such radiation? might this be related to (interpretation of?) radiation 'background' previously interpreted in terms of $\sim 3°K$ 'blackbody radiation'? (please see question 50 for part of an attempt at understanding this)

41. A-Corollary 6.1 and not any information accessible in photon, as consistent with apparent unidirectionality of time? if information were accessible in an entity with velocity $v = c$, would time appear unidirectional anywhere in U?

42. $\overset{(-)}{\nu}$ transport of 'energy' into central region of protostar in region of decreasing $q(r)$ (e.g., of a galaxy) as a means of, or aiding, 'ignition'?

43. Does every observer and/or introspector exist 'for ever' in such entity's own 'time' coordinate?

44. Is *ti* the same coordinate as time?

45. Can 'normal mode' occupation in w state(s) of U (as a whole?) 'spread' 'distance' d (e.g., without 'direction'?), in time $< d/c$? (e.g., in λ_1 w state(s)?)

46. Might Corollary 14.5 step 4 be regarded as related to a type of 'norm' in an 'information representation description space'?

47. Finite linear size R_U of U has to do with being somewhere not representing infinite information, but does it have anything to do with distance of transport of information by a moving system (e.g., a moving 'reporter' and/or 'messenger')? Can information be transported linearly (e.g., by (successive?) moving system(s)), in a single defined 3-space direction, a locally measured distance $>$ or $\geqslant R_U$? If $> R_U$, then by successive communication involving moving system(s), might every observer have access to any information accessible to any observer, in which case finite accessibility of information might imply finite accessible information, the same information for all observers (but not necessarily coincident universe(s)?)? Might this suggest a 3-space 'geometry' with some properties like the 3-surface of a 4-ball? Or, might some 'effect', e.g., 'loss' of definition of direction, make such (a) process undefinable? Might this be related to 'angular momentum'? Might distances add in relation to R_U in a way resembling the way velocities add in relation to c (in domain of applicability of adseparability), so that successive local transport which if added linearly would be $> R_U$ would add to be $< R_U$? Might it be that in all cases sequential transmission would be adequately describable as one observer moving information in its own universe? Might transport by 'messenger' of information in a single defined 3-space direction be like transport by observer between different values of x_0 in 'history keeping' (see also question 5)? Might uniqueness and apparent unidirectionality of x_0 make history keeping by an observer possible, for the same amount of accessible information, over larger range(s) of x_0 than range(s) of (x_1, x_2, x_3) such observer would be able to 'travel' in any (single?) 3-space direction without 'losing track' of its direction (of travel) (and/or reversing direction of change of any or all of x_1, x_2, x_3?)?

48. Is there a difference between 'propagation' as a process adequately describable in terms of transcription through successive (local) regions (but note Cor 10.2 etc.) and some form of communication not involving (such?) propagation? (for communication time interval t and 3-space distance d) and not limited to $t \geqslant d/c$? Is communication time interval $t < d/c$ possible? Might $t < v/c$ imply $t \stackrel{A}{=} 0$?

49. Noting that n_0 as defined in A-Corollary 2.1 step 6 was not specified as integer only, might in some cases n_0 be adequately describable as not changed for $|\Delta n_0| < 2$? Might this appear in description of 'phase space(s)', 'cross sections(s)', 'rates(s)', etc.?

The following 'questions' 50–52 are related to parts of this inquiry which (with others) are, at the time of this writing, at or relatively near the 'edge' of the work in progress reported herein, and, as yet, only partly considered; for this reason, they are presented as subsections of this Section 6. Please note the warning in the Preface, and regard these as offered tentatively.

50. 'Cosmic Background Radiation' and 'Cosmological Theories'. One 'cosmological theory' which has appeared relatively popular prior to and at about the time of and relatively near this writing appears to involve the assumptions that a universe is 'expanding', and that it 'started' as a much denser object which 'exploded' (called the 'big bang theory'). Such hypotheses would seem to have become popular as a way of 'explaining' apparent 'red shifts' of light from astronomical sources, relative abundances of various nuclear species, and 'cosmic background radiation' which has been interpreted in terms of nonrelativistic quantum mechanics (NRQM) 'blackbody' spectrum as approximately corresponding to 'blackbody radiation' at $\sim 3.5°$Kelvin.

With the conceptual and formal structure presented hereinbefore, such 'red shifts' appear for a universe with 'constant size' (Cor 10.2–10.4, pp. 102–105), and objects dense and hot enough for conversion of iron to hydrogen but with mass much less than that of a universe would seem possible (Cor 13.1–13.3, pp. 119–121; see also question 51). Thus, such 'red shifts' and relative abundances do not appear as supporting uniquely such 'big bang theory'.

With respect to such 'cosmic background radiation', one might consider two (crude) numerical estimates:

Using the spectrum of w states in U and Planck's NRQM distribution $(e^{E/kT}-1)^{-1}$ for average occupation number as a function of 'energy' E and 'temperature' T, where k is Boltzmann's constant, one might crudely estimate (a?) flux of 'radiant energy' per unit area, $\mathfrak{F}$, of 'universal' 'starlight' in U:

1. From A-Theorems 1, 3, and 4, and A-Def 7 and 22, noting separability of A-Theorem 1 'wave equations' in 3-space of A-Definition 7 and that there are exactly 2 possible polarizations, the number of w states in 'energy' interval $\Delta E \gg E_1$ is approximately describable by $6\Delta E/E_1$. ('approximately' because only 2 w states for $E = E_1$)
2. From A-Cor 1.5, A-Def 6, 11, and 18, and 1, in universe U with linear size R_U, 'volume' of such w states in U 'filled' from an area of 'emitting surface' with linear size $\ll R_U$ might be crudely describable by $2\pi R_U$ times such area. (linear modes; 'crudely' might comprise, among others, factors $\lesssim (4\pi)^{\pm 1}$, etc.)
3. From A-Theorem 1, A-Cor 1.3, and 2, as in the solid angle and projected area integration previously used in NRQM (subsumed herein) in approximately describing such 'radiant' 'flux' from a surface, $\mathfrak{F}$ might be crudely describable as $c/4$ times the corresponding 'energy density' in U. (factor of $\frac{1}{2}$ from integral of $\cos^2$, factor of $\frac{1}{2}$ from only one 'side' of surface exposed)
4. From quantum mechanics, as subsumed hereinbefore, average occupation number for energy level E in equilibrium with temperature T is approximately describable by $(e^{E/kT}-1)^{-1}$, where k is Boltzmann's constant.
5. From 1–4, noting that photons in energy level E have energy E, for $E \gg \Delta E \gg E_1$, $\Delta\mathfrak{F}(E)\Delta E$ might be crudely describable by

$$\Delta\mathfrak{F}(E)\Delta E \sim \frac{6\Delta E}{E_1}\,\frac{1}{2\pi R_U}\,\frac{c}{4}\,\frac{1}{e^{E/kT}-1}E$$

$$\sim \frac{3c}{4\pi R_U E_1}\,\frac{E}{e^{E/kT}-1}\Delta E.$$

6. From 5, $\mathcal{F}$ might be crudely describable by

$$\mathcal{F} \sim \frac{3c}{4\pi R_U E_1} \int_0^{`\infty'} \frac{E}{e^{E/kT}-1} dE.$$

('∞' in quotes: see Postulate 3)

7. From quantum mechanics (as subsumed hereinbefore), in regions with linear size $\ll R_U$, for a 'black body' at about 'temperature' T_{QM}, flux density F_{QM} is approximately describable by

$$F_{QM} \sim \frac{2\pi^5 k^4}{15h^3c^2} T_{QM}^4.$$

8. From 6 and 7, setting $F_{QM} \approx \mathcal{F}$ and solving for T_{QM}, for such regions, T_{QM} might be crudely describable by

$$T_{QM} \sim \left(\frac{15h^3c^2 3c}{2\pi^5 k^4 4\pi R_U E_1} \int_0^{`\infty'} \frac{E}{e^{E/kT}-1} dE \right)^{1/4};$$

regrouping terms,

$$T_{QM} \sim \left(\frac{45h^2c^2(hc/2\pi R_U)}{4\pi^5 k^4 E_1} \int_0^{`\infty'} \frac{E}{e^{E/kT}-1} dE \right)^{1/4}.$$

9. From A-Def 22, and 8, noting $E_1 = hc/2\pi R_U$, for such regions, T_{QM} might be crudely describable by

$$T_{QM} \sim \left(\frac{45}{4} \right)^{1/4} \frac{h^{1/2}c^{1/2}}{\pi^{5/4}k} \left(\int_0^{`\infty'} \frac{E}{e^{E/kT}-1} dE \right)^{1/4};$$

letting $u = E/kT$; $E = kTu$ and $dE \approx kTdu$;
substituting,

$$T_{QM} \sim 11.25^{1/4} \frac{h^{1/2}c^{1/2}}{\pi^{5/4}k} \left(k^2T^2 \int_0^{`\infty'} \frac{u}{e^u - 1} du \right)^{1/4};$$

collecting terms,

$$T_{QM} \sim \frac{11.25^{1/4}}{\pi^{5/4}} \left(\frac{hc}{k} \right)^{1/2} \left(\int_0^{`\infty'} \frac{u}{e^u - 1} du \right)^{1/4} T^{1/2}.$$

Approximating the integral by a sum of approximate values for the approximate centers of 10^4 steps from $u=0$ to $u=30$ (using the aforementioned Hewlett-Packard model 67 calculator (laziness?)) as $\sim$ $1.644933745(\sim\pi^2/6)$, and substituting (from Cohen and Taylor, ref. 4) $c\approx 299792458$ m/sec, $h\approx 6.626176\times10^{-34}$ J sec, and $k\approx 1.380662\times10^{-23}$ J/°K,

$$T_{QM}\sim\frac{11.25^{1/4}}{\pi^{5/4}}\times 0.1199494191\times 1.644933745^{1/4}T^{1/2}$$

$$T_{QM}\sim 0.059482\ T^{1/2}.$$

Substituting T as approximately the temperature of the earth's sun (from ref. 9, p. F-151)$\sim$5776°K, $T_{QM}\sim$4.52°K. This might crudely describe the temperature of relatively small lumps of matter with 'emissivity' ('absorptivity') nearly independent of frequency over the spectral range involved (e.g., $\sim10^{-6}$ to $\sim$10 eV) absorbing 'energy' from such 'universal' ('starlight') 'radiation' and reemitting it in a region of U with linear size $\ll R_U$ (scale $\ll 1$, 'scale' as in A-Def 10, p. 187) with a ('local') NRQM 'black body' spectrum. For lumps with spectrally varying absorption and/or emission (e.g., because of size and/or composition), and/or other 'source temperature(s)', T_{QM} might appear somewhat different.

Another (crude) numerical estimate might be made by asking, using quantum mechanics (QM) as subsumed hereinbefore, approximately what 'temperature' T_B a 'black body' with radius R_U would have to have, assuming as in QM that the effects of finite linear size of universe may be neglected, to emit as much radiation as would be emitted by all the mass energy M_U in a universe U divided into stars with about the same radius R, mass M, and temperature T as the earth's sun:

$$T_B\sim\left(\frac{M_U}{M}\left(\frac{R}{R_U}\right)^2\right)^{1/4}T;$$

substituting (from Table 2, p. 155) $M_U\sim 4.15508\times10^{53}$ kg and $R_U\sim 3.084568\times10^{26}$ m, and (from ref. 9 p. F-146) $M\sim 1.991\times10^{30}$ kg and $R\sim 6.9595\times10^{8}$ m, $T_B\sim 1.01524\times10^{-3}T$. Substituting $T\sim$5776°K as before, $T_B\sim$5.864°K.

These two crude estimates, although not even 'likeliness' arguments as in Chain 17 and its links, might seem to suggest that it would not be necessary to regard reports of such 'cosmic background radiation' as supporting uniquely such 'big bang theory'.

Such 'big bang theory' appears to have been constructed in terms of prior theories in which space, time, mass energy, relativity, and a single value for 'gravitational constant' G were assumed, and in which 'local' approximate properties of free light propagation (e.g., not any conspicuous apparent wavelength 'change') were assumed to apply 'globally' as exact in (a) 'universe'.

In the physical picture and formal structure presented hereinbefore, space, time, mass energy, relativity, gravity, and apparent 'red shift' of freely propagating light all appear as conceptually and formally derived from the three postulates stated. From A-Definitions 4, 6, and 17, observer O_U's universe U always has adseparable internal and external information. In view of this (and Theorem 11), it would seem that 'change' in 'size' of U might never have any consequence(s) observable by O_U (which is why I put quotation marks around 'change' and 'size': see also A-Cor 1.4 (p. 186) and A-Def 10 (p. 187), and discussion on p. 187 following A-Cor 1.4).

In view of various considerations presented herein, at the time of and near this writing, such 'big bang theory' might seem questionable on conceptual grounds and not uniquely supported by those prior reports, of which I am at the time of this writing aware, of observation.

51. Galaxies and 'Cosmic Rays'. This is intended as an informal suggestion of what might be a way, based on the physical picture and formal structure presented hereinbefore, of thinking about galaxies in (a) universe U with 'constant size'.

51.1 'Lump Gas' in U and 'Cosmic Rays'. Suppose that U were to contain a 'gas' of 'lumps' of 'matter', ranging in 'mass' from particles (e.g., electrons) through objects $\sim 10^6$ times as massive as the earth's sun (1 'solar mass'), many of them in a mass range from large asteroids as found in the solar system up to about 1 solar mass; in which objects much larger than particles would usually, but not always, be moving with velocities substantially less than c. This might be thought of as a 'lump gas', which might have relatively little net (vector) momentum in U as seen by an observer O_U approximately at rest in and relative to U.

Collisions among constituents of such lump gas might accrete relatively small lumps into larger ones, although a suitable sequence of Theorem 13 interactions of a lump with sufficiently many other lumps might accelerate a lump to velocity $\sim c$. (Might this be a source of unusual meteor impacts?)

'Electromagnetic' interaction of 'charged' particles with 'magnetic' and/or 'electrically charged' larger lumps might accelerate such particles to velocities $\sim c$, and energies such as found in (some) 'cosmic rays'.

As noted more fully hereinafter, such 'lump gas' might contain a larger fraction of elements with atomic number >2 than present in luminous matter.

51.2 'Clouds' in Lump Gas. Consider a 'cloud', with 3-space extent $\ll R_U$, of lumps, in U, defined by average density greater than that of the lump gas and average magnitudes of relative velocities among its components less than that of the lump gas. Suppose that the cloud had such density and lump size and distribution that a lump from lump gas would usually be able to traverse the cloud while experiencing only relatively 'distant' interactions approximately describable in terms of Theorem 13 ('gravity') and A-Theorem 1 (etc) ('electromagnetism'); that, during such traversals, effects due to Theorem 13 interaction with cloud lumps were much larger than all effects together due to all other interaction(s) with cloud lumps; and that, for most such interactions, the angle through which each such interaction 'bends' the trajectories, as seen in the respective 'center of mass' 'frame of reference' of each cloud and lump gas pair of interacting lumps, is $\ll \pi/2$ radians ('small angle forward scattering').

In such case, the 'energy' imparted by a lump gas lump to each cloud lump which it passed would be less for faster passage; for relative velocity v substantially less than c, approximately describable as proportional to $1/v^2$. Also, for v substantially less than c, a given (relatively small) energy decrease would correspond to a smaller momentum decrease for a faster moving (lump gas) lump with given 'rest mass'; approximately describable as proportional to $1/v$ times the energy decrease. Thus, the greater the relative velocity between the lump gas lump and the cloud, the less amount of momentum that would be transferred from the lump gas lump to the cloud lump(s); for v substantially less than c, approximately describable as proportional to $1/v^3$. (Note also (possibility of) effect of spectrum of B_0 states in U with $E_j = jE_1$, j finite integer >0, (A-Theorems 3, 4, and 6, A-Cor 2.1 and 4.8, A-Def 12 and 22) on (small angle?) scattering ('partial'?) cross section(s) (and 'phase space').)

After passing through a region of cloud, a lump would have less velocity relative to the cloud, and thus greater ability to transfer momentum during subsequent traversal of other parts of the same cloud, or of other cloud(s) relative to which such lump's velocity had been reduced by such prior interaction(s).

Such decrease of momentum transfer with increase of relative velocity might have several effects.

Consider formation of a cloud with velocity as seen by observer O_U as much less than the average magnitude of lump gas lump velocity: because traversal of one part of the cloud would enhance ability of a passing lump gas lump to transfer momentum to subsequently traversed parts of the cloud, interaction of traversing lump gas lumps with the cloud would tend to tear it apart. Associated with such effect might be a minimum ('threshold') mass for a gravitationally bound cloud to exist for a relatively long time in such environment; and an effect tending to keep clouds apart.

Once such gravitationally bound cloud had begun to form (or formed), the 'probabiiity' of encounters with lump gas lumps approaching from the direction toward which the cloud was then (with velocity Δv) moving would be greater than for encounters with gas lumps approaching from the direction away from which the cloud was then moving; which change in 'probability' might be approximately describable as proportional to $\Delta v/v$. This might seem a 'slower' increase than the $(\sim 1/v^3)$ decrease of momentum transfer with increasing relative velocity. Thus, it might seem that interaction of part or all of the cloud with the lump gas might involve transfer of *more* momentum to the cloud from gas lumps moving in approximately the *same* direction as the cloud than from gas lumps moving in approximately the opposite direction, thus *increasing* the velocity of the (region of the) cloud. If cloud formation involved convergence velocities sufficiently large for this effect to exceed the momentum transfer enhancement effect tending to disperse the cloud, then (further) convergence might be driven by this effect. (Might such adding of momentum to a relatively large entity be thought of as in some sense like a 'momentum space' analog of 'position space' 'condensation' of gravitationally interacting 'gas', e.g., initially nearly uniformly distributed in a large enough 'box'?) (see also questions 53 and 54)

Such gravitationally bound cloud might, by the same effect, be accelerated as seen by O_U (to some range of velocities characteristic of its interaction with, and properties of, the lump gas?), so that, in a set of such gravitationally bound clouds which had existed for a substantial period of time, there might be relatively few with relatively low velocities.

By increasing the velocities of regions of a cloud moving relative to each other, such effect might impart angular momentum to a gravitationally bound cloud, thus causing it to take on a disc shape. Because of effects of collisions among cloud lumps with trajectories tilted with respect to such disc, such disc might become relatively thin and sharply defined.

With gravitational interaction, an approximately uniform distribution of mass in a relatively thin annular region about the disc's center might be unstable with respect to angular distribution of mass, leading to angular dependence with (if symmetric, more than one?) mass concentration(s). Matter in two adjacent such annular regions might orbit with a relation between their respective mass concentrations such that the rotating disc would appear to have 'arms'. As matter moved in or out radially, some of its angular momentum might be gravitationally coupled between matter in such nested annular regions, and the angular offsets of mass concentrations associated with relative 'phases' in such momentum transfer process might lead to 'spiraling' of such 'arms'.

Supplying of more angular momentum from interaction with the lump gas might lead to progressive migration of matter radially outward in such disc, and 'shedding' of lumps from regions near its 'edge'.

After formation of a rotating disc, matter in such disc might partially shield matter in it from passage, parallel to and in the plane of the disc, of gas lumps, and 'orbital crossing' effects might tend to keep matter near the plane of the disc. These effects together might allow smaller 'clusters' of lumps to be gravitationally bound for relatively long periods of time within such (rotating) cloud than would be as likely to exist for such periods of time more nearly isolated in, and subject to disruptive effects of, the lump gas.

Collecting various of these features, there might be a mass threshold for formation of gravitationally bound clouds, which might after formation 'evolve' into 'drifting' rotating discs shedding matter from near their edges, which discs might have spiral arms and might contain cluster(s) of lumps. This might be their form for a period of time relatively long compared to the period of time for their formation, and might last until they hit something. Thus, there might be a positive correlation between drift velocity and angular velocity (disc formation).

Although some of the time periods involved might be longer than R_U/c, they might not seem long compared to $\sim 2^{\sim I_U - 1}\lambda_1/(\sim I_U - 1)c$, which might seem (an approximate?) measure of the longest period of time for keeping a (complete?) historical count (record) of periods of (approximately?) shortest time interval observable in U (the '-1's' are from 1 bit for existence of an observer) (please see also question 5).

51.3 'Converter Bodies', Stellar Composition, Luminosity, and Neutrino Production. During formation of a gravitationally bound cloud, part (e.g., $\sim 1\%$) of the mass in the cloud might form a single body. From Corollary 13.3, accretion of mass onto and/or into such body would not result in 'cutting off' communication of information in either and/or both direction(s) between such body and the rest of U. Thus, relatively large $q(r)$ effects (e.g., $q(r)$ ratios $> 10^6$) might exist in a stable entity. With such large $q(r)$ ratios, starlight might in such region be 'blue shifted' to greater than local electron-positron pair formation 'energy', and the local velocity of light in such region might seem to O_U to be less than a few meters per second. One might think of matter in such (relatively large $q(r)$ effect) region as a 'photon soup', in which there might not be much distinction between 'internal' and 'external' information of 'entities'.

When a lump (e.g., nucleus, neutron star, etc.) 'fell' into such a region, nearly all of its incident mass energy might be available in the form of 'fragments' of its nucle(us,i) by a process of 'tidal' and/or radiation (and/or other) breakup. This might lead to production of mostly electrons and protons, and might involve production of some alpha particles (with some of the incoming 'mass energy' entering the 'soup' to provide 'energy' for the breakup of nucle(us,i) with lower total 'mass energy' per nucleon than the emitted particles?). From Postulates 1 and 2, there would not be any 'loss' of information ('mass energy') in such process(es).

By such process(es), such relatively massive and compact body might act as a 'converter body', interaction with which might convert lumps to mostly hydrogen (gas).

Such converter body might likely be located at or near the center of a gravitationally bound cloud, and, after such cloud had been 'spun up', at or near the center of the resulting disc. Hydrogen from such conversion of lumps might spread and be 'captured' (e.g., by collision and/or gravitationally) by various lumps in the cloud. The 'probability' of a hydrogen atom traveling a distance through a nearly 'uniform' cloud of relatively widely separated lumps might decrease approximately exponentially with increasing distance.

Accumulation of such hydrogen onto lumps might make stars. Thus, the 'dominant' term in approximate description of total (stellar) luminosity per unit area of a gravitationally bound disc cloud might be such exponential decrease with increasing (radial) distance from a converter body near the disc's center, and such luminosity might be distributed substantially differently than matter in such disc (which might be distributed more nearly uniformly per unit disc area).

This notion of stellar formation might lead to somewhat different calculations of stellar composition and evolution than might have seemed likely to be appropriate prior to the time of and near this writing. In particular, iron, hydrogen, helium, beryllium, and boron concentrations and distributions within a star, and stellar interior temperatures, might lead to models (e.g., see ref. 32 (Fred Hoyle)) for nuclear processes in the earth's sun with a neutrino flux and spectrum which might seem consistent with any relatively high energy solar neutrino flux at the earth having not been reported in public reports, of which I am at the time of this writing aware, as significantly different from zero (see ref. 33 (J. N. Bahcall and R. Davis, Jr.)). (See also Appendix 1.)

With such notion of galaxies, one might find ways of addressing various questions relating to sizes, formation, and 'evolution' of galaxies.

51.4 'Life Cycle' of a Galaxy. With this notion of galaxies, one might think of galaxies as likely to have, if no other effects substantially affected galaxy(s), approximately the following 'life cycle':

1. A cloud (e.g., 'due to' 'fluctuation(s)') in lump gas exceeds threshold to become, and becomes, gravitationally bound.

2. A converter body forms, and hydrogen collects on nearby lumps, with star formation. (might be a relatively bright period) ('quasar'?)
3. Interaction with lump gas spins-up cloud, by then a galaxy, thus spreading it into a disc (which might have spiral arms), angular momentum reduces rate of matter 'feed' into converter body, and hydrogen capture distribution changes are followed by changes in luminosity distribution.
4. Interaction with lump gas accelerates the galaxy to nearly some characteristic drift velocity (or range of velocities).
5. The galaxy exists for a relatively long period of time as a drifting rotating disc; with continuing fall of lumps into an approximately central converter body and outflow of hydrogen, with luminosity per unit disc area decreasing approximately exponentially with increasing distance from the disc's center, and more nearly uniform mass per unit disc area; with matter removed from the galaxy by novas, supernovas, as star dust, etc., and shedding of lumps from near its rim as more angular momentum is received by interaction with lump gas. Matter moving radially outward by momentum transfer in the disc might be processed in various ways, for example, in supernovas, 'red giant' stars, etc.
6. The galaxy hits, and/or is hit by, something, for example, another galaxy: if with roughly equal drift velocities and the effects of interaction with lump gas tending to separate two galaxies, such collisions as occurred might usually involve 'center of mass' velocity substantially less than the drift velocity of either galaxy (as in 'head on' rather than 'glancing' collisions). If the converter bodies miss each other, such collision might result in gravitational disruption of one or both galaxies, and/or capture of one galaxy's converter body into the other galaxy (e.g., leaving 'ring' galaxy(s)), and the collision might leave one or both galaxies changed in such way as to be subsequently torn apart by inertial effects and/or interaction with lump gas. If the converter bodies collide, because 'kinetic energy' carried into the collision might more than make up for the relatively small fraction of the rest mass of one converter not 'released' by $q(r)$ effects due to the other, they might form a single converter body (which might not stay with either galaxy) having less 'rest mass' than the more massive one had before collision (about which a 'new' galaxy might form). (might be a relatively bright period) ('quasar'?) (see also Cor 13.2) (if approximately complementary coding of internal information, might be approximately describable in terms of 'mutual annihilation', e.g., transcription to photons and/or other complementary representations)

Various of these processes, such as spin-up and acceleration to nearly (a 'characteristic') drift velocity, might take substantially longer than R_U/c; and, thus, might provide a way to address observationally the hypothesis, found in various prior theory(s), of an 'expanding' universe with a ('Hubble') 'age' $\sim R_U/c$.

51.5 Compositon of Lump Gas. Noting what might be a 'life cycle' for galaxies, after the lump gas had reached approximate 'equilibrium' with galaxies, much of the lump gas might consist of lumps shed from near the edges of galaxies, with a large fraction as iron, neutron stars, and/or (other) residue(s) from nuclear fusion, novas, supernovas, etc. Nearly all of the matter in lump gas might be in the form of lumps containing more than one 'atomic nucleus'. Lump gas might comprise a substantial part of the total information ('mass energy') in U.

51.6 Transcribability of Information. For any notion consistent with Postulates 1, 2, and 3, of any physical process, information must be transcribable from any representation into which it is transcribed in such process.

Converter body(s) might offer a way to transcribe fusion residue(s) into fusible matter (e.g., hydrogen). Might 'momentum accretion' properties of a gravitationally (or other 'long range') interacting (in this case, might be inhomogeneous) lump gas offer a way to transcribe starlight (e.g., emitted from fusion processes) into 'kinetic energy' 'concentrated' enough to participate in transcription of fusion residue(s) into fusible matter? Might this lead to a relation between 'kinetic energy' as a fraction of 'mass energy' of (some?) lumps in lump gas and nuclear 'binding fraction' of fusion residue(s)? Might galactic 'magnetic fields' 'funnel' relatively high velocity ('charged'?) lumps into converter body(s)?

51.7 Some Features Which Might Be Comparable With (Report(s) Of) Observation(s)

1. Sizes of, and size (and mass) threshold for galaxies (partly as partial test of some features of (interaction with) lump gas).
2. Luminosity and mass distributions in galaxy(s) (e.g., in relation to 'converter body(s)' (Cor 13.1–13.3)).
3. Nuclear species distributions in galaxies (as depending on position in galaxies?) and in 'lump gas'.
4. Abundances of various 'types' of galaxies, in relation to time for spin-up, probability of collision, etc.
5. Correlation between rotational velocity and drift velocity.
6. Properties of 'lump gas' (including 'cosmic rays').

7. Behavior of galaxies near each other (partly as partial test of some features of interaction with 'lump gas').
8. Stellar evolution.
9. Solar neutrino production. (See also Appendix 1.)
10. Time scales for formation, spin-up, acquisition of drift speed, reaching luminosity distribution, etc.: if $>R_U/c$, might exclude various prior 'expanding universe' hypothes(i, e)s.

Such approximate luminosity and mass distributions in disc galaxies appear substantially consistent with that part of which I am at the time of this writing aware of prior knowledge (public oral communication, Salpeter, ref. 34).

52. K^0_{Long}, K^0_{Short} 'rest mass difference' Δm: Δm seemed so small a fraction of the approximate 'rest mass' of K^0_L and/or K^0_S, compared to 'd code' "mass splitting"s in $\mu^{\pm}_{L,H}$ and $\pi^{\pm}_{L,H}$, as to suggest that, unlike d code 'splitting' relation to amount of internal information, Δm might be related to coding of external information.

UCLBL 1976 (ref. 8) listed values for $m_{K^0_{L,s}}$ as $\sim(497.70\pm0.13)$ MeV$/c^2$ and Δm as $\sim(5.349\pm0.022)\times10^9\hbar/\text{sec}\ c^2$. From A-Definition 22 (and conditions for Chain 17), $I''=mc^2/E_1$. Substituting $\hbar$ from Cohen and Taylor (CT, ref. 4, listed in Table 1) and E_1 from Table 3 ('intermediate values' listing), $I''_{K^0_{L,S}}\sim7.779863483\times10^{41}$ bits, and $I''_{K^0_L}-I''_{K^0_S}=I''_{\Delta m}$ $\sim5.503592207\times10^{27}$ bits. From this, $I''_{K^0_L}-I''_{K^0_S}$ appears to be within $\sim$ 3.29% of $(I''_{K^0_{L,S}}/2)^{2/3}$; such '2/3' exponent seemed reminiscent of exponents in various 'particle rest information' ('rest mass') approximate calculations appearing hereinbefore. (geometric meaning? factor spaces?)

Note that E_1 enters into this comparison, in conversions between mc^2 and I'', with two different exponents; so, this comparison might offer a (different?) test for, and/or way of obtaining, E_1. (Except in F, $I_U^{1/3}E_1$ does not appear in ratios of 'rest masses' of entities listed in Table 3.) For this reason (among others?), it might seem especially interesting to inquire as to whether such relation might be much more accurate than ~3.29%, and to obtain $I''_{K^0_L}$ and $I''_{K^0_S}$ more nearly formally.

Noting questions raised hereinbefore with respect to 'mass' measurements including assumptions about 'charge' 'interaction strength', if one were to substitute UCLBL's recommended approximate value for $m_{K^0_{L,S}}$, and E_1, as noted hereinbefore, $(m_{K^0_{L,s}}c^2/2E_1)^{2/3}E_1=(m_{K^0_{L,s}}c^2/2)^{2/3}E_1^{1/3}\sim$ $3.408973718\times10^{-6}\text{eV}\sim5.179100510\times10^9\hbar/\text{sec}$. This approximate (suggested?) value is smaller than UCLBL's recommended value by ~3.28048% of this approximate value.

UCLBL stated as 'error' for $\Delta m \sim 0.4112918\%/\sigma$, from which this fractional difference might appear to be $\sim 7.976\sigma$.

With reference to page S76 of reference 8, UCLBL listed 33 reports of Δm, of which they appear to have included only three in their compiled recommended value, which three might seem to have been combined as if 'statistically independent' of each other: M. Cullen+ (ref. 35, 1970), C. Geweniger+ (ref. 36, 1974), and S. Gjesdal+ (ref. 37, 1974).

With reference to Cullen+'s Figure 3 (p. 526 of ref. 35), there might seem to appear in both graphs in such figure a systematic discrepancy between data and calculated 'fit' apparently used in interpretation of such data. This might raise questions in relation thereto, which might not seem completely answered by the statement, appearing in such report, that they had calculated the *exact* shape (italics and wording as appeared on p. 526 of ref. 35).

With reference to Geweniger+ (ref. 36), on page 108 of their report appeared a footnote numbered 1, stating that data analysis included effects not included in the mathematical expression presented therein. With reference to Gjesdal+ (ref. 37), on page 114 of their report appeared a footnote numbered 1, stating that a correct form of the mathematical expression presented in their report had been used in calculation, and that a simplified formula was presented only for illustration. Most of the authors listed for each of these two reports were listed for both.

Footnote 2 on page 111 of Geweniger+'s report (ref. 36) indicated that the experiment(s) reported in both such reports had apparatus in common.

Such reports' statements that analysis used was different from mathematical expressions presented might raise questions as to the use of such reports in formal comparison with this suggested notion relating to Δm, and such commonality of apparatus might raise questions as to whether or not a compilation which treated them as statistically independent might by such treatment have underestimated their combined 'error'.

In view of these questions about the three reports which they included, reliance on UCLBL's compiled 'error' stated for Δm in a formal test might seem unwarranted, and those reports might seem not to formally exclude that $I''_{K^0_L} - I''_{K^0_S}$ might be $\sim (I''_{K^0_{L,S}}/2)^{2/3}$.

53. Might any (inhomogeneous?) 'gas' with 'long range' interaction(s) (with 'strength' approximately independent of velocity?) tend to concentrate 'energy' in (more massive?) components?

54. Might one be able to 'extract' 'energy' from a single 'heat reservoir' using a 'heat engine' having a gravitationally (or other 'long range'?) interacting (inhomogeneous?) 'working' 'medium' ('media'?)?

55. Any entities made of 'neutrino' 'substructures' coupled by 'charge' effects? (rest frame λ_0 for spacings?)

56. Refocusing 'ghost' entities (in U as a whole?)?

57. Might a particle become an antiparticle after traveling distance $2\pi R_U$? and/or existing for longer than longest history in U? (see question 5)

58. Might particle mass be ('slightly'?) less than antiparticle mass? might such a mass difference result in mostly or only particles (and not antiparticles) among those 'predating' longest history in U? (see question 5)

59. 'Scattering' to make particles: entity 'scattered from' as 'source' of 'code' for $I_U^{1/3}$?

60. Particle creation as 'attempt' to code as position information more information than so representable by entity so 'proliferated'?

61. 'Meaning' (e.g., geometric?) for '1/3' and '2/3' exponents ('powers') of information appearing in various 'rest information' ('rest mass') approximate calculations relating to entities ('particles')?

62. One might rewrite the approximate expression in Link 17.4 for $I''_{e\pm}$ in terms of $2I''_{e\pm}$ as a variable, e.g., as $(2I''_{e\pm})^3(\pi \ln(2I''_{e\pm}))^2 \approx (0.9$ (from d code)$)I_U$. Is $2I''_{e\pm}$ 'a' 'natural' variable here?

63. How might one understand 'magnetic moment(s)' of 'particle(s)'? As collective orientation effects on total information of relatively localized system(s) in U? what is relation to 'circulation' of 'currents' (e.g., 'hollow' ('propagating'?) 'modes' of 'electrons' (in 'wires'?)?)?

64. Might 'spin' and 'magnetic moment' be related approximately and/or adequately describable in terms of (an) ensemble(s) with similar 'orientation(s)' in relatively localized region of U?

65. Might 'particle' 'decay(s)' be partially approximately describable in terms of communication (over 3-space distance(s)?) by nonpropagating *ws*? For 'decay(s)' involving transcription of external information to be represented internally by 'decay(s) product(s)', might such distance(s) be related to (linear?) density in U of (such) external information ((as part of?) 'source' for such transcription?)?

66. Might appearance of I_U with 'natural' unit of bits, not scaled by local units, indicate (suggest?) that a (or more than one) concept(s) is (are) missing, such as some (form of) effect 'closing' U in such way that I_U would be independent of choice of (local) units, although R_U, c, and E_1 might be? Might such an effect be in some sense (like) a 'global' 'version' of Theorem 13 interaction? Might with such (a) concept(s) I_U be redundant? (Note that, in information and distance units, $G \approx R_U/I_U$ (e.g., see L 17.10 step 20), $\sim 8.466086265 \times 10^{-97}$ m/bit.)

67. Might every entity be regarded as existing 'simultaneously' at 'both' 'ends' of a 'diameter' of a 4- (or 5- ?) ball? e.g., 'closed' by Theorem 13 type effect(s)? Would I_U (thus?) 'always' apear the same? and thus be redundant?

68. Breakdown of adseparability as breakdown of conservation of momentum? Might this be relevant for consideration of 'reactionless' 'drive(s)'?

69. Is there from '(the?) laws of physics' any distinction between a very 'intelligent' entity and/or a very 'stupid' entity?

70. Is there a set of things an observer can tell an otherwise introspector without changing the otherwise introspector's state? if any, only (the set of?) things which are meaningless? (and, therefore, nonexistent in an information picture?)

71. Might one consider 'polarization' as (a) 'signal encoded for local(ly) propagative communication' bit? Might one consider for this a notion that the 'choice' to make information a transmitted signal is not accessible unless preserved with the signal (polarization?), because 'alternative' not signal is never receivable? (Note Postulate 2 and A-Theorem 1.)

72. 'Hard' walls as implying larger system to contain *ws* to be detected? too big for adseparability? implying momentum not conserved? (for ('locally'?) 'reactionless' drive?)

73. Is there any (formal?) way (criterion by which) an observer seeing an information representation (e.g.?, a collection of atoms?) could (ascertain) distinguish (without 'asking' it) whether (or not?) it is an introspector? What would such 'asking' consist of? Can such asking always be done? Might such a criterion and/or asking suggest and/or appear as (a) 'meaning' of 'alive' and/or 'dead'?

74. Is an introspector an information representation which can exist in ((a) certain) type(s) of 'medium' (and/or combination(s) of media?) (e.g.?, capable of ('supporting') 'information processing'?) is this what is meant by (a) 'soul'? similar questions, but for an observer?

75. What (if any?) is difference between observer and introspector?

76. F in 'strong' $q(r)$ region? apparent R_U as in Cor 14.3, 14.5? spectral line shifts? 'quasars'?

77. Does $\mu \rightarrow e + \nu_e + \bar{\nu}_e$ not $e + \gamma + \overset{(-)}{\gamma}$ suggest that such ν_e and $\bar{\nu}_e$ come into existence spatially separated? and (thus?) that μ decay involves transcription of information by nonpropagating *ws* over nonzero 3-space distance? might this be relevant for decay 'lifetime(s)'?

78. Amplitude in relatively local(ized) $e^{\pm}$ form as number of such relatively local(ized) representations or as (other?) description of amount of information (so?) represented or (as) of information which (would?) be present? compared to $l^{1/2}j^{1/2}A_1$ in W (as in A-Cor 4.7)? Might this affect Corollaries 14.3, 14.5, 14.8, and 14.13–14.16, Theorem 16, (Link 17.4,) and Correspondence Link 17.14? (see also question 38)

79. Might consideration of (various aspects of) adequate description (see also Comment 4.8) be helpful in considering description of atoms and/or molecules? 'solid state'? photons, phonons, and/or photonic and/or phononic 'entities' and/or systems? and/or (local propagative) 'processes'?

80. Might $\hbar^2/8m_0r^2F$ term in Corollary 14.14 step 23 be significant for small enough r? Might this be measurable? Might this account for some of and/or any discrepancy (if any) between approximate calculated value for F and (and/or among) (various) reported value(s) for α^{-1}? and/or appear (with appropriate value for r?) in approximate description as and/or of $e^\pm$ 'magnetic ('dipole'(?)) moment'? Comparing it with $\hbar c/Fr$ term, $(\hbar^2/8m_0r^2F)/(\hbar c/Fr) = (\hbar/8m_0c)/r$, (noting various factors of $2^{\pm 1}$ in Cor 14.3 step 19, Cor 14.8 steps 5 and 11, and Cor 14.14 steps 14–16, 20, and 21) noting that the cancelled F term might be regarded as including various 'effect(s)' of 'charge' 'strength', might this coefficient $\hbar/8m_0c$ of $1/r$ seem reminiscent of 'magnetic moment(s?)' ('strength') of (such) (a) 'particle(s?)'? (note m dependence, unlike 'longer range' term)

ETC...

Section Seven

Conclusion

I have presented herein some concepts related to information, which appear to include a partial formulation of physics in terms of information.

At the time of this writing, this physical picture appears, to the extent tested as presented hereinbefore, substantially allowed by that part of prior knowledge of which I have at the time of this writing both knowledge and awareness, and to offer some insight into space, time, mass, energy, charge, electromagnetic theory, quantum mechanics, special relativity, gravity, 'weakly interacting' particles, 'strong interaction', momentum, inertia, relativity of position, 'spin', 'statistics', virtual states, 'elementary particles', particle decay, conservation of position, and quantum mechanical spreading. In this context were discussed limitations on domains of applicability of special relativity and of the principle of equivalence.

Noting that this report has presented unfinished work in progress, taking all of this material together, I suspect that this physical picture and conceptual structure might merit further thought.

On such basis, I commend to your attention this way of thinking about physical phenomena, calling it

'*information mechanics*'.

Appendix 1

Conjectures on Fast B_0s in Gravity

The following considerations, although not explicitly formal, appear allowed by that part of knowledge prior to the time of this writing of which I am at the time of this writing aware, and sufficiently interesting to offer for your attention. Because two alternative conjectures are presented, both are presented in this appendix, rather than in the main text, pending experimental or theoretical resolution. See also Section 6, question 51.3, in relation to solar neutrino flux and searches for any such flux near the earth.

The question addressed is, what is the behavior of fast B_0s and their emission and/or absorption in a 'gravity well'?

Conjecture 1. *A fast B_0 is emitted into or absorbed from observer O_U's universe U anywhere in U independent of nearness to any relatively localized substantially timelike flow of information in position representation in U.*

This appears consistent with the formal result (Cor 15.1–15.4) that a fast B_0 is not 'attracted' in a Theorem 13 interaction by nearby 'mass' moving in U with $v \ll c$.

From apparent correspondence between fast B_0s and neutrinos, as expressed and tested in various ways in Corollaries 15.11 and 15.12 and Link 17.1–Link 17.14, by comparison, for an observer affected by Theorem 13 $q(r)$, Conjecture 1 might have as a corollary

Conjecture 1.1. *Processes involving making or 'absorbing' neutrinos or antineutrinos are seen by local observer as affected by 'gravity'.*

●

For example, various 'particle' processes in dense objects might be affected.

At the time of this writing, I am not aware of any prior reports of observation such as to exclude Conjectures 1 and 1.1.

Alternatively,

Conjecture 2. *Emission into or absorption from observer O_U's universe U of fast B_0 is subject to Theorem* 13 *$q(r)$ effect(s).*

●

With Conjecture 2, suppose fB_0s were emitted in a $q(r)$ (gravitational) well with red shift somewhat less than $E_2/E_1=2$, referred to O_U. Then, frequency shifted photons associated with such fB_0s might not satisfy spectrum conditions in U in regions with appreciably less red shift, for example, far from the gravitational source.

Consider a fast B_0 propagating in an extended region with slowly spatially varying q, such as a weak gravitational field. Suppose that the fast B_0 were emitted in a region with localization factor q relative to observer O_U. With Conjecture 2, were this B_0 to propagate into a region with localization $q' \neq q$, its momentum would be expected to be seen as if subjected to GS as by $q(r)$. However, allowed momenta for a fast B_0 are relatively widely spaced. Unless q' had a value quite different from q, B_0's associated down-shifted photon could not satisfy $E_j = jE_1$, j integer. Thus, with Conjecture 2, fast B_0 would not propagate into such regions near where it was emitted. The probability of finding a fast B_0 in a region where it would not propagate might be obtained by calculating its 'tunneling' probability. B_0 would have the energy with which it was emitted, but would be detected in a region in which it would be expected to have different energy, for example, 'uphill' from where it was emitted. The 'vertical' distance at which this effect would become important would be where the energy discrepancy equaled the 'quantum mechanical uncertainty' in the energy measurement. When energy discrepancy exceeds energy 'uncertainty', that discrepancy is (statistically) 'quantum mechanically' distinct, and so is (statistically) able to represent information. Such 'penetration' into 'forbidden' regions may be thought of as a property of the nonzero size in phase space of regions belonging to different distinguishable elements by which information may be represented. As discussed in 'Tunneling and Virtual States' (Comment 4.10), information is the conserved quantity here, and any variations which conserve information are allowed.

From this point of view (following notation of Theorem 15 and its corollaries) we can estimate the effect of a weak, nearly uniform and linear, gravitational field on a fast B_0:

Conjecture 2.1. *C3 as in Theorem 15 and its corollaries, with Conjecture 2; in a weak, nearly linear and uniform, 'gravitational field' g, due to source M at distance R; fast B_0 emitted vertically with energy E, B_0's next allowed energies $E' < E$ and $E'' > E$,*

$$E - E' \gg \left(\frac{\hbar g E}{c}\right)^{1/2} \ll E'' - E \quad \textit{and} \quad \frac{E_0^2}{4E} \ll \frac{\hbar c}{R}:$$

fast B_0 is confined to vertical range with characteristic half height

$$|d| \sim \left(\frac{\hbar c^3}{gE}\right)^{1/2}.$$

1. With Conj 2, from A-Cor 2.1, Theorem 13, and A-Def 22, in weak gravitational field g, energy change ΔE expected for fast B_0 in vertical distance d would be approximately describable by $\Delta E \approx (E/c^2) gd$.
2. From A-Cor 2.1, A-Def 22, and Def 13, fast B_0 has velocity $v \approx c$.
3. From 2, observing time t while B_0 moves distance d is $t \approx d/c$.
4. From QM, energy uncertainty conjugate to t is $\sim \hbar c/d$.
5. From QM, if $E - E' \gg \Delta E \ll E'' - E$, observing B_0 at height d above or below where it was emitted has relatively small 'probability' of inducing transition(s) to B_0's next allowed energies E' or E''.
6. From 5, for $E - E' \gg \Delta E \ll E'' - E$, B_0 has energy E with which it was emitted.

7. From 6, for $E-E' \gg \Delta E \ll E''-E$, for information to be conserved, energy $E \pm \Delta E$ which B_0 would be expected to have at height $\mp d$ must be not distinguishable from B_0's energy E. (This may be thought of as the condition for energy 'discrepancy' ΔE to not remove the system from that neighborhood in phase space associated with a distinguishable element which can represent information.)
8. From Postulate 1, and 4, 6, and 7, for $E-E' \gg \Delta E \ll E''-E$, ΔE appears partially approximately describable by $|\Delta E| \lesssim \hbar c/d$.
9. From 1 and 8, for $E-E' \gg \Delta E \ll E''-E$, $|Egd/c^2| \lesssim \hbar c/d$.
10. From 9, for 'statistical' vertical 'range' of B_0 in g, for $E-E' \gg \Delta E \ll E''-E$, $|Egd/c^2| \sim \hbar c/d$.
11. From 10, solving for $|d|$, for $E-E' \gg \Delta E \ll E''-E$, observation of B_0 is (statistically) confined to vertical range with statistical half-height

$$|d| \sim \left(\frac{\hbar c^3}{gE} \right)^{1/2}.$$

12. From Postulates 1 and 2, information exists only as it is represented. (information never 'virtual')
13. From Postulates 1 and 2, and 11 and 12, for $E-E' \gg \Delta E \ll E''-E$, B_0 is so confined to vertical range with statistical half height $|d| \sim (\hbar c^3/gE)^{1/2}$.
14. From 1 and 13, $\Delta E(d) \sim (\hbar gE/c)^{1/2}$.
15. From 13 and 14, for $E-E' \gg (\hbar gE/c)^{1/2} \ll E''-E$, B_0 is confined to vertical range with statistical half-height $|d| \sim (\hbar c^3/gE)^{1/2}$.
16. From Cor 15.1–15.3, and 15, for $E_0^2/4E \ll \hbar c/R$, result in step 15 is consistent with conceptual and formal structure presented hereinbefore. ('principle of equivalence' does not apply)

●

Note that this treatment does not specifically include any effects of motion of B_0's source and/or detector.

This is a remarkable result, and seems quite different from any previous theory of which I am aware. One consequence appears to be that such fast particles might be trapped by gravitational wells which otherwise might have seemed much too weak. In strong gravitational 'fields', restriction of fast B_0s to thin shells about the field source might affect calculations of allowed 'phase space' and rates of emission of such particles.

Note that Conjecture 2.1 restriction of fast B_0s to band with characteristic half-height d appears statistical, and tunneling to greater distances does not appear precluded. d would be 'characteristic half-width' for B_0's vertical distribution.

With Conjecture 2.1, fast B_0s appear constrained by a weak gravitational field to regions with approximately the same gravitational red shift with which they were emitted. If so constrained, B_0s might spread to form a shell about M at the same gravitational potential. If another mass M' were slowly brought near M, such shell about M would be distorted and move out from M, to have the same red shift as seen by observer O_U. The principle of equivalence would not apply.

From this, one would expect that, in a weak gravitational field, a particle P's possible energies after formation would correspond to a set of 'allowed' heights at which P would have those energies. These heights would be allowed locations for P. Were these allowed locations farther apart than P's characteristic vertical half-range d in Conjecture 2.1, one might expect P's behavior to show effects of energy graininess. From this, we can relate interaction 'strength' a_f, E, and gravity g for conditions under which graininess might be seen. (square brackets, as in Section 2, denote portions relating to spectr(um,a) other than $E_j = jE_1$ obtained in axiomatic form (A-Theorems 3 and 4, A-Def 22), retained here for explication) E_1 written for E_{B1}, for $B = U$.

Conjecture 2.2. *C*3; *with Conjecture* 2.1; *in weak gravitational field g due to M at distance R*; *for any vertically moving particle P with energy E and velocity* $v \approx c$, *representing internally amount I of information, interacting differently than photon with energy E*, $(g\hbar/Ec)^{1/2} \ll 1$, *with rest mass* m_0, $m_p > m_0 < m_b$, $m_0^2 c^4/4E \ll \hbar c/R$, *and interaction strength corresponding to rest frame charge magnitude* a_f: *energy graininess affects P's behavior for*

[
$$\left|\frac{a_f}{a_b}\right| \leqslant \sim \frac{(48\pi)^{1/12} E_1^{1/4} E^{7/24}}{(g\hbar)^{1/24} m_b^{1/2} c^{23/24}} \left(\frac{0.9\,Y(3)}{Y(I)}\right)^{1/6}$$

(*least upper bound from finest energy spacings in Corollary* 15.10, *Section* 2)
]

$$\left|\frac{a_f}{a_b}\right| \leqslant \sim \frac{\sqrt{2}\, E_1^{1/4} E^{3/8}}{(g\hbar)^{1/8} m_b^{1/2} c^{7/8}} \left(\frac{0.9\,Y(3)}{Y(I)}\right)^{1/6}$$

[(*greatest upper bound from Section* 3 *spectrum* $E_j = jE_1$)]; *and does not substantially affect P's behavior for*

$$\left|\frac{a_f}{a_b}\right| \gg \frac{\sqrt{2}\,E_1^{1/4}E^{3/8}}{(g\hbar)^{1/8}m_b^{1/2}c^{7/8}}\left(\frac{0.9\,Y(3)}{Y(I)}\right)^{1/6},$$

and I is a finite integer >1.

[1. With Conj 2 and 2.1, as in Conjecture 2.1, in weak gravitational field g, P is expected to lose energy while propagating vertically upward through height L as $\Delta E \approx EgL/c^2$.

2. Call P's next possible energy less than E E'.
3. From 1 and 2, L such that $E-\Delta E=E'$ is an allowed height for P.
4. From A-Def 22 and Conj 2.1, for $m_0^2c^4/4E \ll \hbar c/R$, about each allowed height, P can 'tunnel' vertically within a band with characteristic half-height d.
5. From Postulate 1, and 4, graininess of P's allowed energies would affect P's behavior for $L \gtrsim d$.
6. For specificity, let $L=ud$. ('u' is to help keep track of 'd')
7. From 1 and 6, $\Delta E \approx (E/c^2)\,gud$.
8. From Conj 2.1, $d \sim (\hbar c^3/gE)^{1/2}$.
9. Substituting 8 in 7, $\Delta E \sim (Eg\hbar/c)^{1/2}u$.

[10. From Cor 15.10 and Conj 2.1, for least graininess, using finest energy spacings in Corollary 15.10,

$$E' \lesssim \frac{E}{1+\dfrac{48\pi E_1^3E^3}{m_b^6c^{12}}\left|\dfrac{a_b}{a_f}\right|^{12}\left(\dfrac{0.9\,Y(3)}{Y(I)}\right)^2}.\Bigg]$$

[11. To examine least graininess in P's allowed energies, set E' approximately equal to expression in step 10.]

[12. From 9–11, writing $1+y$ for denominator in step 10,

$$E-\left(\frac{Eg\hbar}{c}\right)^{1/2}u \sim \frac{E}{1+y}.\Bigg]$$

[13. From 12, dividing by E, inverting, transposing 1, and reexpressing y,

$$\frac{48\pi E_1^3 E^3}{m_b^6 c^{12}}\left(\frac{0.9\,Y(3)}{Y(I)}\right)^2\left|\frac{a_b}{a_f}\right|^{12} \sim \frac{1}{1-\left(\frac{g\hbar}{Ec}\right)^{1/2} u} - 1.$$]

14. $(g\hbar/Ec)^{1/2}u \ll 1$, so right side of step 13 is $\approx (g\hbar/Ec)^{1/2}u$.

[15. From 13 and 14, solving 13 for $|a_f/a_b|$, with approximation as in step 14,

$$\left|\frac{a_f}{a_b}\right| \sim \left(\left(\frac{Ec}{g\hbar}\right)^{1/2}\frac{48\pi E_1^3 E^3}{m_b^6 c^{12} u}\left(\frac{0.9\,Y(3)}{Y(I)}\right)^2\right)^{1/12}.$$]

16. From 5 and 6, graininess of P's allowed energies would have greater effect for larger u.

[17. With Conj 2, from 15 and 16, collecting terms and setting $u=1$, for finest energy spacings in Corollary 15.10, graininess would affect P's behavior if

$$\left|\frac{a_f}{a_b}\right| \leqslant\sim \frac{(48\pi)^{1/12} E_1^{1/4} E^{7/24}}{(g\hbar)^{1/24} m_b^{1/2} c^{23/24}}\left(\frac{0.9\,Y(3)}{Y(I)}\right)^{1/6}.$$]

18. From A-Theorems 3 and 4, A-Def 22, Cor 15.10, and Conj 2.1, [for maximum 'graininess', using coarsest energy spacings in Corollary 15.10,]

$$E' \underset{[\gtrsim]}{\approx} \frac{E}{1+\frac{4E_1E}{m_b^2c^4}\left|\frac{a_b}{a_f}\right|^4\left(\frac{0.9\,Y(3)}{Y(I)}\right)^{2/3}}.$$

19. From 9 and 18, as in steps 11 and 12, writing $1+z$ for denominator in step 18,

$$E-\left(\frac{Eg\hbar}{c}\right)^{1/2} u \sim \frac{E}{1+z}.$$

20. From 19, as in step 13, dividing by E, inverting, transposing 1, and reexpressing z

$$\frac{4E_1E}{m_b^2c^4}\left(\frac{0.9\,Y(3)}{Y(I)}\right)^{2/3}\left|\frac{a_b}{a_f}\right|^4 \sim \frac{1}{1-\left(\dfrac{g\hbar}{Ec}\right)^{1/2}u}-1.$$

21. From 14 and 20, solving 20 for $|a_f/a_b|$, with approximation as in step 14,

$$\left|\frac{a_f}{a_b}\right| \sim \left(\left(\frac{Ec}{g\hbar}\right)^{1/2}\frac{4E_1E}{m_b^2c^4u}\left(\frac{0.9\,Y(3)}{Y(I)}\right)^{2/3}\right)^{1/4}.$$

22. With Conj 2, from 16 and 21, collecting terms and setting $u=1$, [for coarsest energy spacings in Corollary 15.10 (corresponding to $E_j = jE_1$ as in Section 3),] graininess would affect P's behavior if

$$\left|\frac{a_f}{a_b}\right| \lesssim \frac{\sqrt{2}\,E_1^{1/4}E^{3/8}}{(g\hbar)^{1/8}m_b^{1/2}c^{7/8}}\left(\frac{0.9\,Y(3)}{Y(I)}\right)^{1/6}$$

[(greatest upper bound)].

23. With Conj 2, from 22, [noting that steps 18–22 treat a greatest upper bound for graininess to affect P's behavior,] graininess would not substantially affect P's behavior if

$$\left|\frac{a_f}{a_b}\right| \gg \frac{\sqrt{2}\,E_1^{1/4}E^{3/8}}{(g\hbar)^{1/8}m_b^{1/2}c^{7/8}}\left(\frac{0.9\,Y(3)}{Y(I)}\right)^{1/6}$$

[where the expression on the right appears here as a greatest lower bound].

24. From Postulates 1 and 2, A-Theorem 1, and A-Def 1, for P to interact differently than photon with energy E, $I>1$.
25. From Postulate 3, I is finite.

With Conjecture 2, Conjecture 2.2 indicates that if a particle with rest mass less than that of an electron were moving nearly as fast as light and interacted weakly enough, it might be observable only in, and trapped into, bands in a gravitational field. This is a remarkable result.

From Conjecture 2.1, $|d| \sim (\hbar c^3/gE)^{1/2}$. For E in electron volts, with $g \sim 9.8$ m/sec^2 as the earth's gravity's 'acceleration' near the earth's surface, using from Cohen and Taylor (ref. 4) 1 eV$\approx 1.6021892 \times 10^{-19}$ J and $c \approx 299792458$ m/sec, $|d| \sim 42540.5/E^{1/2}$ m. For fast B_0 carrying 1 MeV energy, $|d| \sim 42.5$ m. This does not specifically include effects of source and/or detector motion, if any.

With Conjecture 2.2, we can ask how weakly a particle must interact to display energy graininess in the earth's gravity near the earth's surface. [Using the finest energy spacing from Corollary 15.10, from Conjecture 2.2 we obtain, as a least upper bound, with E in joules, $|a_f/a_b| < \sim 3.53153 \times 10^{-5} E^{7/24}/(Y(I_x))^{1/6}$, or, with E in electron volts, $|a_f| < \sim 1.1641 \times 10^{-10} E^{7/24}(Y(I_x))^{-1/6}|e|$, where $|e|$ is magnitude of electron 'charge' strength; for a particle carrying 1 MeV energy, with $I_x = 2$ bits, $|a_f| < \sim 5.8322 \times 10^{-9}|e|$.] Using [coarsest energy spacings in Corollary 15.10,] $E_j = jE_1$ (as in Section 3), from Conjecture 2.2 we obtain, [as a largest upper bound,] for E in joules, $|a_f/a_b| < \sim 0.09379\ E^{3/8}/(Y(I_x))^{1/6}$, or, for E in electron volts, $|a_f| < \sim 8.3931 \times 10^{-9} E^{3/8}(Y(I_x))^{-1/6}|e|$; for a particle carrying 1 MeV energy, with $I_x = 2$, $|a_f| < \sim 1.32969 \times 10^{-6}|e|$. From Conjecture 2.2, [using greatest upper bound] to find a [least] lower bound for E for 'graininess' to appear, setting $|a_f| = |e|$, and noting that for an electron or positron terms in Y cancel, we find that graininess should not substantially affect behavior of an electron or positron propagating substantially vertically in the earth's gravity near the earth's surface for $E \ll 973.12$ J, about 6.07×10^{21} eV.

Conjecture 2.2 suggests that any particle interacting weakly enough and having much less rest frame mass than an electron might have peculiar properties in gravity. I am aware of only one class of particles which might satisfy Conjecture 2.2's conditions in laboratory experiments on earth: neutrinos. If neutrinos correspond to fast B_0s or fastest B_0s, as might be suggested by correspondences tabulated in Table 3, Conjecture 2.2 might suggest that neutrinos emitted in the sun would not be observable on earth. This appears allowed by those measurements prior to the time of this writing of which I am at the time of this writing aware (ref. 33). (See also question 51.3, Section 6.) Even more striking, with Conjectures 2.1 and 2.2, effects of energy graininess might be observable in experiments on earth: from Conjecture 2.1, as noted, d might be not unreasonable for such experiments.

Conjectures 2.1 and 2.2 represent a qualitative change from various prior theories about particle motion. If neutrinos are described by these conjectures, such effects might be observable on earth. Various prior theories say that neutrino flux near a nuclear reactor on earth should be described by an inverse square law, and that 'gravity' would cause at most only substantially negligible anisotropy of such flux. If neutrinos are described by Conjectures 2.1 and 2.2, and if effects of source and/or detector motion are negligible, a nuclear reactor's neutrino flux should begin to depart from an inverse square law in a vertical (and, with 'trapping', horizontal) distance $\sim$50 m, and should be anisotropic. Such an effect might be sought near a large nuclear reactor, for example, in a power plant. Tunnels might house movable detector(s) below the reactor, and detector(s) might be placed above and around the reactor. Prudence might suggest an experiment sensitive enough to respond to an effect somewhat smaller than expected, for example, by a factor of $\sim 2\pi$.

Observation of such an effect might have many consequences.

Were such an effect observed, one would expect neutrinos to form shells around massive objects. This might change astrophysical theories in ways discussed more fully hereinafter. More urgently, in such case one would like to know in detail what happens to neutrinos produced in nuclear power plants.

1. With Conjectures 2.1 and 2.2, fast B_0s would be (statistically) confined to gravitational equipotential shells about massive objects. From Conjecture 2.1, if 4 MeV neutrinos from nuclear reactors correspond to fast B_0s, they would form a shell $\sim$42.5 m thick about the earth, at nearly the gravitational potential at which they were emitted. From Corollaries 15.1–15.3, the principle of equivalence would seem substantially irrelevant in defining location of such gravitational equipotential shells. For this reason, other bodies in the solar system might have much larger effect on the positions of such shells than is seen in usual tidal effects.

Two principal effects might be expected:

(1) Shells would be further from the center of the earth on the side facing another massive body. As the earth rotated, a nuclear reactor would effectively be raised and lowered through a belt of equipotential shells, with belt thickness depending on the reactor's lattitude, with neutrinos trapped in successive shells at the gravitational potential at which the reactor emitted them.

(2) As the earth and another massive object approached toward or receded from each other, equipotential shells would spread from or converge towards the earth, respectively. The magnitude of effect (1) would also increase as the bodies approached each other.

Both these effects would tend to shift the height of a shell, and so would cause neutrinos from a steady source to be trapped in a belt with thickness greater than $2d$ (d as in Conjecture 2.1).

Starting with a gravitational potential $V \approx -Gm/r$, one can estimate the effect of a mass m, starting at distance r_1 from the earth and moving to distance $r_2 < r_1$, on a shell starting near the earth's surface and moving out to height x above its starting height. Writing the earth's mass as M and radius as R, for the shell to stay at the same red shift,

$$-\frac{Gm}{r_1} - \frac{GM}{R} \approx -\frac{Gm}{r_2} - \frac{GM}{R+x}.$$

Factoring out $-G$ and grouping terms,

$$\frac{1}{R+x} \approx \frac{m}{M}\left(\frac{1}{r_1} - \frac{1}{r_2}\right) + \frac{1}{R}.$$

Note that $r_1 > r_2$, so x is positive. For calculating shell offset on near and far sides of planet, r_1 and r_2 can be written as $r_1 + R$ and $r_1 - R$. For $r_1 \gg R$, $1/r_1 - 1/r_2 \approx -2R/r_1^2$, and

$$\frac{1}{R+x} \approx -\frac{m}{M}\frac{2R}{r_1^2} + \frac{1}{R}.$$

(Note that 'tidal' ΔR of the earth's surface would be approximately symmetric on sides toward and away from other body, and for this reason does not need to be included in this approximate description.)

As a massive body receded from the earth, a shell would contract into the earth to some new radius r. With a $1/r$ approximation for gravity, the field in a homogeneous solid with density ρ would be approximately describable as

$$F(r) \approx -\frac{4}{3}\pi r \rho G.$$

Integrating $F dr$ between R and $r < R$, and adding the potential at R,

$$V(r) \approx -\frac{GM}{R} - \frac{2}{3}\pi \rho G(R^2 - r^2).$$

Thus, for constant red shift, a shell would move from R to r. This constancy can be expressed as

$$-\frac{Gm}{r_1} - \frac{GM}{R} \approx -\frac{Gm}{r_2} - \frac{GM}{R} - G\frac{2}{3}\pi\rho(R^2 - r^2).$$

Solving for r,

$$r \approx \left(R^2 - \frac{m\left(\frac{1}{r_1} - \frac{1}{r_2}\right)}{\frac{2}{3}\pi\rho} \right)^{1/2}$$

Note that $r_1 \leqslant r_2$, and $r \leqslant R$. The earth is denser near its center than near its surface, so this approximation based on uniform density would tend to underestimate r.

Using these formulae, I estimated effects which might be expected with Conjectures 2–2.2 of other masses in our solar system on the height of a shell starting near the earth's surface and being shifted outward. For these estimates, I took as R the earth's equatorial radius. Data were from astronomical tables on pages F146 and F147 of the *Handbook of Chemistry and Physics*, 52nd edition (1971, ref. 9). Intermediate results were carried to 10 decimal digits.

As the earth turns, a shell would appear to an observer on the earth's surface to rise on the side toward the sun, moon, and Jupiter by approximately:

1. Sun ~7.735 km
2. Moon ~43 m
3. Jupiter ~0.42 m
4. As the moon and earth orbit each other, the earth 'wobbles' toward and away from the sun, and a shell which was at the earth's surface when the earth was farthest from the sun would rise to a maximum height of ~5.662 km.
5. In orbiting the earth, the moon approaches and recedes, and (neglecting other effects) a shell at the earth's surface at lunar apogee would rise to a height of ~143 m at lunar perigee.
6. The earth's orbit about the sun is eccentric, and a shell at the earth's surface at aphelion would rise to a maximum height of ~5758 km at perhelion. Using the formula for a uniform solid, a shell at the earth's surface at perhelion would move in to radius ~1487 km at aphelion, or a depth ~4892 km below the earth's surface at the earth's equator.

Approaching from positions of the earth in its orbit farthest to closest to each of the following planets would raise a shell from the earth's surface to a height of approximately:

7. Jupiter ~6.6428 km
8. Venus ~672.6 m
9. Saturn ~574.6 m
10. Mars ~44.2 m
11. Uranus ~21.5 m
12. Mercury ~13.2 m
13. Neptune ~10.3 m

(Pluto's mass has been reported (ref. 9) as roughly 1% of Neptune's mass, but would seem to in that report be so roughly stated that, in view of relative smallness of its reported mass, detailed estimate of any such effect from Pluto's orbital eccentricity seemed unnecessary here.)

Gravitational sources outside our solar system:

14. A 'gravitational field' from a source far from our solar system might affect shells near the earth. The effect due to the mass of our galaxy may be estimated as follows: Our solar system is presumed to be in nearly circular orbit about the galactic center, at distance ~30000 light years. The mass of the galaxy is taken as $\sim 10^{11}$ times our sun's mass. The earth's solar orbit provides an excursion in this 'field' of approximately its orbital diameter. This excursion would change the height of a shell near the earth's surface by ~5 m, about 10^{-6} of the effect due to the earth's solar orbit's eccentricity. Note that, if our solar system were approaching or receding from the galactic center at a substantial velocity, effect(s) due to galactic mass might be substantially larger.
15. One might crudely estimate a rough order of magnitude for effects due to sources outside our galaxy, by taking (from Gérard de Vaucouleurs, ref. 38) an approximate velocity v for the earth relative to other relatively nearby galaxies of ~330 km/sec, and a rough distance r ~20 'megaparsecs' (1 parsec $\approx 3.08374 \times 10^{16}$ m (ref. 9)); and estimating the change x in height of a 'gravitational equipotential' near the earth's surface if the earth were approaching a cluster of ~100 galaxies (total mass M) at such distance and with such relative velocity.

    In the Theorem 13 weak field case, for relatively small displacements, the amount of displacement x in the earth's gravity needed to compensate for displacement in time interval Δt in the more remote source's gravity is given approximately by $x \approx v\,\Delta t\, MR^2/mr^2$, with 'average' R, and m as before in this Appendix 1, where MR^2/mr^2 is just the ratio of approximate 'field strengths', appearing as a scale factor relating displacement in one field to compensating displacement in the other, with sign determined by choice of x as positive for increasing distance from the earth and v as positive for approach to the other source.

Substituting (various data from ref. 9),
$R \sim 6371315$ m. ('average'),
$r \sim 20 \times 3.08374 \times 10^{22}$ m,
$m \sim 5.979 \times 10^{24}$ kg,
$M \sim 10^2 \times 10^{11} \times 1.991 \times 10^{30}$ kg ($\sim 10^{13}$ 'solar mass(es)'),
$v \sim 330000$ m/sec,
$\Delta t \sim 31536000$ sec (~ 1 year),
I found $x \sim 3.697 \times 10^{-3}$ m.

Values for M and r do not seem to have been accurately obtained yet by direct observation, and, in view of Corollaries 10.2–10.4, Theorem 14–Corollary 14.16, and A-Theorems 3 and 4, one might wish to reexamine interpretation of data in various reports relating to M, r, and v.

16. One can similarly estimate effect due to a 'wandering' object, not substantially comoving with the local galactic velocity field, such as a dark star with about the same mass as the earth's sun, about 1 light year away, with radial (component) of relative velocity ~ 500 km/sec. Substituting $M \sim 1.991 \times 10^{30}$ kg, $r \sim 9.454 \times 10^{15}$ m, and $v \sim 5 \times 10^5$ m/sec, with R, m, and t as before, I found $x \sim 2.37$ m for 1 year.

I have included in this list one indirect effect, the moon's 'wobbling' of the earth's solar orbit. Many other effects would be expected in the many body trajectories in the solar system, such as the motion of the sun as Jupiter orbits it. Detailed behavior of a shell about the earth might, if such study seems appropriate, reasonably be studied with iterative numerical solution for positions of masses in the solar system, using a computer.

It seems remarkable that relatively distant clusters of galaxies and/or wandering objects might have so large a cumulative effect on the height of a 'gravitational equipotential' near the earth. A rapidly approaching more massive object might produce substantially larger rates of shift of such equipotentials, and, because orbital changes would depend on 'tidal forces' rather than potential changes, such equipotential shifts might be relatively more sensitive to approach (or recession) of a massive object.

With the cautionary note that this is exploration of one of two alternative conjectures, and that there may be effects from outside our solar system which exceed the annual shift estimated from the earth's orbit's eccentricity in the sun's gravity, by, for example, relatively rapidly 'raising' or 'lowering' equipotentials so that they would move only monotonically inward or outward over periods of time much longer than 1 year, one might ask what might happen if such an effect were to be present and then disappear (e.g., from passage of a relatively massive and nearby 'wanderer'). Part of the reason for considering such question is that the amount and behavior of nonluminous matter does not seem well known, but such matter might seem relatively abundant (see Table 2; 'seem' as used in this Appendix 1 not referencing 'introspection' as in Comment 4.5); with Corollary 13.3, stable, relatively massive and compact, nonluminous objects might exist in significant number; and closest approach distances of less than 1 light year for relatively rapid passage of such objects might lead to relatively large effects which could change magnitude and/or sign in a relatively short period of time.

If Conjectures 2–2.2 approximately describe behavior of neutrinos in gravity, then, absent or on ending of effects from outside our solar system, one might see the following effects:

Of items 1–13, it might seem that the sun would have the largest effect on shells about the earth. Each day, the earth's rotation would sweep a source at the equator through $\sim$7.7 km of trapping belt thickness. Each year, a belt $\sim$5758 km thick would expand and contract through the earth's surface, passing sources. During an annual cycle, the lower portion of the belt would spend most of its time in the earth, in which neutrinos from a nuclear reactor would have a half-life $\sim$5 years. The portion of the belt near the top would spend almost none of its time in the earth (only mountains would reach into it) and most of its time above the earth's atmosphere, and so would be cleared of neutrinos only very slowly, over thousands of years. Neutrinos from nuclear explosions might be trapped as shells and swept radially for years, with decay time depending on time averaged density of matter in such swept shell.

At equilibrium, a shell containing neutrinos would lose energy as fast as it received energy. Assuming that all matter is roughly equally effective for absorbing neutrinos, the rate at which energy would be absorbed from the shell per unit mass in the shell would be inversely proportional to the time average of the mass in the shell and proportional to the time averaged power input. 1 Rem (Roentgen equivalent in man) $=100$ ergs/g. Natural background radiation near the earth's surface is $\sim 10^{-5}$ Rem/hr., or about 2.77×10^{-14} W/g.

With this, one can estimate the effect which neutrino trapping might have on the level of background radiation to which people might be subjected. The earth's rotation would spread flux over a belt of height $\sim$7.7 km. Were that belt filled with earth's crustal solid, density $\sim$2.5 g/cm^3, it would contain $\sim 9.82\times 10^{24}$ g. Thus, to raise background by a factor of 2 would require $\sim 2.73\times 10^{11}$ W of neutrino flux, average. Half-life would be about 5 years. At the top of the annual belt, buildup would be at roughly the same rate. That is, $\sim 2.73\times 10^{11}$ W flux would roughly double background in 5 years of total input time, adding input time as the belt's upper part came near the ground and entered the region of injection. The region within about 7.7 km below the top would spend $\sim(2\times 365/2\pi)\cos^{-1}(1-7.7\times 2/5758)\sim 8.5$ days each year in the daily spread belt, so 2.73×10^{11} W input would take about 214 years to double background.

Suppose, however, that a society of 5 billion people on earth were to use as much power per capita as the United States does at the time of this writing, and that all such power were to be obtained from nuclear power plants near the earth's surface. For 50% power plant efficiency, this would release roughly 10^{13} W neutrino flux. Such a nuclear power 'economy' would inject enough neutrinos into the region near the top of the belt to raise background radiation by about 17% of present background during that 8.5 days in each year when the top of the belt was near the ground, *cumulatively*. That seems quite likely unacceptable. Also, such flux might take thousands of years to substantially clear.

Consider next a 10^{13} W neutrino flux injected near the bottom of the belt. This part would spend all the rest of the year in the earth, and so would build up to $\sim(10^{13}/2.73\times 10^{11})(8.5/365)\times$background, or about 85% above present background, in equilibrium. This estimate assumes a density of about 2.5 g/cm^3, but at the depths at which this part of the belt would spend much of the year, the earth's density is considerably greater, so at equilibrium this part of the belt might contain somewhat less neutrino flux. Also, this flux would be substantially cleared from the belt in a few decades.

Near the middle of the annual belt, the rate of motion of the belt would sweep it through a 7.7 km band in $\sim 7.7\times 365\times 2/(2\pi\times 5758)$ day twice each year, for a total of about 0.31 day/year. This portion of the belt would spend about half the time in the earth. Thus, 10^{13} W of neutrino flux would increase background by $\sim(10^{13}/2.73\times 10^{11})(2\times 0.31/365)\times$ background, or about 6.2% above present background, and would have a half life of about 10 years. This might seem less ugly, but seems clearly not desired.

Detailed study of belt behavior might properly include analysis of all effects which might raise the top of the belt. Were the top of the belt to be raised far enough, it might then merge with a shell about the sun, and thus have a much longer time constant for accumulation of neutrinos. However, estimates of direct effects of other objects in our solar system do not indicate that such 'dumping' into a solar shell would occur.

From these estimates of the behavior of gravitational equipotentials near the earth, it appears that one might draw the following conclusions:

1.1 It would seem important to examine neutrino flux for dependence on direction (and energy) in vertical and horizontal planes about a nuclear reactor, for anisotropy, at distances of 0 to about 200 m, to observe whether neutrinos might be trapped by the earth's gravity. Restriction of only observation of neutrinos would be indicated by decrease 'faster' than $1/r^2$ for vertical distance from the reactor and $\sim 1/r^2$ for horizontal distance; trapping would be indicated by decrease 'faster' than $1/r^2$ for vertical distance, and 'slower' than $1/r^2$ for horizontal distance.

1.2 If neutrinos are trapped, it would be important to study in detail the behavior of neutrino shells near the earth.

1.3 If neutrinos are trapped into a rising and falling belt, it would seem essential that possible desirability of not using substantial amounts (compared to absorption) of neutrino or antineutrino emitting nuclear reactions near the earth be seriously considered, and that any large-scale use of nuclear power include some form of energy storage which would allow substantially all nuclear power production to stop for about 1 month while the top of the belt was near the ground. Were this not done, a particularly ugly and persistent form of radiation pollution might occur.

Although one relatively convenient and economical way to accommodate timed use of reactors might be to use an intermediate fuel, for example, in a hydrogen 'secondary energy' 'economy', for storage and distribution of 'primary' nuclear energy, which storage might also facilitate reducing nuclear output while the bottom of the belt was near the earth's surface, if neutrinos are trapped into a rising and falling belt, the substantial use (compared to absorption) on earth of nuclear reactions which emit neutrinos or antineutrinos, whether fission or fusion, would seem in various ways substantially less desirable than use of nuclear reactions far enough from earth to avoid such neutrinos, for example, far enough from earth so that a belt not intersecting the earth would be formed about the sun. Solar power might, prior to the time of and near this writing, seem to be an example of nuclear processes in such a location.

2. With Conjectures 2.1 and 2.2, in stars, neutrinos would not be very effective for radial energy transport. Instead, neutrino shells might be the strongest horizontal energy transport mechanism in the nuclear burning region. By horizontal energy transport, they might enhance stellar symmetry, perhaps even during supernova. They would also store energy.

3. With Conjectures 2.1 and 2.2, that neutrinos would not transport energy long distances radially might affect various detailed aspects of stellar evolution calculations.

4. With Conjectures 2.1 and 2.2, intense neutrino shells might remain about neutron stars, left there by decay in transit of heavy elements formed in the supernova which left the neutron star, and decaying slowly by interaction with matter (including $\overset{(-)}{\nu}$'s?) present at much lower density after formation.

5. With Conjectures 2.1 and 2.2, transverse spread of a horizontally directed neutrino beam might be greater horizontally than vertically, and response of such beam to changes in 'aim' might be greater horizontally than vertically. Dependence of such spread and aim effects on neutrino energy might help distinguish such effects (if they exist) from 'background'. Also, with Conjectures 2.1 and 2.2, beam 'aim' might vary somewhat as the earth turned relative to the sun, which effect might (if it existed) depend on beam orientation.

These results appear substantially different from all previous theories of which I am at the time of this writing aware, and might be testable. There seem to be sufficient physical, environmental, and economic reasons for performing relevant experiments, such as measuring neutrino flux about a nuclear reactor to see if neutrinos propagate anisotropically in the earth's gravity.

If such experiments show substantially isotropic propagation, then one might examine theoretical models of dense objects for possibly observable consequences of effects of variations, as seen by local observer, in 'particle' processes involving neutrinos and/or antineutrinos.

If such experiments show substantially anisotropic propagation, prudence would suggest serious study of trapping and/or subsequent behavior, to provide a basis for consideration of various practical consequences mentioned in this appendix.

In view of the extent to which various results presented in Sections 2 and 3 of the main text appear compatible with knowledge prior to the time of this writing, and reports[33] prior to the time of this writing of nonobservation of solar neutrinos in experiments on earth (but see also Section 6 question 51.3), experimental resolution of questions raised on conjectures appearing in this appendix might appear warranted.

Appendix Two

Reference Lists

For your convenience in checking various references, I have included the following lists briefly noting each referenced postulate, condition, lemma, theorem, corollary, chain, link, definition, principle, interpretation, and conjecture. I found such lists helpful in writing this material. These notes are not intended to be self explanatory, and likely will be nearly incomprehensible before you read the text. (Note order specified in Subsection 1.1 (p. 3) of performing mathematical operations.)

To such lists which I used, I have added in this appendix a list of abbreviations, for convenience in interrupted and/or incomplete reading.

You might wish to copy these lists so that you can keep them handy while you read.

APPEARING IN SECTION 1

Interpretation (p. 2) interpretation of references to prior knowledge and/or theory(s).

APPEARING IN SECTION 2

Conditions ('C(*integer*)'), Lemmas ('L'), Theorems ('T'), and Corollaries ('C(*integer*).(*integer*)')

	page	
T1	39	NRQM; information conserved in isolated finite QM system
C1.1	40	NRQM; I conservation if not energy exchange
C1.2	41	NRQM; equal amounts of information in complementary representations
T2	41	B, photons; I by photon occupation number
T3	42	B_1 in B_2; I depends on E_2 and B_2, not (details of) B_1's photons
C3.1	43	photons and boxes; I depends on E and B
L4.1	44	B_0 in B; $\lambda_b \approx c^2 h / E_0 v$, NRA
T4	45	B_0 and P states 1:1 iff $E_0 \approx m_P c^2$, $v \neq 0$.
T4′	72	relativistically B_0 corresponds to P, $v \neq 0$.
C4.1	46	NRA, $E_0 \approx m_P c^2$, external I of B_0 and P the same in B_2
C4.2	47	NRA, $E_0 \approx m_P c^2$, I same in B_2 for B_0 and P
C4.3	47	NRA; I of photons $= I$ of mass particle P iff $E \approx m_P c^2$
C4.4	74	some collected results for B_0s
C4.5	77	some collected results for B_0s in finite enclosure
T5	48	NRPPR; I depends on E in box
C5.1	49	NRPPR; I of B_1 or P in B_2 depends on E and B_2
C5.2	49	NRPPR; finite universe, I of B_1 or P depends only on E
C5.3	50	NRPPR; I of any electromagnetic and/or mass configuration depends only on E
T6	50	NRPPR; in box, I substantially proportional to E
C6.1	51	NRPPR; I substantially proportional to E
C6.2	51	NRPPR; in box, I conservation implies substantially E conservation
C6.3	51	NRQM and NRPPR; I conservation substantially equivalent to E conservation
C6.4	51	NRQM and NRPPR; co-possible modes: A_j substantially proportional to $1/I_j^{1/2}$
T7	53	NRA; NRPPR applies to particles and photons in B_1, $v \ll c$

	page	
C7.1	54	GA of B_1 preserves equivalence of boundary conditions for P and B_0
	55	restriction of L8.1–C8.16 to electromagnetic energy and/or mass particles, only
L8.1	55	NRQM, NRPPR, NRA; GA to $v \ll c$ preserves amount and representation of internal I
L8.2	57	NRPPR; E conservation implies sum rule on photon occupation numbers in 'stationary' $B, \Sigma_i M_i E_{Bi} =$ constant
L8.3	58	NRPPR, NRA; E conservation implies same sum rule in B, $v \ll c$
L8.4	59	NRPPR; $\Sigma_i n_i E_{Bi} =$constant implies I conservation in B
L8.5	59	NRPPR; $\Sigma_i n_i E_{Bi} =$constant for photon representation in 'stationary' B implies E substantially conserved for any system in B
L8.6	60	NRPPR, NRA; E conservation in B moving $v \ll c$ is substantially equivalent to I conservation in B moving
L8.7	61	NRQM, NRPPR, NRA; all physical processes in B moving $v \ll c$ conserve I
C1	61	condition referenced in formal statements of L8.8, 8.10, and 8.11, T8, C8.1–8.3, conforms to formally neglecting finiteness of enclosure
L8.8	61	GS of B_1 does not change amount or representation of contained I_1
L8.9	62	NRPPR applies in B' with size changed by GS the same as in B
L8.10	63	NRPPR; all physical processes in B deformed by GS conserve E and I
L8.11	63	NRPPR; all physical processes in B deformed and/or moving with $v \ll c$ conserve E and I
T8	63	NRPPR, NRA; separability of internal and external information
C8.1	65	NRPPR, NRA; B closed, uniform velocity $v \ll c$, I separable
C8.2	66	NRA, NRPPR; laws of physics and speed of light unchanged to observer in B, $v \ll c$
C8.3	66	NRA, NRPPR; T1–C8.3 apply in closed system, uniform $v \ll c$

	page	
C8.4	67	infinite enclosure, PPR; laws, c, and T1–C8.4 unchanged $v<c$
	67	constructive reference to C8.4
C8.5	78	PPR; in finite B, PODIC applies for B_0, or two lowest mode photons in box, perpendicular photon polarizations, or corresponding P, with $v<v_m$
C1, C2	82	conditions referenced in formal statements of C8.6–8.16
C8.6	82	$E_0(v)=2E_{Bj}(1+v/c)/(1-v^2/c^2)^{1/2}$
C8.7	83	fastest B_0, $E_0(v)\sim 2E_{B1}(1+v/c)/(1-v^2/c^2)^{1/2}$ (see text for cautionary note)
C8.8	84	$E(v)=2E_{Bj}/(1-v/c)$
C8.9	84	$v(E)=c(1-2E_{Bj}/E)$
C8.10	84	reference frame $v<c(1-\epsilon)$, $\epsilon\gg 2E_{Bj}/E$, B_0 v looks like $\approx c$
C8.11	85	fast B_0; $E_0\approx 2\sqrt{EE_{Bj}}$
C8.12	85	for $\tau_0\approx 2h/E_0:\tau\approx h/2E_{Bj}$, fB_0 $\tau\sim\pi R_B/c$ (see text for cautionary note)
C8.13	86	fB_0 lags behind photon by about one wavelength in time τ
C8.14	87	on forming fast B_0, $\Delta E/E=E_{B1}/E\ll 1$
C8.15	87	already formed B_0, $E'<E$,

$$\left[\frac{1}{\dfrac{4}{3\pi}\left(\dfrac{E_{Bj}}{E_{B1}}\right)^3+1}\lesssim\right]\delta \underset{[\lesssim]}{\approx} \frac{1}{\dfrac{E_{Bj}}{E_{B1}}+1}=\frac{1}{j+1}.$$

C8.16	90	PODIC applies for $v<\approx c(1-(R_3/R_B)^2)/(1+(R_3/R_B)^2)$, separability fails for $v\geqslant\approx c(1-(R_3/R_B)^2)/(1+(R_3/R_B)^2)$
T9	97	B_2's photon states spread and raised for UDA by D
C9.1	98	energy carried by moving object attributable to information transport
	100	statement of condition for T10–C13.5
C1	100	condition for T10–C11.5
T10	100	total $I=I_1L_2/L_1$, external $I'=I_1(L_2/L_1-1)$
C10.1	101	detecting B's external information without detecting B's internal information reduces B's localization

	page	
C 10.2	102	in B constant size R_B, for $d \ll R_B$, $\lambda(d) \approx \lambda_0(1 + d/R_B)$
C 10.3	103	like C 10.2, fictitious $v_r(d) \approx Hd$, $H \approx c/R_B$
C 10.4	105	C 10.2 and 10.3 appear allowed by prior observation, with H corresponding to local Hubble's constant
T 11	105	GS of B, laws and constants of physics appear unchanged to observer in B
C 11.1	106	laws and constants as seen by observer in box B_1 in B are independent of B_1's localization
C 11.2	106	after GS of B by factor q, outside observer sees internal units M as $\approx M/q$, L as $\approx Lq$, and T as $\approx Tq$
	106	restriction of QMT to same representation
C 11.3	107	$W_1 = E_1(q_1/q_2 - 1)$
C 11.4	108	O_1 sees just fast enough box as shortened by q_2/q_1
C 11.5	108	O_0 sees B unchanged length parallel motion.
C2	112	condition for T 12–C 12.2
T 12	112	$\Delta I \approx (I_1 + I_1')R_B I_M / I_B r$
C 12.1	114	$\Delta I \approx I_T R_B M / M_B r$
C 12.2	114	$q(r) \approx 1 + R_B M / M_B r$
(cond)	116	condition for L 13.1–T 13
L 13.1	116	B_1 freely QMT toward M seen locally with nominal
L 13.2	116	width $R_B M / M_B r \ll 1$, B_1 freely QMT toward M locally shorter by $1/q(r) \approx 1 - R_B M / M_B r$
L 13.3	117	$R_B M / M_B r \ll 1$, clocks near M slower by $q(r) \approx 1 + R_B M / M_B r$
L 13.4	117	$R_B M / M_B r \ll 1$, $\Delta E \approx (m_1 c^2 R_B M / M_B)(1/r_2 - 1/r_1)$
T 13	118	$R_B M / M_B r \ll 1$, gravity about M, $G \approx c^2 R_B / M_B$
C 13.1	119	strong field $q(r) \approx \dfrac{1 \pm (1 + 4R_B M / M_B r)^{1/2}}{2}$
C 13.2	119	released $M_r(r) \approx 2M\left(1 - \dfrac{2}{1 + \left(1 + \dfrac{4R_B M}{M_B r}\right)^{1/2}}\right)$
C 13.3	121	nearness to mass does not isolate from information exchange

	page	
C 13.4	121	T 13 $q(r)$ effect due to M at t if observable as locally propagated to $\approx r$ observable there at $t' \approx t + r/c$
C 13.5	123	correspondence: gravity as interaction of time–like I flows in position representation
	124	statement of condition for T 14–C 14.16
T 14	124	for every form of photon interaction, lower bound on energy to form bound B_0 in B
C 14.1	125	bound B_0 in B; lowest momentum state $v \neq 0$, same as for QM mass particle
C 14.2	125	$\sim E(N)$
C 14.3	127	upper bound for least m_0 for B_0 in B bound by Theorem 13 interaction, N bits internal $m_0\left(32\pi^2(\ln(R_B 2m_0c/\hbar))^2\right)^{1/3} \approx \left(\frac{\hbar^2 M_B Y(N)}{c^2 R_B^2}\right)^{1/3}$
C 14.4	131	$1/4$ cycle lead and lag
C 14.5	131	with 1 bit of N bits represented internally in B_0 by $\pm$ relative phase $\pi/2$, upper bound as in C 14.3 becomes $m_0\left(32\pi^2(\ln(R_B 2m_0c/\hbar))^2\right)^{1/3} \approx \left(\frac{\hbar^2 M_B Y(N)}{c^2 R_B^2\left(1+\frac{1}{N^2}\right)}\right)^{1/3}$
C 14.6	133	two relative polarizations
C 14.7	133	$N \geqslant 3$ bits
C 14.8	134	$F \approx (32\pi^2(\ln(R_B 2m_0c/\hbar))^2)^{1/3}$
C 14.9	136	likes repel, unlikes attract
C 14.10	137	assembly of like B_0s at energy $\ll mc^2$, Fermi–Dirac statistics
C 14.11	139	unlike equal mass B_0s; annihilation
C 14.12	141	B_0 coding complementarity
C 14.13	141	$1/F^{1/2}$ localized amplitude, phase $\pm 1/4$ cycle
C 14.14	141	for $r \gg \hbar/m_0c$, $\Delta E(r) \approx \pm \hbar c/Fr$
C 14.15	144	collected properties of T 14–C 14.14 3 bit T 13 B_0s in B
C 14.16	152	electron and positron correspondence with least mass T 13 P_0, $\alpha^{-3}\exp(\alpha^{-3}/32\pi^2)^{1/2} \approx 7.2\hbar c/m_e^2 G$
C1, C2, C3	157	conditions used in formal statements in T 15–C 15.11

	page	
T 15	157	principle of equivalence, B_0 within S linear size r if $E_{Bj} \geqslant \hbar c/r$ and $E - 2E_{Bj} \geqslant \hbar c/r$; for fast B_0 and $E_0 \gg \hbar c/r$, if $E \leqslant \approx E_0^2 r/4\hbar c$
C 15.1	158	M at distance R, 'principle of equivalence' substantially irrelevant if $E_{Bj} \ll \hbar c/R$; for fast B_0, if $E \gg E_0^2 R/4\hbar c$
C 15.2	159	fB_0 does not have any position information in B
C 15.3	160	B_0 and M substantially unattracted by T 13 interaction if $E_{Bj} \ll \hbar c/R$; for fast B_0, if $E \gg E_0^2 R/4\hbar c$
C 15.4	160	fB_0 and M do not attract each other by T 13 interaction
C 15.5	161	$F_f \approx \dfrac{1}{E_0}\left(\dfrac{M_B Y(N)\hbar^2 c^4}{R_B^2}\right)^{1/3}$
C 15.6	162	$F_f \approx \dfrac{1}{2\sqrt{EE_{Bj}}}\left(\dfrac{M_B Y(N)\hbar^2 c^4}{R_B^2}\right)^{1/3}$
C 15.7	163	$\lvert a_f\rvert = \lvert a_b\rvert \left(\dfrac{F_b}{F_f}\right)^{1/2} \approx \lvert a_b\rvert \sqrt{2}\, E^{1/4} \dfrac{E_{Bj}^{1/4}}{m_b^{1/2} c}\left(\dfrac{0.9Y(3)}{Y(N)}\right)^{1/6}$
C 15.8	164	$m_p > E_0/c^2 < m_b$; $N \geqslant 2$ bits
C 15.9	165	fB_0; $N = 2$ bits
C 15.10	165	$m_p > m_0 < m_b$, $N \geqslant 2$,

$$\frac{E}{1 + \dfrac{4E_{B1}E}{m_b^2 c^4}\left|\dfrac{a_b}{a_f}\right|^4 \left(\dfrac{0.9\,Y(3)}{Y(N)}\right)^{2/3}} \approx E' \; [\lesssim]$$

$$\left[\lesssim \frac{E}{1 + \dfrac{48\pi E_{B1}^3 E^3}{m_b^6 c^{12}}\left|\dfrac{a_b}{a_f}\right|^{12} \left(\dfrac{0.9\,Y(3)}{Y(N)}\right)^{2}}\right]$$

C 15.11	168	T 13 and C 14.15 3 bit B_0 with 2 bit fast B_0
C 15.12	172	correspondence between fast B_0s and neutrinos
T 16	173	correspondence, inertia in terms of information

Definitions

1	37	$I = \log_2 J$
2	38	complementary representations and codes

	page	
3	42	internal information of B_1
4	42	external information of B_1 in B_2
5	43	soft box
6	43	B_0
7	48	NRPPR
8	49	attribute energy to B_1 or P in B_2
9	49	attribute information to B_1 or P in B_2
10	54	gentle acceleration (GA)
11	61	gentle scaling (GS)
12	63	gentle conversion (GC)
13	82	fast B_0
14	83	fastest B_0 (fB_0)
15	99	classical transcription (CT)
16	99	quantum mechanical transcription (QMT)
17	101	localization = (total information)/(internal information)
18	126	$Y(N)=\Sigma_{i=1}^{N} E_{Bi}/E_{B1}$
19	156	photicle
20	163	charge magnitude$=(\hbar c/F)^{1/2}$

Principles

1	68	Information proportional to energy
2	68	PODIC
3	71	changing observer's external state without changing internal state iff laws and constants of physics preserved
4	107	if observer can not have internal history that would tell it its length unit is changed, its report of local length measurement is honored

APPEARING IN SECTION 3

Postulates

	page	
1	179	information is conserved
2	179	information is communicable
3	179	information is finitely accessible

'A-' Prefix Conditions ('C(*integer*)'), Lemmas ('L(<17).(*integer*)') Theorems ('T'), and Corollaries ('C(*integer*).(*integer*)')

	page	
A-T 1	179	4-space and equations for (local propagative) communication
A-C 1.1	182	observed event 4-space as in A-T 1
A-C 1.2	182	distinct homogeneous apparently unidirectional x_0 and isotropic homogeneous 3-space
A-C 1.3	185	A-C 1.2 in terms of A-Def 7 of space, time, and space-time: O's homogeneous apparently unidirectional time and homogeneous isotropic space
A-C 1.4	186	R_U of U; as scale calibration
A-C 1.5	187	finite maximum spatial period for solution(s) of WE with amplitude >0 ('λ_1')
A-L 2.1	188	$A'_{z,k}(lI_z)=l^{1/2}A_z(I_z)$
A-T 2	191	N_1 GA1, N_2 GS1; amount and representation of internal and external information separable and separately conserved if $\Sigma_{i=1}^{N_1+N_2}s_i^2<1$, s_i taken before each step ('adseparability')
A-C 2.1	194	some B_0 properties in domain of adseparability
A-C 2.2	202	adseparability for $v<v_{\max}$
A-T 3	206	all $w_{j,z}$ possible in U have $\lambda_j=\lambda_1/j$, j finite integer >0
A-T 4	207	$I_j=j$ bit(s), j as in A-T 3
A-C 4.1	210	consequences of an 'adequate' description are not observably, representably, and/or communicably different from those of an 'exact' description
A-C 4.2	211	A-D 12 and 13, and A-T 2–A-C 4.1 inclusive, are consistent with respect to B_0s
A-C 4.3	211	$I(f_j)=(\lambda_1/c)f_j=I_j=j$
A-C 4.4	213	co-possible modes, A_y proportional to $1/I_y^{1/2}$
A-C 4.5	213	$I=\Sigma_j l_j j$ represented by ws
A-C 4.6	213	$I=\Sigma_z l_z(n_{+,z}+n_{-,z})$ represented by B_0s
C1	214	$w_{j,z}$ orthonormal, for B_0s A-D 12 used, for A-C 4.7–4.8
A-C 4.7	214	$W=\Sigma_{j,z}l_{j,z}^{1/2}j^{1/2}w_{j,z}$ representing $I=\lvert W\rvert^2=\Sigma_{j,z}l_{j,z}\,j$ bit(s)
A-C 4.8	216	$\psi=\Sigma_z l_z^{1/2}(n_{+_z}^{1/2}+in_{-_z}^{1/2})w_{n_{+z},z}w_{n_{-z},z}$ representing $I=\lvert\psi\rvert^2=\Sigma_z l_z(n_{+_z}+n_{-_z})$ bits
A-T 5	217	representations for adseparable I_0 and I_1
A-C 5.1	218	describing B_0 with I_0 bits of internal information by 2^{I_0} ψs as in A-C 4.8

	page	
A-T6	220	adseparability of system S in B_1 with scale s in U iff $v < c\dfrac{1-s^2}{1+s^2}$
A-C6.1	225	adseparable system S's existence bit not accessible in S
A-T7	226	complementary representations represent equal amounts of information
A-C7.1	226	complementary representations iff even number of bits
A-C7.2	226	irad S fully transcribable alone iff even number of bits
A-C7.3	227	adseparable S, information accessible in S, or if S required to exist, fully transcribable alone iff I_S as measurable by O_U is odd number of bits
A-T8	229	sufficiency of Postulates 1, 2, and 3, with lists of substituted referents for merging Sections 2 and 3

A-Definitions

1	179	$I = \log_2$(number of possibilities) in bits
2	179	'adequately', 'irad'
3	181	'w', 'w_j', 'w_z', '$w_{j,k}$', '$w_{j,z}$', '$w_{z,k}$', $k = 0, 1$; 'WE'
4	181	'observer'
5	182	'observation'
6	182	'universe' ('U')
7	185	'time', 'space', 'spacetime'
8	186	globally flat coordinates as in A-D7
9	187	R_U linear size of U
10	187	'scale' $s = R/R_U$
11	188	λ_1
	189	usual frequency, velocity, acceleration, defined in terms of A-D7 and 8
12	189	'B_0', 'w_+', 'w_-', 'λ_+', 'λ_-', 'λ_0', 'λ_b', 'f_+', 'f_-', 'f_0', 'f_d', 'velocity of B_0', O', scale s_0 of B_0, operations on B_0
13	190	'transitions among allowed states of B_0'
14	190	'GA1', 'step'
15	190	'GS1', 'step'
16	190	internal and external information of B_1 containing S_1 in B_2
17	193	'adseparability'
18	212	$R_U = \lambda_1/2\pi$
19	213	'occupation number' ('l')

	page	
20	226	'complementary' (representations and codes)
21	227	for I in bits: (a) 'energy' $E=(\hbar c/R_U)I$; (b) 'mass' $m=(\hbar/R_U c)I$; (c) 'momentum' $p=(\hbar/R_U c)vI$, $v \leqslant c$
22	228	$E_1=\hbar c/R_U$: $E=E_1 I$; $m=(E_1/c^2)I$; $p=(E_1/c^2)vI$, $v \leqslant c$
of 'finite'	231	for other than Postulate 3, 'finite' may occasionally be taken as meaning 'consistent with amount of accessible information'.

In Chain 17 and its Links ('L 17.(*integer*)')

Conditions for Chain 17	233	definitions of terms for Chain 17
L 17.1	234	w (photon) states and multiplicities used in Chain 17
L 17.2	235	photon, $I_x=1$
L 17.3	236	$\nu_e, \bar{\nu}_e$, likely $I_x=2$
L 17.4	238	$e^{\pm}$, likely $I_x=3$
L 17.5	240	$\mu^{\pm}_{L,H}$, likely $I_x=7$
L 17.6	244	$\nu_{\mu(L,H)}, \bar{\nu}_{\mu(L,H)}$, likely $I_x=6$
L 17.7	246	$\pi^{\pm}_{L,H}$, likely $I_x=13$
L 17.8	256	likely if $\sim$C3, I_x, b,d,f and $I_x<12$, much more rapid decay than $\pi^{\pm}_{L,H}$
L 17.9	256	π^0, likely $I_x=12$
L 17.10	258	L 17.10 interaction, subsumption of I_x by T 13 interaction with y
L 17.11	263	likely least I_y'' for L 17.10 interaction with various x
CL 17.12	294	correspondence link, neutrinos and 'weak interaction'
CL 17.13	296	correspondence link, 'strong interaction'
CL 17.14	296	apparent substantial compatibility, to extent tested as reported herein before CL 17.14, with knowledge prior to the time of this writing.

Chain 17 Related Definitions

21	254	various 'particle' notations
22	255	$\delta 3,2,I$

In Appendix 1, Conjectures

1	350	fast B_0 emission or absorption independent of nearness
1.1	351	processes involving making or 'absorbing' fast B_0 appear locally as affected by T 13 $q(r)$ effects

page

2 351 Emission or absorption of fast B_0 in U is subject to Theorem 13 $q(r)$ effect(s)

2.1 352 fast B_0 emitted vertically in 'gravity'; if

$$E-E' \gg \left(\frac{\hbar g E}{c}\right)^{1/2} \ll E''-E$$

and

$$\frac{E_0^2}{4E} \ll \frac{\hbar c}{R},$$

fast B_0 confined to vertical range with characteristic half height

$$|d| \sim \left(\frac{\hbar c^3}{gE}\right)^{1/2}$$

2.2 354 particle P moving vertically in 'gravity'; graininess if

$$\left[\left|\frac{a_f}{a_b}\right| \leqslant \sim \frac{(48\pi)^{1/12} E_1^{1/4} E^{7/24}}{(g\hbar)^{1/24} m_b^{1/2} c^{23/24}} \left(\frac{0.9\,Y(3)}{Y(I)}\right)^{1/6} \quad \text{(NRQM spectrum)}\right]$$

$$\left|\frac{a_f}{a_b}\right| \leqslant \sim \frac{\sqrt{2}\, E_1^{1/4} E^{3/8}}{(g\hbar)^{1/8} m_b^{1/2} c^{7/8}} \left(\frac{0.9\,Y(3)}{Y(I)}\right)^{1/6} = \mathcal{X}$$

[(Section 3 spectrum $E_j = jE_1$)]; not substantially grainy if $\left|\frac{a_f}{a_b}\right| \gg \mathcal{X}$; and I is a finite integer >1

Abbreviations

[ ] used to enclose portions of text and expressions relating to spectra other than $E_j = jE_1$ derived in Section 3

$\overset{(-)}{x}$ used to indicate that both entity x and anti-entity $\bar{x}$ are designated

A above relational symbol, denotes 'is adequately describable as', (see A-Def 2, p. 179, and A-Cor 4.1, p. 210, with discussion following)

A-	prefix distinguishing formal statements in 'axiomatic' form in Section 3
adequately	has special meaning herein, see A-Def 2, p. 179, and A-Cor 4.1, p. 210
BCS	Bardeen, Cooper, and Schrieffer (ref. 7)
B_0	see Def 6, p. 43, and A-Def 12, p. 189
box	in some places, used with general concept of containment and/or belonging, as in and/or to a 'system', with respect to one (or more) criterion(s)
CD	Cohen and Du Mond, ref. 6
CT	Cohen and Taylor, ref. 4; classical transcription, Def 15, p. 99
drtas	deemed reference to as sufficient
EMT	electromagnetic theory
exspace	external (observation) space as in Comment, 4.5, p. 305
fB_0	fastest B_0, Def 14, p. 83
GA	gentle acceleration, Def 10, p. 54
GA1	gentle acceleration for system containing 1 w or B_0 as in A-Def 14, p. 190
GC	gentle conversion, Def 12, p. 63
GS	gentle scaling, Def 11, p. 61
GS1	gentle scaling for system containing 1 w or B_0 as in A-Def 15, p. 190
hB_0	hypothetical B_0
iff	if and only if
inspace	introspection space as in Comment 4.5, p. 305
irad	(an) information representation adequately describable (as)(by)(in such way), A-Def 2, p. 179
likely	it seems to me reasonably likely that, p. 233
NRA	nonrelativistic approximation
NRCM	nonrelativistic classical mechanics
NRPPR	nonrelativistic photon particle representation
NRQM	nonrelativistic quantum mechanics
oie	or its equivalent
PODIC	principle of detailed information conservation
PPR	photon particle representation
QED	quantum electrodynamics
QM	quantum mechanics
QMT	quantum mechanical transcription, Def 16, p. 99
sasd	simplest approximate substantial description
SR	(Einstein's) special theory of relativity
SRCM	special relativistic classical mechanics
ti	inspace 'time'

TPL	Taylor, Parker, and Langenberg, ref. 5
UCLBL	University of California Lawrence Berkeley Laboratories, ref. 8
w	'wave', solution of WE (see next entry)
WE	'wave equations' as in A-Theorem 1, p. 179, so designated in A-Def 3, p. 181
xi	inspace coordinate other than *ti*

Appendix Three

Map

Interpretation p. 2

Start of Nonaxiomatic Form

↓	page →		page →		page →		page →		page
D1	x 37	C4.2	x 47	D10	x 54	L8.11	x 63	Post 1	179
D2	x 38	C4.3	x 47	C7.1	x 54	D12	x 63	Post 2	179
T1	x 39	D7	x 48	condtn	x 55	T8	x 63	Post 3	179
C1.1	x 40	T5	x 48	L8.1	x 55	C8.1	x 65	A-D2	179
C1.2	x 41	D8	x 49	L8.2	x 57	C8.2	x 66	A-T1	179
T2	x 41	D9	x 49	L8.3	x 58	C8.3	x 66	A-D3	181
D3	x 42	C5.1	x 49	L8.4	x 59	C8.4	x 67	A-D4	181
D4	x 42	C5.2	x 49	L8.5	x 59	ref to		A-D5	182
T3	x 42	C5.3	x 50	L8.6	x 60	C8.4	x 67	A-C1.1	182
C3.1	x 43	T6	x 50	L8.7	x 61	Pr1	x 68	A-D6	182
D5	x 43	C6.1	x 51	D11	x 61	Pr2 (PODIC)		A-C1.2	182
D6	x 43	C6.2	x 51	C1	x 61		x 68	A-D7	185
L4.1	x 44	C6.3	x 51	L8.8	x 61	(GO TO 'MERGE')		A-C1.3	185
T4	x 45	C6.4	x 51	L8.9	x 62	Start of Axiomatic Form		A-D8	186
C4.1	x 46	T7	x 53	L8.10	x 63	A-D1	179	A-C1.4	186

Start of Nonaxiomatic Form

↓	page →		page →		page →		page →		page
A-D9	187	A-D22	228	D17	101	C14.8	134	Lnk 17.6	244
A-D10	187	A-T8	229	C10.1	101	C14.9	136	Lnk 17.7	245
A-C1.5	187	Merge of Non-		C10.2	102	C14.10	137	D21	254
A-D11	188	axiomatic and		C10.3	103	C14.11	139	D22	255
A-L2.1	188	Axiomatic		C10.4	105	C14.12	141	Lnk 17.8	256
f, *v*, acc.	189	Forms Here	68	T11	105	C14.13	141	Lnk 17.9	256
A-D12	189	exmpl 1	69	C11.1	106	C14.14	141	L17.10	258
A-D13	190	exmpl 2	69	C11.2	106	C14.15	144	L17.11	263
A-D14	190	exmpl 3	69	condtn	106	Tbl 1	150	Tbl 3	268
A-D15	190	exmpl 4	69	Pr 4	107	CC14.16	152	Tbl 4	285
A-D16	190	exmpl 5	x 70	C11.3	107	Tbl 2	155	Tbl 5	289
A-T2	191	Pr 3	71	C11.4	108	D19	156	CL17.12	294
A-D17	193	T4′	x 72	C11.5	108	C1	157	CL17.13	296
A-C2.1	194	C4.4	x 74	C2	112	C2	157	CL17.14	296
A-C2.2	202	C4.5	x 77	T12	112	C3	157	Break of Main	
A-T3	206	C8.5	x 78	C12.1	114	T15	157	Development	
A-T4	207	D13	82	C12.2	114	C15.1	158	Tbl 6	322
A-C4.1	210	C1	82	condtn	116	C15.2	159	Through	325
A-C4.2	211	C2	82	L13.1	116	C15.3	160	Conj 1	350
A-C4.3	211	C8.6	82	L13.2	116	C15.4	160	Conj 1.1	351
A-D18	212	C8.7	83	L13.3	117	C15.5	161	Conj 2	351
A-C4.4	213	D14	83	L13.4	117	C15.6	162	Conj 2.1	352
A-D19	213	C8.8	84	T13	118	D20	163	Conj 2.2	354
A-C4.5	213	C8.9	84	C13.1	119	C15.7	163	End of Map so	
A-C4.6	213	C8.10	84	C13.2	119	C15.8	164	far on 1977	
C1	214	C8.11	85	C13.3	121	C15.9	165	June 6	
A-C4.7	214	C8.12	85	C13.4	121	C15.10	165		
A-C4.8	216	C8.13	86	CC13.5	123	C15.11	168		
A-T5	217	C8.14	87	condtn	124	CC15.12	172		
A-C5.1	218	C8.15	87	T14	124	CT16	173		
A-T6	220	C8.16	90	C14.1	125	Chain 17	233		
A-C6.1	225	T9	97	C14.2	125	condtns for			
A-D20	226	C9.1	98	D18	126	Chain 17	233		
A-T7	226	D15	99	C14.3	127	Lnk 17.1	234		
A-C7.1	226	D16	99	C14.4	131	Lnk 17.2	235		
A-C7.2	226	condtn	100	C14.5	131	Lnk 17.3	236		
A-C7.3	227	C1	100	C14.6	133	Lnk 17.4	238		
A-D21	227	T10	100	C14.7	133	Lnk 17.5	240		

'x' denotes items not included in axiomatic form

REFERENCES AND NOTE

Following the '/' after each reference or note is a list of pages on which it is referenced.

1. See, for example, Albert Einstein, 'Die Grundlage der Allgemeinen Relativitatstheorie', *Annalen der Physik*, 49, 1916/118,.
2. Allan R. Sandage, Cosmology: A search for two numbers, *Physics Today*, **23** (2) Feb. 1970, p. 34/15, 123, 145, 151.
3. Estimates of density of luminous matter, as by Oort, depend on H, and would seem to lie in the range of roughly 10^{-31} to 10^{-30} gm/cm^3, for H of $\sim$75 km/sec·megalightyear. However, other contributions, for example dark matter, radiation, cosmic rays, etc., vary differently with H. I have not examined each source of estimated mass to determine this dependence. I am not, at the time of this writing, aware of any report of a direct accurate measurement of the average density of mass (energy) in our universe/15, 123.
4. E. Richard Cohen and B. N. Taylor, The 1973 least-squares adjustment of the fundamental constants *J. Phys. Chem. Ref. Data* **2** (4), 663–734 (1973)/18, 145, 147, 149, 151, 155, 231, 265, 271, 290, 291, 331, 336.
5. B. N. Taylor, W. H. Parker, and D. N. Langenberg, Determination of e/h, QED, and the fundamental constants, *Rev. Mod. Phys.* **41**, 375 (1969)/18, 148, 149.
6. E. R. Cohen and J. W. M. Du Mond, Our knowledge of the fundamental constants of physics and chemistry in 1965, *Rev. Mod. Phys.* **37**, 537–594 (1965)/18, 149, 291.
7. J. Bardeen, L. N. Cooper, and J. R. Schrieffer, Theory of superconductivity, *Phys. Rev.* **108**, 1175–1204 (1957)/151.
8. N. Barash-Schmidt, A. Barbaro-Galtieri, C. Bricman, V. Chaloupka, R. J. Hemingway, R. L. Kelly, M. J. Losty, A. Rittenberg, M. Roos, A. H. Rosenfeld, T. G. Trippe, G. P. Yost, and B. Armstrong, Review of particle properties, *Rev. Mod. Phys.* **48** (2) Part II, 21, 23, 60–63 (1976) ('UCLBL 1976')/172, 233, 265, 267, 280, 281, 283, 284, 285, 288, 289, 290, 292, 293, 295, 344, 345.
9. *Handbook of Chemistry and Physics*, 52nd edition, Chemical Rubber Co., Cleveland, Ohio, 1971/336, 361, 362, 363.
10. Robert E. Shafer, Pion-mass measurement by crystal diffraction of mesonic X rays, *Phys. Rev.* **163**, 1451–1461 (1967)/267, 272, 281.
11. G. Backenstoss, Pionic atoms, *Annu. Rev. Nucl. Sci.* **20**, 467–508 (pp. 479–484) (1970)/267, 272, 282.
12. G. Backenstoss, H. Daniel, H. Koch, Ch. Von der Malsburg, G. Poelz, H. P. Povel, H. Schmitt, K. Springer, and L. Tauscher, New determination of the π^- and ν_μ masses, *Phys. Lett.* **36B**, 403–408 (1971)/267, 272, 282.
13. G. Backenstoss, H. Daniel, H. Koch, Ch. Von der Malsburg, G. Poelz, H. P. Povel, H. Schmitt, and L. Tauscher, π^--mass and vacuum polarization. A re-evaluation, *Phys. Lett.* **43B**, 539–541 (1973)/267, 272, 282.
14. Walter H. Barkas, Wallace Birnbaum, and Frances M. Smith, Mass-ratio method applied to the measurement of L-meson masses and the energy balance in pion decay, *Phys. Rev.* **101**, 778–795 (1956)/267, 272, 276, 277.
15. Kenneth M. Crowe and Robert H. Phillips, Precision measurement of the negative pion mass from its radiative absorption in hydrogen, *Phys. Rev.* **96**, 470–483 (1954)/267, 271, 272, 276.

16. Roy P. Haddock, Alexander Abashian, Kenneth M. Crowe, and John B. Czirr, Pion mass measurements using neutron time-of-flight techniques, *Phys. Rev. Lett.* **3**, 478–480 (1959)/272, 276, 283, 285, 287.

17. John B. Czirr, Determination of $\pi-$meson masses by neutron time of flight, *Phys. Rev.* **130**, 341–350 (1963)/271, 272, 275, 277, 279, 281, 285, 286.

18. Wolfgang K. H. Panofsky, R. Lee Aamodt, and James Hadley, The gamma-ray spectrum resulting from capture of negative π-mesons in hydrogen and deuterium, *Phys. Rev.* **81**, 565–574 (1951)/272, 276.

19. William Chinowsky and Jack Steinberger, The mass difference of neutral and negative π mesons*, *Phys. Rev.* **93**, 586–589 (1954)/272, 276.

20. M. Cini, R. Gatto, E. L. Goldwasser, and M. Ruderman, A note on the apparent inconsistency among measurements of threshold pion interactions, *Nuovo Cimento* **X**, 243–258 (1958)/272, 277.

21. J. M. Cassels, D. P. Jones, P. G. Murphy, and P. L. O'Neill, Neutral pions, *Proc. Phys. Soc. (Lond.)* **74**, 92–96 (1959)/272, 277.

22. M. Gettner, L. Holloway, D. Kraus, K. Lande, E. Leboy, and W. Selove, Measurement of the π^0 mass, and search for $A\pi_0{}^0$, *Phys. Rev. Lett.* **2**, 471–473 (1959)/272, 277, 287.

23. P. Hillman, W. C. Middelkoop, T. Yamagata, and E. Zavattini, The $\pi^- - \pi^0$ mass difference, *Nuovo Cimento* **XIV**, 885–894 (1959)/272, 277, 288.

24. N. P. Samios, $\pi^0 - \pi$ mass difference as determined from double Dalitz pairs, *Nuovo Cimento* **XVIII**, 154–159 (1960)/272, 277.

25. W. Selove and M. Gettner, Search for a second π^0, *Phys. Rev.* **120**, 593–598 (1960)/272, 277, 288.

26. I. M. Vasilevsky, V. V. Vishnyakov, A. F. Dunaitsev, Yu. D. Prokoshkin, V. I. Rykalin, and A. A. Tyapkin, (in Russian), *12th Int. Conf. High Energy Physics*, Vol. I, 883–884 (1964)/272, 279, 280.

27. I. M. Valilevsky, V. V. Vishnyakov, A. F. Dunaitsev, Yu. D. Prokoshkin, V. I. Rykalin, and A. A. Tyapkin, Mass difference on the negative and neutral pions, *Phys. Lett.* **23**, 281–282 (1966)/272, 279, 280, 281, 285, 286.

28. L. G. Hyman, J. Loken, E. G. Pewitt, M. Derrick, T. Fields, J. McKenzie, I. T. Wang, J. Fetkovich, and G. Keyes, Mass measurements using the range-energy relation in a helium bubble chamber, *Phys. Lett.* **25B**, 376–380 (1967)/272, 281.

29. P. S. L. Booth, R. G. Johnson, E. G. H. Williams, and J. R. Wormald, Measurement of momentum of muons from pion decay upper limit for muon neutrino mass, *Phys. Lett.* **26B**, 39–40 (1967)/272, 281.

30. P. S. L. Booth, R. G. Johnson, E. G. H. Williams, and J. R. Wormald, Mass of the positive pion, *Phys. Lett.* **32B** 723–724 (1970)/272, 282.

31. Halton Arp reviewed some related experimental data in his article, Observational paradoxes in extragalactic astronomy, *Science* **174**, 1189 (1971)/321.

32. Fred Hoyle, A solar model with low neutrino emission, *Astrophys. J.* **197**, L127–L131, 1975 May 1/341.

33. J. N. Bahcall and R. Davis, Jr., Solar neutrinos: A scientific puzzle, *Science* **191**, 264 (1976)/358, 367.

34. Edwin H. Salpeter, colloquium at Columbia University, New York, N.Y. room 428 Pupin Physical Laboratories, 1976 November 17, ~16:30/344.

35. M. Cullen, P. Darriulat, J. Deutsch, H. Foeth, C. Grosso, M. Holder, K. Kleinknecht, E. Radermacher, C. Rubbia, D. Shambroom, M. Sciré, A. Staude, and K. Tittel, A precision

determination of the K_L-K_S mass difference, *Phys. Lett.* **32B** (6) 523–527, 1970 August 17/345.

36. C. Geweniger, S. Gjesdal, G. Presser, P. Steffen, J. Steinberger, F. Vannucci, H. Wahl, F. Eisele, V. Lüth, G. Zech, K. Kleinknecht, Measurement of the kaon mass difference m_L-m_S by the two regenerator method, *Phys. Lett.* **52B** (1) 108–112, 1974 September 16/345.
37. S. Gjesdal, G. Presser, T. Kamae, P. Steffen, J. Steinberger, F. Vannucci, H. Wahl, F. Eisele, H. Filthuth, V. Lüth, G. Zech, K. Kleinknecht, A measurement of the K_L-K_S mass difference from the charge assymmetry in semi-leptonic kaon decays, *Phys. Lett.* **52B** (1) 113–118, 1974 September 16/345.
38. G. de Vaucouleurs, The velocity-distance relation and the Hubble constant for nearby groups of galaxies, in *International Astronomical Union Symposium No. 44*, David S. Evans, Ed., D. Reidel Publishing Co., Dordrecht, Holland, 1972, pp. 353–366/362.

Index

This is a partial alphabetical index. Words indexed herein have been selected partly in an attempt to match way(s) in which a person who has *not* read the main text might try to find various things. Please excuse any imprecision and/or informality in this index. For more nearly formal use, you might wish to use lists presented in Appendix 2, as I have.

Index material other than this index, with page numbers, may be found in:

Table of Contents, by headings;
Appendix 2, lists of statement numbers with reminders (I found these helpful), and a list of abbreviations;
Appendix 3, map of logical sequence of statements; and
References and Note, in which page(s) on which each entry is referenced is (are) listed after the entry.

Please note that material presented in different sections hereinbefore may have been offered as being of different kind and/or quality: in using this material, please note the descriptions, near the beginnings of Sections 1-6 and Appendix 1, of material appearing therein. You might find Appendices 2 and 3 helpful in tracing logical context of a result located using this index. (the contents of Section 6 ('Questions') are not fully indexed in this index; hereinafter 'q' is used occasionally to separate page and question numbers) ('&' denotes that the closest following subheading is of the same 'level' as the closest preceding subheading)

Acceleration, axiomatic definition, 189
  gentle, 7, 25
    Def 10, 54
    GA1 A-Def 14, 190
    effect on boundary conditions, (for $v << c$ in NRA), 54
Acknowledgements, viii
Adequate, 23, 27
  A-Def 2, 179
  description, 309–313
    and B_0s, 330q27, 333q49
    and cross-sections, 333q49
    good enough, 210, 211
      A-Cor 4.1, 210
    helpful for atoms and/or molecules, 347q79
    and rates, 333q49
  interpretation of (mathematical) statements, 211
Adseparability, 26, 188
  domain of applicability,
    for B_0 or specified w in box, 25, 26, 202
    for systems, 28, 217, 219, 220
      A-Theorem 6, 220
      discussion and examples, 224

Alternatives, number of observable, only one present, 231
Amplitudes squared, 177, 188
  A-Lemma 2.1, 188
Aristotle's (two valued) logic, 312
Atom, diffraction of (example 5), 219
Axiomatic form, 30, 230
  axiomatization, 22, 173, 174–297
  constants, 231
  entry in Section 2, 68
  partial test (*see also* Particles) 31, 232
  postulates (*see also* Postulates) 179, 230
    looking for, 174–179
    testing, 179–230
    sufficiency, 229

B_0(*see also* Particles; w), 5, 43
  algebraic expressions, some, 74, 77, 194
  allowed states, transitions among (A-Def 13), 190
  axiomatic, 25, 189
    A-Def 12, 189
  consistency of treatment (A-Cor 4.2), 211
  coupling in, 15, 16
    two photon, for representing internally N bits, 125, 126
  and 'elementary particles' (*see also* Particles), 15, 124
  domain of applicability of adseparability for (A-Cor 2.2), 202
    of PODIC for, 78–81
  energies allowed 86–89
    after formation, with constant E_0, 87
    on formation, 87
  fast, 8, 9, 82, 156
    calculated values, some approximate, 171, 172
    correspondence with neutrinos (*see also* Particles, neutrino and antineutrio), 21, 172
    charge magnitude (*see also* B_0, fast, interaction strength), 20, 163
    energies, allowed, 20, 87
      in relation to charge, 20, 165
    energy in 'rest frame', 85
    fastest, 8, 19, 83
      calculated values, some approximate, 171, 172
      energies (momenta), allowed, *see* B_0, fast, energies, allowed
      internal information of unbound, 165
      lifetime, 8, 9, 85
        lag behind photon in, 86
      position information, not having in $U(B)$, 159
      production of 2 bit, with 3 bit B_0, 20, 21, 168, 172, 236
    in gravity, conjectures on, 33, 350–367
      if emission and absorption seen by local observers as affected by Theorem 13 effects (Conj. 1-1.1), 350–351
      if emission and absorption subject to Theorem 13 effects (Conj. 2), 351
        trapping (Conj. 2.1), 352
          consideration of tests and conjectured consequences, 358–367
          dependence on interaction strength and 'gravity' strength, 354
          numerical estimates, approximate, for fast B_0, 358 shielding by, 331q32
          and solar neutrinos, 358
      'principle of equivalence,' as applicable, 157
        as not applicable, 158
    interaction strength (*see also* B_0, fast, charge magnitude), 20, 161–164
      as charge, 163
    internal information of, unbound, 164
    lifetime of unbound, 8, 85
    numerical and other results, some approximate, 21
      also for fastest B_0, 171, 172
    two bit, production with 3 bit B_0, 20, 21, 168, 172, 236
    unattraction, by Theorem 13 interaction with mass (*see also* B_0, fast, in gravity, conjectures) 160, 161
    velocity of, 84
  mass(es), least, 124
  'mechanics' of, in domain of adseparability, 194, 227–229
    A-Cor 2.1, 194
    A-Def 21 and 22, 227, 228
  momentum, lowest, 125
  operations on (A-Def 12), 189, 190
  states in box, 44, 72
  velocity of (A-Def 12), 189, 190
    maximum (*see also* B_0, fast (fastest)), 78, 83, 202

Ball, 'soft', dielectric, 329q18
Bit, 36, 311, 312
Boundary conditions, 212, 347q72
for U (or B) (*see also* Universe, boundary conditions), 104, 105, 212, 313, 332q47
Brackets, square, use of herein, 4
Break of development, 297, 326

Cautionary notes and warnings, vii, 1, 65, 83, 85, 86, 89, 113, 114, 130, 236, 270, 347q78
Chain 17 (*see also* Axiomatic form, Partial test; L 17.10 interaction; particles; 'Weak interaction'), 233
conditions for, 233, 234
Channel capacity, 36
Charge (*see also* B_0, fast, charge magnitude & interaction strength; Particle, electron and positron, charge)
magnitude, more nearly general definition of, 163
sign, 16, 17, 131, 136
coding, conservation of, and particle stability, 258
strength (*see also* Charge magnitude)
conservation or not, of, 258
electromagnetic interaction, different particles, 151, 331q38, 348q80
prediction as a possibility (14), 320
Code(s), coding, 2, 29, 36, 226,
'*b*', 234, 236 (Step 4) 331q37
'book', 2, 38
charge sign, *see* Charge, sign
complementary, 2, 38
A-Def 20, 226
'*d*', 234, 330q23
and parity, 227, 236, 246, 293, 294, 329q20
polarization, *see* Particle, photon, polorization
'spin ½', *see* Particle, electron and position, 'spin ½'
Coherence, 52, 125, 215–217
Comments (*see also separate entries this index*), 33, 298
Communication, describing, 23, 24, 177–181
if local propagative by gravity, 121
local propagative, and polarization, 177–181, 347q71
in strong gravity, 121
time associated with, 181, 303, 332q45, 333q49
Comparison, of some particle approximate descriptions with prior reports of observations, 145–152, 170–172, 265–297
Table 1 (electron charge via 'fine structure constant'), 150
Table 3 (various), 268
to various prior theories, 227–229, 232, 233, 268, 296
of units (*see also* Acceleration, gentle; gravity, metric effects; units), 12, 107, 298–300
Compatibility with prior knowledge, 4, 21, 22, 32, 33, 105, 123, 152, 172, 173, 296, 297
Compilations and co-adjustments by others, 147–152, especially 147–149
Conclusion, 349
Conjectures on fast B_0 in gravity (*see also* B_0, fast, in gravity, conjectures), 350
Conservation of information, *see* Information, conservation & separability
Constants, 231, 232
postulates and, 314
various in U, with (approx.) expressions for, Table 2, 155
Construction, proof by, 6
Conventions, 3
Co-possible modes, 6, 52, 53, (A-Cor 4.4) 212, 213
Correspondence, apparent, 296, 297
Cosmic background electromagnetic radiation, 321, 332q40
and 'cosmological theories', 333–337q50
'Cosmological' red shift (*see also* Hubble's constant; transcription of information between position and momentum representations; Universe, radius, 11, 12, 102, 329q11, 332q47
'CPT' breakdown, 328q7

Date(s) hereof, viii, 270, 297, Appendix 3
Description with finite information, with examples, 309–311
Detector, classical mechanical, 36
quantum mechanical, 36
Detection (*see also* transportation), 38, 104, 105
classical, 99
quantum mechanical, 99

Dirac's equation, 326q1
spinors (examples 3, 4), 219
Distance, red shift, *see* 'cosmological' red shift; Hubble's constant; Universe, 'radius'

Einstein, Albert, *see* gravity; special relativity; 'Statistics', Bose-Einstein
Electromagnetic field (example 2), 219
Electromagnetic theory, quantized, 30, 227–229
Electron *see* Particle, electron and positron
Electrostatic interaction (*see also* Charge magnitude & strength; Particle, electron and positron, charge & F & 'fine structure constant'), of e^+e^- and/or p^+e^-, 151
'Elementary' particles *see* Particles
Elements observable present together, number of, 231
Energy, 98, 227, 228
carried by moving object, 10, 11, 39, 96, 98
conservation of (example 1), 69
'kinetic' *see* Energy carried by moving object
'potential' in gravity, 111
Entity, in 'closed' U, 346q67
possibility of reconstruction, 331q36
'Entropy', NOT reasoning as for various prior notions of, 2, 4, 42, 43, 338q51, 339q51, 343q51, 345q53, 54
example, 2, 4
'Equipartition', 52, 312, 313
Experiments, notes and/or suggestions for, 33, 321
related to galaxies, 343q51.7, 344q51.7
Table 6, of, 322–325

Finite accessiblity of information (*see also* Axiomatic form, Description with finite information; Information; Postulates), 175, 176, 179,
comment on, 312
'Finite', meaning of, in axiomatic form, 231
Free will, 306–307
Frequency, and amount of information exchangeable with w, 211
axiomatic definition of, 211

Galaxies and cosmic rays, 337–344q51
clouds in lump gas, 338q51.2
and cosmic rays, 337q51.1
comparable with (report(s) of) observation(s), some features which might be, 343q51.7
converter bodies, 340q51.3
'life cycle' of, 341q51.4
luminosity distribution in, 341q51.3
star formation and neutrino production, 341q51.3
lump gas, clouds in, 338q51.2
and cosmic rays, 337q51.1
composition of, 343q51.5
mass distribution in, 341q51.3
star formation and neutrino production, 341q51.3
statistics of, as test of various theories of 'age' of universe, 343q51.4, 344q51.7
transcribability of information, and processes related to, 343q51.6
General relativity, *see* gravity . . . ; Theorem 13 interaction
Gravity (Theorem 13 interaction herein), 11, 13, 105, 123
absence of, between fastest B_0 and mass, 19, 158–161
communication by, if local propagative, time associated with, 121
effect on (*see also* Gravity, metric effects of) G, h, $\bar{\alpha}$ m_e, etc.?, 331q34, F?, 347q76
'equivalence, principle of', 11, 110, 124, 157
breakdown of, 19, 121, 157, 158
not for fast B_0s, 158
field energy, not, 124
geometry in, 111
'graviton'?, 330q26
metric effects of localized mass energy, 13, 14, 105–123, 304
not isolating any regions in U, 121
prediction (2), 319
locally, 110, 111
'strong field', 14, 15, 118–121
not isolating any region, 121
two equal masses, choice of coordinates, 118, 119
'energy' released, 119
$q(r)$, 119
'weak field' near mass, 116–118
on clocks, 117
'energy' released, 117
gravitational 'constant' G, 14, 118
experiment, note for, 322

on length, parallel to radius, 116
perpendicular to radius, 116
neutrino and/or antineutrino, in, *see* B_0, fast, in gravity, conjectures
'potential' energy, 111
q(r) near localized mass energy, 114–116
shielding by, (*see also* B_0, fast, in gravity, Conjectures 2-2.2), 331q32
theories of, prior, Einstein's comment on, 304
Newton's, comment on, 303
treatment herein, comment on, 304

'Heat engine' with long range interacting 'working fluid'? 345q53, 54
History, 328q5
internal (*see also* Adseparability; Information, separability of internal and external), 298–303
How doing this has felt, comment on, 315
Hubble's constant (*see also* 'Cosmological' red shift; Transcription of information between position and momentum representations; Universe, 'radius', 12, 18, 19, 103–105, 122, 123, 313
approximate calculated value, 151–155
and electron and positron, 145
experiment, note for, 324
expressions for, 155, 232
prediction (4), 319

Impedances, parallel, 52
Inertia (*see also* Energy, carried by moving object; Particle, electron and positron, charge & *F* & 'spin ½' & 'statistics'; Position; Transportation), 10, 39, 96, 98, 172, 227–229, 304
Informal introduction, 1
Information, additivity of measure, 36
aspects of some, 311, 312
bit, 36
as measure of, 311, 312, 346q66
communicability (*see also* Communication)
in gravity (Theorem 13 interaction), 121
postulate 2 (*see also* Postulates), 177, 179
conservation, 4, 39
mass energy interconversion and, 5, 45, 68, 227, 228
postulate 1 (*see also* Postulates), 175, 179
Def 1, 37
A-Def 1, 179
energy proportional to, 6, 50, 51
A-Def 21 and 22, 227, 228
principle 1, 68
example of, 2
external, 25, 190
amount of, 13, 100, 101
A-Def 16, 190
representation of (*see also* Momentum; Position), examples 1–4, 69, 70
subsumption by nearness, 13, 112–114
effect on localization, 114–116
and inertia, 21, 22, 97, 98, 172, 173
transport of energy by moving object, 98
internal, 4, 25, 190
apparent amount of, inside and/or outside adseparable system (A-Cor 6.1), 225
A-Def 16, 190
'pure', 2, 38
representation, 35–38
adseparable, describing, (A-Theorem 5), 217
examples, 219
using ψ (A-Cor 5.1), 218
complementary, 4, 38, 40, 41, 226
describing by, relative phase, of B_0s, 216, 217
of *ws*, 215
set of B_0s, 213
of *ws*, 213
w s, 207–217
wave, 214
wave function ψ, 216
externally by mass particle (*see also* B_0 Information, external; Particle), 5, 44, 72
'space' 'norm'?, 332q46
separability of internal and external (*see also* Adseparability; PODIC; Principle 2; Special relativity), 7–10, 25, 26, 28, 29, 52, 63, 188, 191, 193, 219–225
breakdown of, 6, 7, 9, 10, 81, 90, 95, 219–225
comparison of derivations, velocity (Cor 8.16), acceleration (A-Theorem 6), 225, 298–303
in neutron decay, approx. calculated transcription rate from external to internal, 172

transcribability and amounts, 226, 227
transformation of conserved amount, describing, 218, 219
'Information mechanics', 349
Intelligence, 347q69
alien, reality seen by, 305
artificial, entity, reality seen by, 305
Interaction, *see* Electromagnetic theory, quantized; Gravity (Theorem 13 interaction herein); L 17.10 interaction; Particle, electron and positron, charge & electrostatic interaction neutrino and antineutrino; 'Spin'; 'Statistics'; 'Strong interaction'; 'Weak interaction'
Interpretation of, (mathematical) statements, 211
references to prior knowledge and/or theory(s), 2
Introspection, 305
postulates for axiomatic description of, (?) 305
Introspector, as an information representation (?), 347q74
test for, 347q73
'alive', 347q73
time, subjective, of, 305, 306, 332q43, 44
Irad, 23, 179, 210, 211

Kaon, *see* Particle, kaon
Knowledge, prior, interpretation of reference(s) to, 2

L 17.10 interaction, 258
'mixing' of internal information, 262, 263
rest information of some 'other' entities for, 263
'strong' interaction as, 296
comparison with prior reports of observations, 268, 294–296
Labeling, conventions for, 3, 4
Length and time units, comparisons, *see* Units, comparison(s) of
'Lepton number', 233, 234, 268, 294
Light, speed of, 181
special role (nonaxiomatic form), 48
'Likely', 233
Lists, 33, 34, Appendix 2
Localization, 101
contained observer, 106
detection of external information, effect on, 101
Location, 98

Magnetic moment, of electron and positron (and other(s)?), 348q80
and spin, 346q63, 64
Map of logical sequence, 33, 34, Appendix 3
Mass, A-Def 21 and 22, 227, 228
energy interconversion, *see* Information, conservation, mass energy interconversion
'rest', 98, 227, 228, 233, 234
Maxwell's equations, not, for $d << R_U$, 103
source free, 228
in vacuum, 23, 24, 179–181, 228
Mechanics, Newtonian, 118, 227–229
quantum, 227–229
relativistic, 227–229
'Mind' and 'body' (*see also* 'Soul'), 306
Momentum, 10, 11, 39, 96–98, 227, 228
conservation, 'breakdown' of (*see also* Transportation), 347q68, 72
as PODIC example 1, 69
interaction, 329q8
Motion, of material particle, 39
energy of, 39, 98
in region with spatial q dependence, 107–111
Moving systems, *see* Adseparability for systems; Information representation, separability of internal and external; Special relativity
Muon, *see* Particle, muon

Nearness, 11, 111
amount of information subsumed by, 111, 112
Neutrino, *see* B_0, fast; Particle, neutrino and antineutrino; 'Weak' interaction
Neutron, 329q17
decay, 172
'Noise', 36–38
Noncommutativity of observables, and inertia, 39
Notation, conventions for herein, 4
Notice, patent, v

Observation, 182, 304
Observer, 181, 328q5
absence of internal history of change, report(s) honored, 12, 107, 298–300
as an information representation, 347q74
subjective time (*see also* Time), 68, 90, 106, 182, 220, 227–229
as 'for ever', 332q43

Occupation number, 213
Order of mathematical operations, convention for, 3

'Parity' conservation and nonconservation (*see also* Particle, neutrino and antineutrino; 'Weak' interation), 227, 332q39
Particle (*see also* B_0; Electromagnetic theory, quantized; Gravity . . . ; (Information representation, L 17.10 interaction; Mechanics)
  and antiparticle, conversion between, 328q5, 346q57
    mass difference (*see also* Particle, kaon) and 'History' in *U*, 328q5, 346q58
  classes and masses, some, 254, 255
  coding of internal information (*see also* Coding), 233, 234, 329q13, 14
  conversion from or to antiparticle, 328q5, 346q57
  decay (*see also* Particle (name)),
    and interaction 'strength', 292–294
    and lifetimes, 327q2–4, 346q65
    spatial separation of products in, 346q65, 347q77
    transcription of external to internal information in neutron decay (?), 172
  electron and positron, 18, 19, 31, 126–156, 168–172, 238, 268, 290–292
    annihilation, mutual, 139
      and complementary coding, 141
    calculational results, approximate, collected, 144, 145, 150, 155, 238, 268
      coding, 238
      numerical, 145, 150, 155, 268
    cautionary notes, 130, 347q78
    charge (*see also* Charge; Particle, electron and positron, extended interaction)
      coding, 131, 136, 139, 141, 144, 238, 268
        in mass approx. calculation, 17, 131–133
        and stability (?), 258
      magnitude, 17, 134, 330q25
        approx. calculated value, 153–155
        and 'fine structure constant', 145–152, 268
        Table 1, various comparisons with prior compilations, 150
      sign, 131
    comparisons with prior reports of observations (*see also* Particle, neutrino and antineutrino, electron; Particle, muon, decay; Particle, pion, charged, decay; (Particle) Neutron, decay), 145–152, 268, 290–292
      and Hubble's constant, as input, 145
        obtained approximately (with other inputs), 146, 151, 155
        prediction (4), 319
    complementarity of internal codes, 141
      and conservation of charge sign, 'lepton number', and 'spin', 156
    correspondence, apparent, as entity representing internally 3 bits, 152
    description, 126–144, 238, 346q62
    electrostatic interaction (*see also* Particle, electron and positron, charge)., attraction and repulsion, 17, 136
    extended interaction, 16, 127–131
      charge coding, effect in, 131–133
      in other 'geometries', 330q25
    F, 17, 134, 141–143
      and 'fine structure' constant', 17, 18, 145–152
        comparison with prior reports of observations, 145–152, 268, 290–292
        Table 1 of *F* and α^{-1}, 150
        experiment, suggestion for, 322
    gravity due to and effect on, 155
      prediction (3), 319
    localized form (*see also* Particle, muon), 17, 134, 135
      amplitude in, 141, 347q78
    magnetic moment, 348q80
    making, with electron neutrino *or* antineutrino, 168, 172
      with electron neutrino *and* antineutrino, 240–244
    mass, 16, 126–133, 144
      with Hubble's constant as input, 145
      as input, 150, 268
      relation to α, *c*, *G*, and *h*, 146, 147
        comparison with prior reports of observations, 268, 290–292

'spin ½', 17, 133, 238
approx. Fermi-Dirac 'statistics', 17, 137
breakdown?, prediction (6), 319
approx. 'Pauli exclusion', 17, 137
'statistics' at external energy $<< m_e c^2$, 137
expression for partial comparison δ 3,2,*I*, 255
kaon, neutral (m_{K_L}-m_{K_S}), and E_1, 312, 344–345q52
experiment, suggestion for, 325
making (*see also* Particle, electron and positron, complementarity of coding & making with. . . ; Particle, muon; Particle, neutrino and antineutrino), 346q 59, 60
mass, motion of (*see also* B_0, mechanics), 39
masses (*see also* Particle (name)), predictions (9), 319
Table 3, comparisons of approx. calculated values with prior reports of observations, 268
at maximum velocity (*see also* B_0, fast, fastest), 330q29
meson masses and lifetimes, 331q35
modeled on, 'diametric modes' in *U*, 329q17
multineutrinos, 345q55
muon, 31, 240–244, 268
comparisons of calculations and approx. values with prior reports
of observation, 268, 288–290
literature review, *see* Particle, pion, comparisons with...; partial review of literature
Table 5, 289
decay, 240, 244, 329q21, 347q77
prediction (8), 319
as a possibility (11), 319
heavy, mass, rest, 240, 268
suggestions for experiments, 322, 323
internal information and (partial) coding, 240
light, mass, rest, 240, 268
prediction as a possibility (12), 320
suggestions for experiments, 322
neutrino and antineutrino (*see also* B_0, fast (fastest); 'Weak' interaction;), 327q3-3.4, 330q22
electron, 31, 168–172, 236, 268, 292–294
cautionary note, 236
comparison with prior reports of observations, 268, 292–294
internal information and coding, 236
making both with electron, positron, 240–244
making one with electron, positron, 168–172, 236
'mass', 268
two component, 236, 293, 294
and 'weak' interaction, 168–172, 292–294
in gravity, 158–164, 171, Appendix 1
suggested experiments, 325
muon, 31, 244, 268, 290, 292–294
cautionary note, 246
comparison with prior reports of observations, 268, 290, 292–294
internal information and (partial) coding, 244
'odd' bit and parity, 246
making one with muon, 245, 247, 292–294
'mass', 268
and 'weak' interaction, 292–294
stellar (*see also* B_0, fast, fastest), 330q28, 340–341q51.3, Appendix 1
suggestions for experiments, 325
and stellar ignition, 332q42, Appendix 1
neutron, 329q17
decay, 172
photon (*see also* *w*), 28, 31, 179–181, 235, 268, 294
polarization, 176, 179–181, 347q71
spectrum of states in *U*, 9, 27, 105, 206–210, 234
comparison (including multiplicities) with prior reports of
observations, 32, 287, 288, 294
and cosmic background electromagnetic radiation, 332q40, 333–337q50
and 'cosmological theories' 333–337q50
counting lowest few, 234
energy of lowest, approx. calculated value, 154, 155
experiment, suggestion for, 324
prediction (7), 319
photon particle representation, 6, 19, 48, 67
pion, charged, 31, 246, 254
comparisons with prior reports of observations, 267–288
amount of internal information, 283
partial review of literature, 276–282
Table 3, 266–270
Table 4, 285

decay, 31, 32, 247, 292–294
experiments, suggestions for (light and heavy), 323, 324
heavy, mass, 246, 268
internal information and (partial) coding, 246
light, evidence for existence, 272, 273
mass, 246, 268
prediction as a possibility (13) 320
neutral, 31, 256, 268, 285
adseparability, applicability of? 257, 258
decay, 257
experiment, suggestion for, 323
internal information and (partial) coding, 256
mass, 256, 268
'spin 0', 248 steps 9–14
'virtual', in L 17.10 ('strong') interaction 295, 296
pionlike, lifetime dependence on amount of internal information, 256
prediction (9), 319
proton (*see also* L 17.10 interaction) 172, 329q17
Patent notice, v
Phase, relative, addition without, 188
of B_0s, 216, 217
and 'statistics', of photons, 188
'spin 0', 248 step 11
'spin ½', 137
of *w*s, 215
Phase space (*see also* B_0, fast (fastest)), 39, 52, 330q27, 333q49
Phononic and photonic systems, 313
Photon, *see* Electromagnetic theory, quantized; Particle, photon; Time, apparent unidirectionality; *w*
Planck's constant and, (informal, nonaxiomatic), 307
representation for particles, 6, 19, 48, 67, 156
(convenience, nonaxiomatic, 48)
Photonic particle (*see also* B_0; Particle), 19, 156
photicle, 156
Photonic, phononic and, systems, 313
Pion, 'virtual', in L 17.10 ('strong') interaction, 295, 296
Place, 98
Planck's constant, 211, 329q9, 10
expression for, 232
and photons (informal, nonaxiomatic), 307
PODIC (*see also* Adseparability; Information, separability . . .), 7, 68
domain of, for B_0 or two specified photons in box, 78
for systems, 90–96
examples, 69–71
Polarization, photon, *see* Particle, photon, polarization; *w*
Position, 10, 11, 98, 100, 328q6
conservation of (localizability of), example 2, 69
information, amount of, 100
fastest B_0 not having any in U, 159
relativity of, example 4, 69
Positron, *see* Particle, electron and positron
Postulates, 22, 179, 304, 329q16
and constants, 230–232, 314
comment on, 314
equivalencies, 23, 174, 175
introspection, for (approximate?) description of, 305
looking for, 22, 174
sufficiency of, 30, 229, 230
and reference substitutions for merging Sections 2 and 3, 229, 230
testing, 179–230
'Predetermination', 307
'prediction', 318
Predictions, 33, 318
and experiments which might be interesting, some, 318
some implications of, 320, 321
Table 6, suggestions for experiments, 322–325
Principle 1 (I proportional to E), 7, 68
2 (PODIC, *see also* Adseparability; Information, separability . . . ; PODIC), 7, 68
3 (devoid of history of change), 8, 71, 299
4 (report of local length measurements), 12, 107, 300
'Principle of equivalence', *see* Gravity, 'principle of equivalence'
Proton (*see also* L 17.10 interaction), 329q17

Quantization, 25, 176, 177, 188, 206–210, 227–229
informal, nonaxiomatic, 307
Quantum mechanics, 227–229
operators, 228, 229
prior formalism, counterexample (internal inconsistency), 3

relativistic, 30, 227–229
spreading, 70, 80 (steps 29–32), 206
prediction (1), 319
Quasars, 320, 321, 342q51.4 (2,6)
Questions (not fully indexed in this index, but some various separate entries), 33, 326
Quotation marks use herein, 4

Reality, objective, 304–305
subjective, 305–306
Relativistic and nonrelativistic theories (*see also* Theories . . .), 298
Relativity, general, *see* Gravity...; Theorem 13 interaction
special, *see Special relativity; see also* Adseparability; Information, separability...; PODIC
'Resonances', 330q24
'Representing measurements (*see also* Description with finite information; Information, separability...), 309

Sanity, 305
Scale, 186, 187
Scaling, gentle (*see also* Gravity, metric effects; Localization), 7, 25, 61
gentle scaling 1, 190
observer, contained, effect on, 105
units of, 106
'Scattering' (*see also* Phase space), 329q15
Signal, 2, 36, 38
'Soul' (*see also* 'Mind'...), 347q74
Space, 24, 25, 181, 184–187
choice of coordinates, 24, 25, 186
'external' and 'internal', 305–306
homogeneity and isotropy, 182, 185
3-dimensionality of, 179–185
Spacetime, 23–25, 185
choice of coordinates, 24, 25, 186
4-dimensionality of, 179–185
Special relativistic covariance, hereof, 71, 227–229
Special relativity, 5, 6, 26, 47, 48, 55–57, 66. 67
from adseparability, 193, 227–229
domain of, for B_0 or specified w in box, 202
for systems, 220–225
comment on, 298–303
a limitation on domain of applicability of, 7–9, 96, 220–225
from PODIC, 69
domain of, for B_0 or specified ws in box, 78
for systems, 90–96
'Spin', '0', 248 (steps 8–11)
'½', *see* Particle, electron and positron, 'spin ½' & 'statistics'
'Spinors', Dirac's examples 3 and 4, 219
Square brackets, use herein of, 4
'Statistics', Bose-Einstein, *see* Particle, photon, Particle, pion, 'spin 0'
Fermi-Dirac, *see* Particle, electron and positron, 'spin ½' & 'statistics'
'Strong' interaction (*see also* L 17.10 interaction) 32, 294–296
prediction (10), 319

Tables: 1 F α^{-2}, 150
2 Some (approx.?) expressions and value in U, 155
3 Particles, 268
4 Pions, 285
5 Muons, 289
6 Experiments, 322
Theorem 13 interaction (*see also* Gravity (Theorem 13 interaction herein)), 100–123
Theorem 13, 118
'weak field' apparent correspondence with general relativity, 118, 123
Theories, comments on, 298–304
cosmological, and cosmic background electromagnetic radiation, 333–337q50
multi-observer, 298–300
prior, interpretation of reference herein to, 2
relativistic and nonrelativistic, 298–303
'thermodynamic', 2
'second law of thermodynamics', not, 43
Ti, 306
and time, 306, 332q44
Time, 24, 25, 181, 185
apparent undirectionality of, 24, 70, 182, 185, 332q41
choice of coordinates, 185, 186
homogeneity, 182, 185, 186
Transcription, between adjoining regions with different localization (*see also* Gravity, geometry in & metric effects), 107
between position and momentum representations (*see also*

'Cosmological' red shift; Hubble's constant; Universe, 'radius', 102–105
detection and, 11, 37–39, 99
classical, 11, 99
quantum mechanical, 11, 99
of external to internal information in neutron decay, approx. calculated rate, 172
Transmitter (*see also* Communication; Transportation)
classical mechanical, 37
quantum mechanical, 37
Transportation, viii, 96, 98, 99, 172, 173, 181, 328q6
drive, 347q68, 72
local propagative in gravity, 107
time (?) associated with, 121, 181, 303, 332q45, 333q48
Truth-telling, 305
Tunneling and 'virtual states', 312
Two, looking for a factor of, 317

Units, comparison(s) of (*see also* Adseparability; Gravity, metric effects; Information, separability...; PODIC, examples), 12, 107, 298–300
during gentle scaling, 105
and localization, 106
Universe, 182
'age' of, 328q5, 331q30,31, 343q51.4, 344q51,7,10
apparent finiteness of accessibility, 175, 176, 186, 187, 312, 313, 332q47
different observers, universe(s) of, 114, 130, 175, 312, 313, 332q47
(re-)focusing in, 346q56, 57
information in, total amount of,
approx. calculated value, 154, 155
as a constant, 231
is it necessary?, 315, 346q66
approx. expression for, 154, 155
mass energy in, 'average' density of, 15, 19, 122, 123
approx. calculated value, 153, 155
experiment, note for, 324
(approx.?) expression for, 155
prediction (5), 319
total, approx. calculated value, 154, 155
(approx.?) expression for, 155
expression for, 232
'radius' or 'size' of, 102–105, 186, 187, 212, 332q47
approx. calculated value for, 154, 155
as a constant, 231
expression (approx.?) for, 155
$R_U \rho_U$ (*B* as *U*), 154, 155
'size', as not 'variable', 104, 187, 231
Up, height of (how high is,?), 316, 332q47

Velocity (*see also* Momentum), 10
axiomatic definition, 189
'Virtual states', 295, 296, 312
w (*see also* Electromagnetic theory, quantized; Information; Particle, photon), 181
amplitudes, 188
describing information representation, by, 27, 207
by set of, 213
least amount of information exchangeable with irad *w*, 207
orthonormality, 179, 181, 185, 186, 214
states in *U*, spectrum of (*see also* Particle, photon, spectrum...), 27, 28, 206
wavelengths, 206
maximum for amplitude >0 (λ_1), 187, 188

Warnings (*see also* Cautionary notes and warnings) vii, 1
Wave, 179, 181, 214
exchange of information with, 207, 211
informal, nonaxiomatic, 307
standing, example 1, 219
Wave function, 216
factoring, 217, 218
Wavelength, of B_0 and particle, 5, 44–46, 72, 74, 194
maximum for *w*('s) with amplitude > 0 in *U* (λ_1), 187, 188
'Weak' interaction (*see also* Particle, neutrino and antineutrino), 32, 292–294
and two component prior theory of electron neutrino and antineutrino, 236–238, 293, 294
various prior theories, prediction (10), 319, 321
'Why do things fall?', 315

Xi, 306